Sustainable Technologies for the Oil Palm Industry

Dominic C. Y. Foo ·
Mustafa Kamal Tun Abdul Aziz · Suzana Yusup
Editors

Sustainable Technologies for the Oil Palm Industry

Latest Advances and Case Studies

Springer

Editors
Dominic C. Y. Foo
Centre of Excellence for Green Technologies
University of Nottingham Malaysia
Semenyih, Selangor, Malaysia

Mustafa Kamal Tun Abdul Aziz
Department of Chemical and Environmental Engineering
University of Notthingham Malaysia
Semenyih, Selangor, Malaysia

Suzana Yusup
Department of Generation and Environment
Tenaga National Berhad Research
Kajang, Selangor, Malaysia

ISBN 978-981-19-4846-6 ISBN 978-981-19-4847-3 (eBook)
https://doi.org/10.1007/978-981-19-4847-3

This Springer imprint is published by the registered company Springer Nature Singapore Pte Ltd.
The registered company address is: 152 Beach Road, #21-01/04 Gateway East, Singapore 189721, Singapore

Dominic C. Y. Foo would like to dedicate this book to his wife Cecilia and their daughters Irene, Jessica and Helena. They are his driving force in developing various cutting-end technologies for a sustainable future.

Mustafa Kamal Tun Abdul Aziz would like to dedicate this book to his late father Tun (Dr.) Abdul Aziz Abdul Majid and mother Toh Puan Raja Teh Zaitun Kamarulzaman, wife Noor Azian, sons Abdul Muhaimin and Abdul Aziz, his late student Dr. Mohd Halim Syah Ismail as well as the well-respected the late Royal Professor Ungku Aziz bin Ungku Abdul Hamid.

Suzana Yusup would like to dedicate this book to her parents Yusup Abd Rasip and Salmah Baba, her husband Mohamad Zaki Yusof, her sons Mohamad Akmal, Mohamad Hazim , Mohamad Harith and daughter Nur Atikah. Their understanding, support and encouragement are the key to her journey towards advancing sustainable greener processes and technologies.

Foreword

I am delighted to congratulate the editors and authors of this excellent book *Sustainable Technologies for the Oil Palm Industry—Latest Advances and Case Studies.* It encompassed the latest work of both industrial players and academia for the sustainable oil palm industry, with emphasis on various novel disruptive technologies for the oil palm business and radical innovative processes. Certainly, the book is in line with the commitment of Council of Palm Oil Producing Countries (CPOPC) in furthering the United Nations Sustainable Development Goals (UNSDG).

I am aware that the same group of editors and authors produced the earlier book *Green Technologies for the Oil Palm Industry* in 2019. It serves as a good guide for many practitioners in seeking for latest innovations in green technologies for the oil palm industries. So I hope this second book with a focus on sustainable technologies is equally beneficial to the oil palm industrial players. It is a sign of the strength and continuity of sustainable oil palm research and development, as well as Malaysia commitments to scientific and evidence-based research.

Let me conclude by wishing the editors and authors a big success for this book. Judging from its outstanding content, I am sure it will contribute to the primary roles of palm oil and palm biomass to the economic benefits of the producers, while putting environmental sustainability as a primary focus for the benefits of global consumers. This will ensure a long and stable low-carbon economy and global food security future for all.

Jakarta, Indonesia

Tan Sri Datuk Dr. Yusof Basiron
Executive Director of Council of Palm
Oil Producing Countries (CPOPC)

Preface

Palm oil is the most widely used edible oil in the world. Being the second-largest world producer and exporter of palm oil, Malaysia exported 16.2 million metric tonnes (MT) of palm oil and palm-based products in the year 2020, valued at approximately 73.3 billion Malaysian ringgit (approximately 17.5 billion USD). In the same year, the oil palm industry contributed about 2.7% or 36.87 billion Malaysian ringgit (approximately 8.78 billion USD) to the gross domestic product (GDP) of Malaysia (www.statista.com).

In recent years, major efforts have been put to enhance sustainability aspects of palm oil products. The Roundtable on Sustainable Palm Oil was established since the year 2004 to develop and implement environmental-friendly standards in palm oil production. As an effort to promote sustainability, various new technologies were reported in our last book published in 2019—*Green Technologies for the Oil Palm Industry* with Springer Nature. Since its publication, the book has received an overwhelming response in the oil palm industrial sector as it serves as a guide for promoting the use of green technologies. As a follow-up, this book is meant to promote some recent developed sustainable/green technologies for the oil palm industry. It comprises 14 chapters outlining the state-of-the-art advances in various aspects of palm biomass and palm oil processing from renowned scholars and industrial practitioners. These chapters may be read independently of each other without a particular sequence. Synopses of the chapters are given as follows.

Synopses of Chapters in Part One—Disruptive Technologies for Palm Biomass

This part consists of eight chapters on various disruptive technologies developed for palm biomass-based processes in recent years, ranging from experimental research, computational tools and industrial practices. The first chapter by Law and Foo entitled "Maximizing Valorisation Opportunities from Palm Biomass" provides an overview of various innovations involving the conversion of biomasses into value-added products. In Chapter Two entitled "Aviation Biofuels: Conversion Routes and Challenges", Chong and co-workers discuss the opportunities and challenges of sustainable production of jet biofuel. In Chapter Three entitled "Oil Palm-Based Nanocellulose: From Extraction to Applications", a comprehensive review was reported for palm-based nanocellulose from its initial isolation, characterisations to the final applications. Chapter Four by Perera and co-workers, entitled "Effect of Nanolignin and Nanocrystalline Cellulose on Thermal, Mechanical, and Water Barrier Properties of Starch Composites", discusses a ternary biopolymer composite that was obtained by reinforcing thermoplastic starch with nanocrystalline cellulose and nanolignin from oil palm empty fruit bunch.

The few following chapters reported some computational tools for various biomass processes. Chapter Five entitled "Process Simulation and Scheduling of Bio-succinic Acid Production from Palm Biomass" by Foo and co-worker reported process modelling and scheduling of bio-succinic acid production from palm biomass. In Chapter Six entitled "Stochastic Modeling for Palm Biomass Supply Chain", How and co-worker presented an overview of key supply chain uncertainties that should be incorporated into the biomass supply chain model. In Foo's Chapter Seven entitled "Process Integration Tools for Optimal Allocation of Palm Biomass", two process integration tools, i.e. graphical pinch diagram and mathematical programming model, were presented to determine the optimal allocation of palm biomass for bioenergy production. In the final chapter of this part entitled "Innovations to a Palm Biomass–Fueled Power Plant", industrial practitioners Lakshmanan and co-workers described some changes made for the water tube boiler to improve its fuel handling and combustion efficiency, which allowed handling of fibre with higher moisture content.

Synopses of Chapters in Part Two—Innovations in Palm Oil Processing

This part consists of six chapters that have their focus on various innovations in palm oil processing. Chapter Nine by Lim and co-worker entitled "Mathematical Modelling and Optimisation for Fresh Fruit Bunch Harvesting and Evacuation" proposed a potential solution to the labour-intensive fresh fruit bunch harvesting and evacuation process using mathematical optimisation modelling approach. In Chapter Ten entitled "Zero-Waste Technologies for the Sustainable Development of Oil Palm Mills", Teow and co-workers described an integrated technology for zero-waste oil palm processing in addressing the challenges of highly polluting effluent, greenhouse gases emission and solid waste generation. Two following chapters focus on the treatment of palm oil mill effluent. In Chapter Eleven entitled "Advancement in Various Stages of Palm Oil Mill Effluent (POME) Treatment Process", three established methods to modernise POME treatment are described by Khadaroo and co-worker. Chapter Twelve entitled "Performance and Stability of Pre-commercialized Integrated Anaerobic–Aerobic Bioreactor (IAAB) for the Treatment of Palm Oil Mill Effluent (POME)" next reported the performance of a pre-commercialised integrated anaerobic–aerobic bioreactor under variable organic loadings and environmental conditions with respect to effluent quality and methane yield. In Chapter Thirteen entitled "3-MCPDE in Palm Oil Processing: Formation Factors, Transference to Food and Mitigation Approaches" by Chew and co-workers, the formation factors and mitigation strategies of process contaminant, i.e. 3-monochloro-1, 2-propanediol esters (3-MCPDE), are discussed. In the final chapter entitled "Sustainable Practices of IOI Palm Oil and Palm Kernel Processing Complex in Sabah", industrial practitioners reported various sustainable practices carried out in a palm oil and kernel processing complex, resulted in 85% recovery of treated effluent and 40% reduction of power consumption, apart from improved product quality and lower treatment chemical usage.

Together, these 14 chapters present some latest advancements of sustainable technologies in the oil palm industry. It is hoped that the idea and innovations in these chapters will motivate readers in nurturing a sustainable palm oil industry.

Selangor, Malaysia

Dominic C. Y. Foo
Mustafa Kamal Tun Abdul Aziz
Suzana Yusup

Contents

Disruptive Technologies for Palm Biomass

Maximizing Valorisation Opportunities from Palm Biomass

Albert Yuen Hong Law and Dominic C. Y. Foo

Abstract Malaysia is the second largest world palm oil producer. The oil palm industry has been the main contributor of biomasses to Malaysia. These biomasses can be turned into various value-added bio-based products that are in high demand, such as bioenergy, bio-agriculture, eco-products, and bio-chemicals. In this chapter, the availability of various palm biomasses in Malaysia, and its current industry practices were reviewed. The innovations involving the conversion of biomasses into value-added products were outlined. Challenges to unlock the potential of the underutilized biomasses are further discussed.

Keywords Value-added products · Empty fruit bunch · Palm kernel shell · Oil palm trunk · Oil palm frond · Mesocarp fiber

1 Introduction

Crude palm oil (CPO) is the main product of the oil palm industry. Fresh fruit bunch harvested from the oil palm plantation is sent to a typical palm oil mill (POM), where CPO is produced through a series of pressing actions to the fruit. The leftover wastes from the CPO production process include the empty fruit bunches (EFBs), mesocarp fiber (MF), palm kernel shell (PKS), and palm oil mill effluent (POME). While these are typically viewed as wastes, it is a fact that the multiple trails of biomasses left behind the CPO production should be treated as "opportunities". In the past decades, various technologies and innovations have transformed the biomasses into feedstock and/or bioenergy. It has been proven that the oil palm biomass can be converted into solid biofuel in pellets, briquettes, or shell forms, which can replace coal for energy

A. Y. H. Law (✉)
ENLPRO Solutions (M) Sdn. Bhd, Level 28, The Gardens South Tower, Mid Valley City, Lingkaran Syed Putra, 59200 Kuala Lumpur, Malaysia
e-mail: albertlaw@enlpro.net

D. C. Y. Foo
Center for Green Technologies/Department of Chemical and Environmental Engineering, University of Nottingham Malaysia, Broga Road, 43500 Semenyih, Selangor, Malaysia

D. C. Y. Foo et al. (eds.), *Sustainable Technologies for the Oil Palm Industry*,
https://doi.org/10.1007/978-981-19-4847-3_1

generation. Besides, through bio-chemical extraction, it can be converted to bio-ethanol, medical grade alcohol, bio-diesel, or even sugar, and many other possibilities that are waiting to be harnessed, i.e., bio-fertilizer, biogas power generation, eco-fiber matt, raw material for mattresses and bio-composite, etc. In other words, treating biomass as waste is an outdated view.

The National Biomass Strategy 2020 (Agensi Inovasi Malaysia (AIM), 2013) blueprint projected that Malaysia to produce up to 80 million tons of biomass annually, with a big majority from the oil palm sector. The biomass industry itself has an enormous potential to drive a sustainable and circular economy in the country.

The demand for bioenergy is rising, especially in East Asian countries such as Japan, China, South Korea, and those in the Europe. Some of these countries are already importing biofuel pellets (Northport (Malaysia) Bhd., 2021; Department of Statistics Malaysia, 2021a), and palm kernel shells (Department of Statistics Malaysia, 2021b) from Malaysia, Indonesia, and Vietnam. Although some might argue that this highly policy-dependence industry has no longevity in its current form, it has been reported that more biomasses are now replacing fossil fuels in the energy industry in the above-mentioned countries.

At present, the traceability of the palm biomass is being scrutinized. The main principle is to ensure that these biomasses come from the waste streams. According to MIGHT 2019 (Malaysia Industry-Government Group for High Technology (MIGHT), 2019) MITI 2018/2019 review (Ministry of International Trade and Industry, 2019), as well as the Roundtable on Sustainable Palm Oil (2021) certification, sustainability and chain of custody are the requirements of the importing countries. Nevertheless, the fact remains that substantial portion of these biomasses is still underutilized. Hence, the enormous potential should be explored, and its unexpected challenges should be addressed.

The palm biomass may be categorized into those that originated from the plantation activities, and those from the milling operations. These are discussed in the following sections. This chapter provides an overview of the following:

- Types of biomasses produced from both plantations and mills,
- Characteristics of each type of raw biomass, and
- Market potential and industrial utilization.

The information herein was based on personal interviews and experiences accumulated over the years with various industry players. The chapter summarizes those understandings, views, and recommendations of those experiences and interviews.

2 Oil Palm Plantations in Malaysia

Malaysia is blessed with a humid tropical climate that comes with ample of sunshine and evenly distributed annual rainfall. This promotes the growth of oil palm trees and its CPO production under this optimal condition, which has led to tremendous contribution to gross domestic product of Malaysia in the past four decades.

To date, Indonesia and Malaysia have become the epicenter of the booming palm oil production, which resulted in the extensive expansion of palm oil plantation area. Since 1960s, Malaysia government has identified oil palm as a key agricultural crop apart from rubber. The Malaysian Palm Oil Board (MPOB) reported that there are about 5.9 million ha (MPOB 2021) of oil palm planted areas nationwide at the end of 2020, which is equivalent to about 25% of Malaysia's total land area. This includes 2.8 million ha of plantation in Peninsular Malaysia, 1.5 million ha in Sabah, and 1.6 million ha in Sarawak (MPOB 2021). Figure 1 shows the locations of oil palm plantations in Malaysia.

The palm plantations are owned by large conglomerates (e.g., IOI, Sime Darby, KLK), governmental agencies (i.e.: FELDA and FELCRA) (Teoh, 2002) and independent smallholders (see Fig. 2). Back in the 1990s, under the encouragement from the government to promote growth, urbanization, and infrastructure developments in rural areas, a surge of oil palm plantation and industry could be seen through various government programs such as 2nd–11th Malaysia Plan, National Key Economic Areas (NKEA) policies (World Bank Group, 2019), etc. The industry is export-oriented and hence serves as one of the major contributors to Malaysia's national revenue.

Oil palm yield is about ten times higher than that of soybean (Singh et al., 2021), and usually needs 26–33 months for the planted sapling to bear mature fresh fruit bunch (FFB) ready for harvest. On average, a hectare of mature palm trees can produce 1.3–1.6 tons of FFB monthly (dan Komoditi, 2019). Each of these FFB weighs 10–25 kg (Ismail & Mamat, 2002) and it bears fruits all year long. The production of FFB for year 2019 is reported to be 99.3 million tons (see Fig. 3).

The oil palm tree can grow up to 6 m and have a life span of 30 years, MPOB recommends 25 years replanting cycle for the oil palm plantations (Eco Synergy Solutions Sdn, 2017). Normally, one hectare of palm plantation is generally planted with to 132–150 oil palm trees (Malaysian Palm Oil Board (MPOB), 2011), depending on terrain and other factors.

Fig. 1 Locations of oil palm plantation in Malaysia (EU-Malaysia Chamber of Commerce and Industry (EUMCCI), 2017)

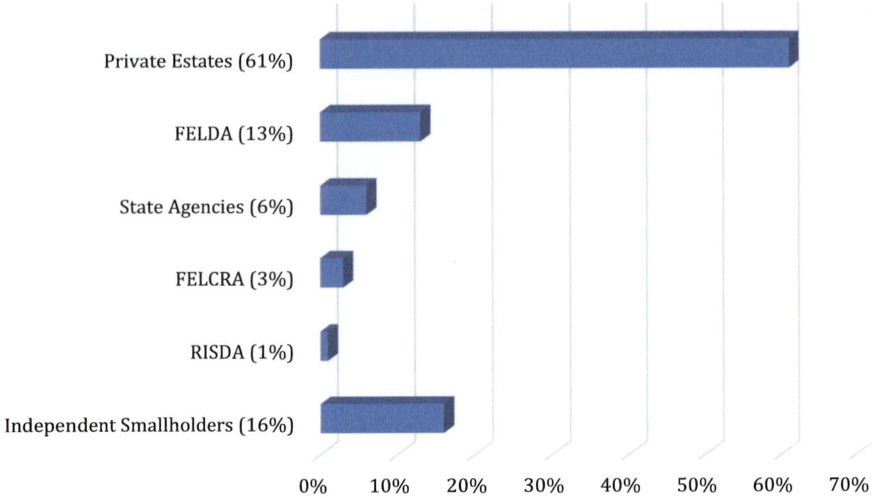

Fig. 2 Stakeholders distribution of oil palm plantation (Hasan et al., 2019)

Fig. 3 Overview of biomasses from Malaysia oil palm Plantation of 2019 (World Bank Group, 2019)

3 Biomass from Plantations

3.1 Oil Palm Trunks (OPT)

Oil palm trunks (OPT, Fig. 4) become available at the end of a plantation's life cycle, i.e., every 25–30 years. As a rule of thumb, it is estimated that 74.48 (Zulkifli et al., 2010) tons of OPT biomass could be generated of each hectare of palm oil plantation during a replanting operation. For year 2018, a total of 6.63 million tons of OPT was generated (see Fig. 3).

Note that the replanting schedule and quantity of available OPT are not directly correlated with the age of oil palm trees that is to be replanted. Instead, it depends on the CPO market price and owner discretion. MPOB recommended replantation to take place when trees grow above the age of 26 (Nambiappan et al., 2018). However, the palm replanting effort has not been encouraging despise encouragement of the government agencies. It is estimated that one-sixth of matured plantation in Malaysia has surpassed its optimum fruit yield period. Hence, this led to reduced CPO and hence biomass production.

There are assorted reasons for the hesitation of replanting among the conglomerate's management. These include factors such as the nature of palm oil production cycle, volatility of CPO market price, invariably association between rising costs of production and falling profit (due to deteriorating fruit yield), and sudden changes of international purchasing policy. The smallholders however face different obstacles on the other hand. These include difficulties to do proper replanting (for those with crops are in the forest areas), excessive cost for difficult terrain replanting, incompletion

Fig. 4 Freshly harvested OPT stacked and waiting to be transported out of the replanting field during replanting operation (photo taken in Kluang, Johor)

Fig. 5 Tree felling, chipping of OPT into chunks and pieces for natural decomposing process for nutrient recovery to the plantation

of smallholders' legal documents, lack of other sources of income during the grace period, and limited operational budget for replanting, etc. Furthermore, smallholders have difficulty accessing loan from banks due to the lack of third-party guarantors for small replanting projects. These unstable reasons halted OPT in becoming a stable supply of biomass feedstock for industrial-scale downstream projects. Hence, oil palm trunks are left to decay in the plantation area (Bukhari et al., 2019). This is indeed a major loss of business opportunity.

3.1.1 Opportunities of OPT Biomass as Natural Fertilizer

Current practice of the industrial players is to retain the OPT in the plantation as fertilizer. Trunks are either felled or chipped to allow for quicker return of nutrients to the soil. Alternatively, they are killed with poison and left standing to decompose naturally in the field (Fig. 5). A small percentage of trunks are currently used in the wood industry, e.g., plywood, medium-density fiberboard (MDF) (Rosli et al., 2016), and energy generation (Ibrahim et al., 2014a)

3.1.2 Opportunities of OPT as Raw Material for Plywood

The wood from the oil palm tree cannot be used as timber directly, as it is softwood with high moisture content. Extensive research was conducted to convert OPT into plywood and medium-density fiberboard. It is a mature technology that has been commercialized and deployed (Ibrahim et al., 2014b) (see Fig. 6). However, in author's opinion, there are challenges that need to be addressed and overcome before these technologies are ready for commercialization:

- Inconsistency of raw material supply because its availability is dependent on decision of plantation owners.
- For plantations in the rural area, harvesting, collection, and transportation costs for the site may be higher than the cost–benefit curve.

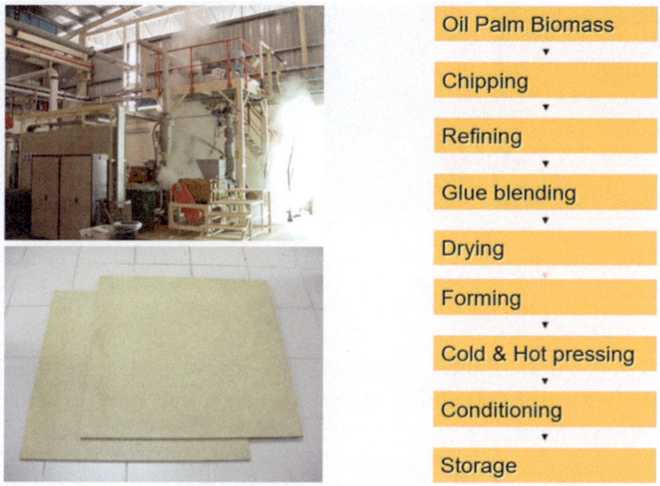

(a)

(b)

Fig. 6 Steps to produce **a** MDF; **b** MDF production (photos are courtesy of Dongwha MDF Fiberboard Commercialization trial plant)

Fig. 7 Process of plywood production (*Courtesy picture from MPOB Palm Plywood processing research)

- Due to its soft nature, the final product does not have the same quality as other fiberboards in terms of durability to heat, pressure, thus have lower acceptability in the existing market.

As a rule of thumb, an OPT of 18 ft long can be used to produce about 40 pieces of wood veneers (with size of 2 ft × 8 ft) of 4 mm of thickness. Alternatively, the 18 ft long OPT can also be made into seven pieces of veneers, each with thickness of 12 mm.

Even though OPT is a good candidate for the deprived plywood market. However, due to its weaker durability, most harvested OPT is not being used. This left behind some waste in the plywood-making process. Besides, OPT is not woody material anatomically. Therefore, the produced plywood (Fig. 7) can only meet the minimum strength requirement according to the Japanese Standard Method (JAS 233:2003), which resulted in less attractive profit margin. That might be the reason why OPT plywood does not have large-scale production, even though it is technically viable and achievable. It is also worth mentioning that to achieve desired productivity in operations and product quality, some modifications are necessary for the processing techniques and equipment, such as blade and the intake size of the chipping machine.

3.1.3 Opportunities of OPT as Raw Material for Solid Biofuel

In recent years, it has been a global trend in seeking alternative energy for the replacement of fossil fuels. As such, palm pellet (Fig. 8) has been identified as favorite for global energy storage. It is a natural entry point a biomass owner can capitalize almost

Fig. 8 Palm pellet made from OPT

immediately, as the technology is reasonably mature, while the cost of developing infrastructure is relatively low. Hence, the project has short payback period.

Apart from its obvious function as an energy deliverance for the energy sector, pellet is also a very cost-effective way to mobilize the huge, bulky, highly moist biomasses for long distances. It is arguably that large-scale palletization of OPT is an efficient way of energy reservoirs for regional energy security.

Palm pellets in Malaysia have been shipped to Japan and Korea as a renewable energy source in recent years. There is also market demand for pellets from European energy utility companies where European renewable energy targets are to be met. Similarly, Japan is sourcing biomass for co-firing to ensure sufficient energy supply given its reduction in nuclear energy capacity. Unfortunately, the current Malaysian pellet is unable to meet these needs in terms of production volume, and standardization of its quality requirement, as well as its policy compliances.

Even though the palm pellet industry has a bright future, it does face some challenges. Malaysia pellet industry is currently shaping its self-regulating standardization to meet the quality requirement of the importing countries. Besides, there is no clear longevity policy and demand on Malaysian biomass pellet from importing countries like Japan and Korea. In other words, the absence of a ready market is still the major obstacle for the biomass pellet business currently.

3.1.4 Opportunities of OPT as Raw Material for Bioproduct and Biofuel

Due to its relatively high sugar content in the trunk sap, OPT may be considered for bioethanol production, which in turn can be used as biofuel. Production of bioethanol via fermentation of sugars and to be blended with gasoline has always been seen as a wise strategy. Apart from being able to reduce dependency on fossil fuels, this sustainable alternative fuel helps to reduce carbon footprints. There is on-going effort is trying to unlock the potential of OPT sap (Jalil & Hossain, 2015) for bioethanol production. This pilot plant scale effort has successfully optimized the usage of sugar content within OPT sap as carbon source. One could now conclude that OPT sap provides major advantages for bioconversion due to the following reasons:

- the process is inexpensive,
- no additional chemical or enzymatic treatment is needed,
- excellent properties in terms of sugars, nitrogen, amino acids, vitamins, and minerals, hence good yield of desired products,
- low quantity of by-products from the process which makes the fermentation process economically viable.

There are numerous flickering sparks and initial successful pilot biofuel projects in several areas with high density of palm plantation. A full commercial-scale bioconversion complex is also expected in the coming future.

3.2 Oil Palm Fronds (OPF)

Oil palm fronds (OPFs) are obtained during replanting activities, pruning of oil palm trees, and harvesting of FFBs, with the latter accounting for the greatest share of volume. It has been estimated that about 14.47 tons/ha of OPF can be gathered from pruning activities, while 15 tons/ha of OPF could be gathered when a plantation is being replanted. It is logical to assume that the OPF availability is depending on FFB harvesting activities (Onoja et al., 2018). It was reported that a total of 46.5 million tons of OPF was generated for year 2019 (see Fig. 3).

The OPF is approximately 2–3 m long and weighs about 10 kg (wet weight; see Fig. 9). It consists of two main parts, i.e., the petiole (stem) and many long leaflets on either side of the stem. The top two-thirds of the frond contains most of the nutrients, while the basal (lower) third is rich in cellulosic materials and sugars, which are needed in the production of biofuels and bio-based chemicals.

Frond pruning of oil palm trees significantly affects fruit bunch production. Trees without frond pruning produce lesser and lighter bunches comparing to those with excessive pruning below 32 fronds (Agensi Inovasi Malaysia (AIM), 2013). This leads to the abundant availability of OPF all year round following the regular industry frond pruning practices.

3.2.1 Opportunities of OPF as Natural Fertilizer Through Natural Decomposition

Most OPFs are left in the oil palm plantation in the current practice. Even if there is any collection activity of OPF from the field, only the basal portion, i.e., one-third of the fronds are being collected for downstream uses. The remaining parts of the

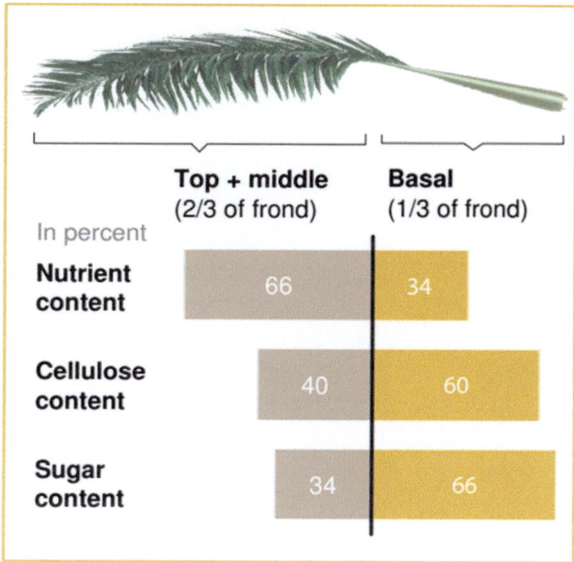

Fig. 9 Structure and contents of OPF (Agensi Inovasi Malaysia (AIM), 2013)

frond (top and middle portion) which contain most of the nutrients (mulch or whole) are left as topsoil replacement and natural fertilizer. The basal portion may be made available for fibers production, feeds for livestock, biofuels, or bio-based chemicals.

During the replanting activity, the teared down midrib is arranged in the soil to become line marker for the field work and as organic fertilizer. They also help to maintain soil moisture and avoid erosion to certain extent.

Due to its bulky nature, the OPF collection methods range from simple manual collection with a wheelbarrow, animal cart, or motor-cart, to advanced mechanization, which varies with collection cost. Terrain problem (e.g., elevation, spacing of trees, rivers, and holes), labor constraints, and economies of scale are among reasons why OPF have yet to reach its maturity as raw material for downstream innovation projects at industrial scale.

3.2.2 Opportunities of OPF as Animal Feeds for Livestock

OPFF contains high fibrous materials which can be further processed as ruminant animals' feed (Fig. 10). OPF has been widely utilized as feedstock in the form of freshly chopped, silage, pellets, or cubes form. Mixture made from freshly chopped OPF and palm kernel cake has been extensively used by local farmers for feeding beef and dairy cattle in Malaysia. Many studies have been reported to raise feed conversion efficiency (FCE) and average daily gain (ADG), with an acceptable ratio of 20% OPF: 80% PKC-based mixture diet. Note however that this does not show promising economics (Zahari et al., 2002). A mixture diet containing OPF and PKC with ratio of 40:60 is most economical as indicated by cost per weight gain value

Fig. 10 OPF as livestock feeds

[26]. Hence, the optimum OPF inclusion level for beef and dairy animals feed is at about 30% in their diet from the economical perspective, i.e., feed cost per weight gain value.

3.2.3 Other Commercial Potential of OPF

There are other potential applications for OPF. It has been reported that OPF could also be used as a starting material to produce bio-composites for high-performance applications (Suhaily et al., 2012; Rasat et al., 2014). The substitution of petroleum-based materials with this natural fiber would reduce the dependency on imported oil thus contributing greatly to overall cost-effectiveness.

There is also potential to make use of OPF juice in producing biofuels (Lim and Ibrahim, 2012; Kumneadklang et al., 2015) (bio-ethanol primarily) and bio-based chemicals such as bio-plastics (Zahari et al., 2015), nutraceuticals (Ofori-Boateng, 2013), etc. Note however that the available technology is yet to reach its mature stage, and hence the utilization of OPF on large industrial scale is yet to be seen.

4 Palm Biomass from Processing Activities in POM

In 2020, there were 457 POMs in operation in Malaysia (Malaysia Palm Oil Board MPOB, 2020) (see Fig. 11). The main product from the POMs is CPO which is mainly used as cooking oil and many other palm products. A total of 19,140,613 tons of CPO was produced in year 2020 (Malaysia Palm Oil Board MPOB, 2020), which made Malaysia the second world CPO producer.

In the POMs, CPO is produced from FFB after undergoing a series of purification processes. The milling operations also generate a significant amount of biomass. Figure 12 shows the biomass from the milling process include FFB, mesocarp fiber, palm kernel shell (PKS), as well as the liquid effluent known as POME.

Although large amount of palm biomass is available in Malaysia, there are no official statistics announced by the authority. This is because biomass has no official custodian ministry or agency to date. Hence, the palm industry normally adopts "guesstimation" approach in obtaining biomass data. Figure 13 shows amount of solid biomass produced for every ton of FFB processed. As shown, approximately 22% of the processed FFB becomes EFB, 12% turns into mesocarp fiber, while 5.5% is as palm kernel shell. Besides, about 60.75 litter of POME is produced. Note however that these figures are based on general rules of estimation in the oil palm industry. As each POM has different operating conditions (e.g., energy consumption practices, automation-oriented or labor dependency, management principles, etc.), their biomass yield may differ from these figures.

Fig. 11 Number of palm oil mills in Malaysia by states, 2020 (Malaysia Palm Oil Board MPOB, 2020)

Fig. 12 Different types of biomasses produced from a POM

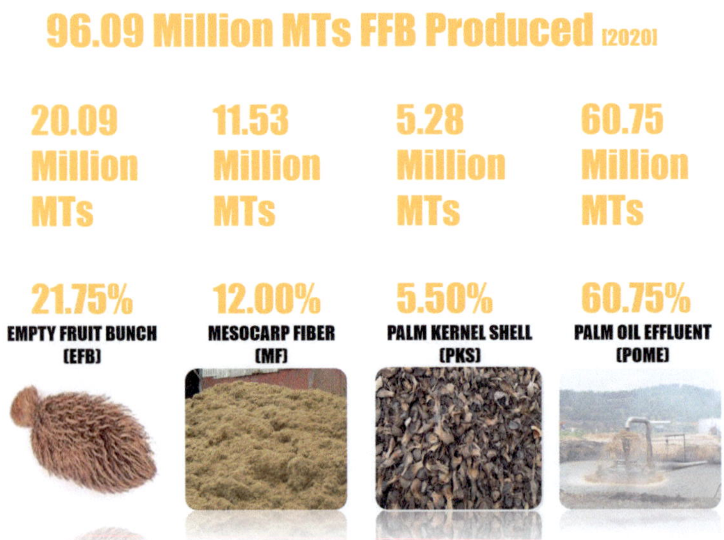

Fig. 13 Different types of biomasses generated from POM, calculative estimation to industrial best knowledge

4.1 Empty Fruit Bunch (EFB)

EFB is the remains after removal of oil from the FFB. The upstream process of separating the palm fruits from the bunch involves cooking of bunch with steam under high pressure, which is followed by threshing operation. These operations lead to separation of fruits from the bunches, with disposal of the "empty" fruit bunch (see Fig. 14).

It is estimated that for each ton of FFB processed, 21% of EFB is generated. For year 2020 where 96 million tons of FFB that have being processed, it was estimated that a slightly over 20 million tons of EFB was produced from the entire Malaysia palm industry (Fig. 13).

In current industrial practices, some EFBs are utilized for steam and power generation in the mill, or other niche downstream applications (see discussion in the later section), note however that a large portion (estimated three quarters) of EFB is used for nutrient replacement and mulching purposes in the plantations, or worse landfilling within the POM premise. These are considered as loss of opportunity, as EFB can be converted to other value-added products.

However, there are technical hurdles that need to be overcome before the EFB can be better utilized. Firstly, EFB is bulky. Its length may be as 50 cm and weighs up to 20 kg per bunch. This increases the cost of transportation significantly and preventing them for downstream utilization. Furthermore, these highly moist EFBs must be processed within 3 days after its disposal from the mill. Failing to do so leads to fermentation process where the EFBs will start rotting. Note that the abovementioned issues do not occur in other biomass types, such as wood waste. Fortunately, technological breakthrough has enabled EFB to be utilized in various forms of consumer products. These are discussed in the following sub-sections.

4.1.1 EFB as Natural Fertilizer

For many decades, EFBs were burned in an open air, and its ashes are used as soil-treatment agents or fertilizer. However, the rising awareness of environmental

Fig. 14 Fresh fruit bunch (FFB) is processed in the palm oil mill, which resulted with palm fruit and empty fruit bunch (EFB)

sustainability and air pollution has led to the enforcement of legislation that prohibited this traditional practice.

However, utilization of EFB to field mulching and composting is a long-standing practice of the industry and is still widely being carried out (Fig. 15). Other benefits of this practice include better weed control, prevention of soil erosion, and moisture retention in the soil. Note however that there are opposing views on this practice, arguing that it is merely a cost-effective way of disposing of the EFBs, as compared to the alternative such as landfilling within the mill area, which is land consuming, hazardous, and costly. Note however that in the recent years, the main problem encountered with mulching practices is the rising cost of labor and transportation. These factors have led to the increasing cost of the mulching process.

On the other hand, EFB compost treatment is a more cost-effective and environmental-friendly option. It turns EFB into a usable compost. This process reduces the bulkiness of EFB so that it is more cost-effective for transportation and distribution work at the plantation fields. It also retains the benefit of protecting the soil and crops by reducing the risk of weed seeds, parasites, and pathogens spreading, while enjoying lower transportation costs.

EFB may also be composted along with POME. Apart from cost reduction (due to co-treatment of both by-products), EFB compost treatment also provides another product for the POM with saleable value (Fig. 16). Generally, there are two methods practiced by the industry, i.e., indoor and outdoor composting processes. Each method uses EFB, other organic materials and mostly couples those with POME as part of the ingredient. In both processes, this compost will be mixed and turned and introduced

Fig. 15 EFB mulching in the plantation

Fig. 16 Indoor composting process and outdoor composting process, turning EFB and POME into sellable compost

with its own "cocktails recipe" of enzyme to increase its yield. However, as the name suggested, the outdoor composting process is fully exposed to nature, hence the composting process is harder to manage and control. Hence, in general, the indoor composting process is much well-managed and its outcome is also more controllable.

4.1.2 Fiberized EFB and Its Opportunities

Technological breakthroughs for EFB happened in the 1990s, where successful mechanical shredding and fiberizing of EFB were reported. The fiberization of EFB enables various value-added products to be introduced in the market. Among them, the production of EFB pellets as solid biofuels has attracted many small and medium enterprises (SMEs) to venture into the business. These fuel pellets and briquettes made from palm EFB are meant for the export market (Fig. 17a). These solid fuels are used for both small-scale home heating and large-scale power generation. Besides the EFBs are made into dried long fiber, to be used as raw material for consumable mattresses (Fig. 17b). These are primarily aimed for China market as a replacement of coil fibers for mattresses manufacturing. The fibrous nature of EFB with a lower price (as compared to coil fibers) allows the mattresses industry to reduce its production cost. Besides, fiber mats for soil erosion prevention or for agriculture seed bedding purposes can also be made from EFB (Fig. 17c).

Yet another area that attracted good attention is eco-products. There are several companies that venture into the production of pulp and paper, as well as stationaries products from EFB fibers in Malaysia. Another promising sub-sector that shows

Fig. 17 **a** Different types of solid biofuel produced from EFB; **b** dried long fiber made from EFB which is then weaved into mattresses; **c** soil erosion control fiber mat (courtesy of Global Green Synergy Sdn. Bhd. and Fibromat (M) Sdn. Bhd.)

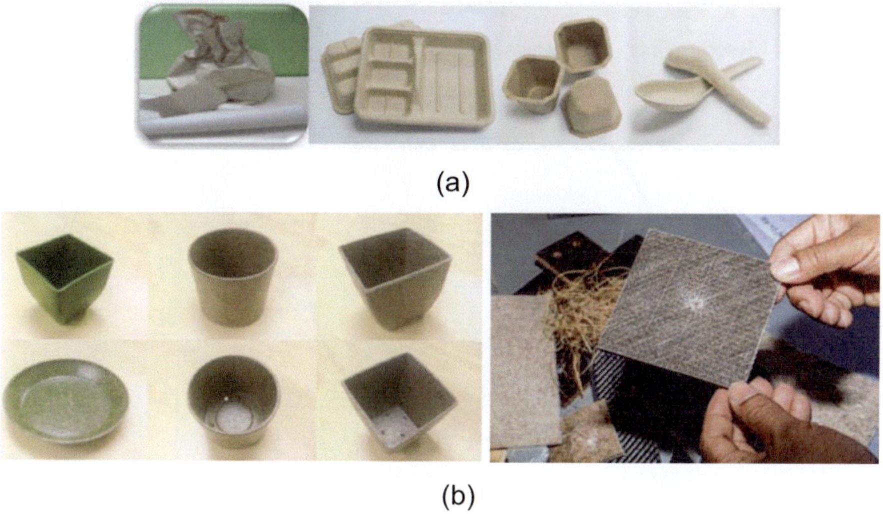

Fig. 18 **a** Pulp, paper and green product; **b** EFB-based bio-composite (courtesy of Malaysia Palm Oil Board)

growth potential is the use of EFB fiber and recycled plastics to manufacture high-end bio-composite products for building and automobile accessory industry (Fig. 18).

Currently, most biomass ventures in Malaysia (operational or under planning) are singular product and feedstock-focused. While it is still too early to judge its advantage or disadvantage, it is clear that EFB biomass resources are abundant, and their full opportunity is yet to be explored. However, due to lack of data in production, sales, and consumption, it is difficult to gauge which of these new exploratories will venture into full-scale production soon.

4.2 Mesocarp Fiber (MF) and Its Opportunities

Mesocarp fibers (MFs) are by-products from the extraction of CPO and palm kernel oil. Of every kg of FFB processed, 12% of its weight is MFs (Fig. 13). It is estimated that 11.53 million tons of MFs existed in year 2020 (see Fig. 13). In most cases, the MFs are used as raw materials as fuel within the mill itself as it is an excellent biofuel for steam and power generation. As very few POMs are willing to sell their MF, hence it is not easily found in the market, which makes it one of the more expensive biofuels for domestic use. Note however that there are some small amounts of MFs being sold as biofuel to CPO refineries and steam manufacturing plants. Figure 19 exhibits the inner component of a palm fruit which consists of mesocarp fiber, palm kernel shell, and palm kernel.

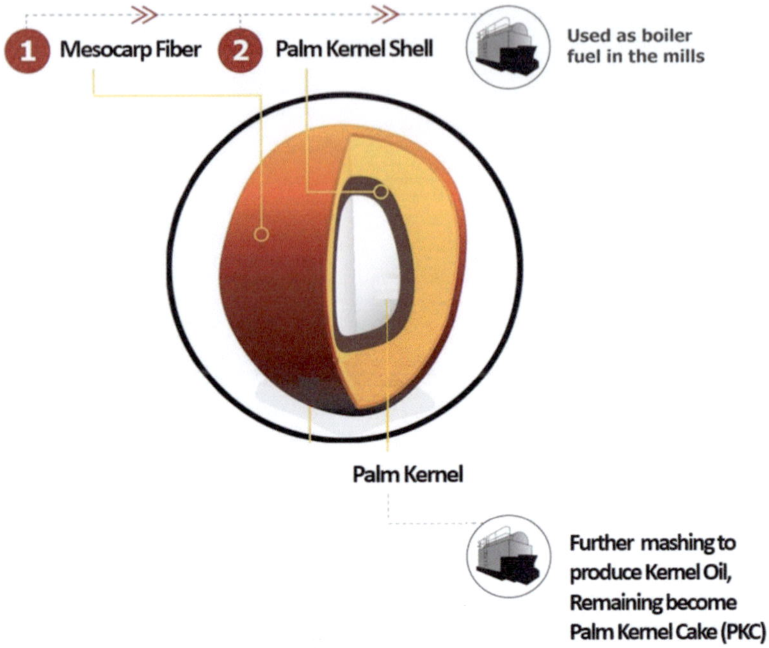

Fig. 19 Showing a palm fruit, the inner component, mesocarp fiber, palm kernel shell, and palm kernel and their respective further process or usage

Table 1 Calorific value and moisture content of palm-based biomass (Subramaniam and Chin, 2004)

Biomass	Calorific value (GJ/ton)	Moisture content (%)
PKS	20.01	23
MF	19.00	35–37
EFB Fiber	18.88	65–67
POME	20 MJ/m^3	98–99

Table 1 clearly shows that apart from PKS, MFs have a higher calorific value compared to EFB. Since most POM would sell their PKS (to domestic users or for export), and it is unsuitable to use 100% EFB to fuel the boiler for steam and power generation (Alaw & Sulaiman, 2019), MF is the only viable option for biofuel in the POM.

4.3 Palm Kernel Shell (PKS) and Its Opportunities

Kernel is found within each palm fruit and is "protected" by the PKS (see Fig. 19). Many POMs have a kernel processing plant where the shell is cracked before the

Fig. 20 Palm kernel shell in its raw form

kernel can be processed into kernel oil. The cracked shell is then separated from the kernel seed using a clay bath or hydro-cyclone methods. The collected PKS is then dried off (Fig. 20).

Like coconut shells, PKS contains high calorific value which can be used as fuel for energy generation in replacement of coal. The incineration of PKS results in ashes that can be used as fertilizers for crop production. It has been reported that roughly 5.5% of PKS is produced of each ton of FFB being processed. For year 2020, it is estimated that 5.28 million tons of PKS is being produced as biomass (Fig. 13).

PKS contains high calorific value which makes it a good fuel source for the generation of steam and energy. The POMs may sometime reserve some PKS for own consumption, when their boilers are commissioned, or when they need a boost of temperature.

PKSs have a high acquisition need and their selling price is predominantly due to domestic demand as supplementary fuel in high-energy industries such as the manufacturing of rubber gloves, bricks, cements, and ceramics. Besides, PKS are highly demanded by foreign users, particularly Japanese biomass power plants (Levinson, 2020). PKS is used as coal substitution in Japan power sector because of its uniform size distribution, easy handling, easy crushing, and limited biological activity due to low moisture content. Hence, it serves as a *good quality biomass fuel* option.

Nevertheless, there are challenges faced by PKS. It was recently reported that all exporting PKS would require a sustainable traceability system in place to secure long-term purchases from Japan (Levinson, 2021). This might impose a tough hurdle for Malaysian biomass exporters. Nevertheless, it is believed that the high demand of PKS will be unshaken for at least the next decade. Figure 21 shows the imports of PKS to Japan from 2014 to 2019. To date, there are 21 biomass power plants in Japan (some are still under construction). Surely this would suggest the logical increase in PKS demand in the near future.

Figure 22 shows some common sections of PKS accumulating and exporting sites commonly found in Malaysia and Indonesia. The pictures show that extra processes are needed to ensure the exported PKS meets certain quality mutually agreed.

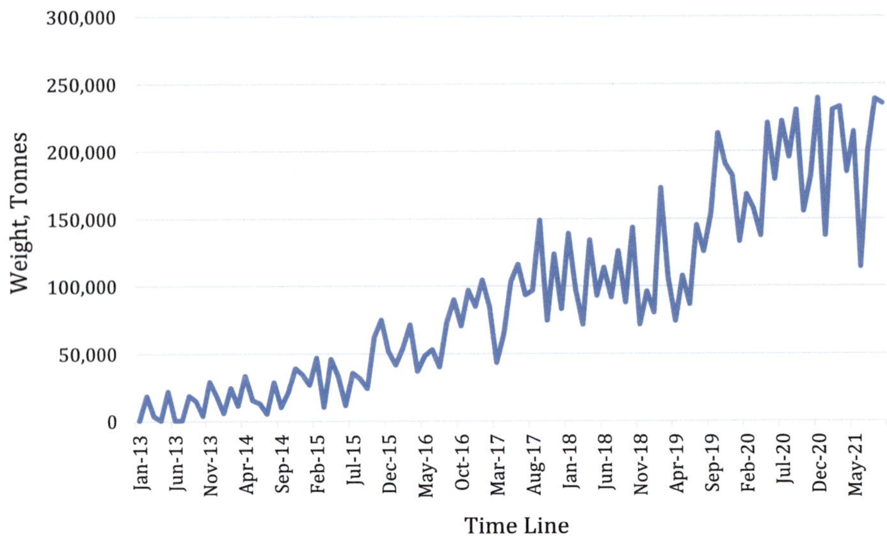

Fig. 21 Japan Imports of PKS, 12 month moving total from 2013 to 2021. *Source* Japanese custom data on H.S. Code 230,660

Fig. 22 Showing PKS outdoor storage, PKS processing facility, PKS weighing bridge, PKS being loaded unto truck and vessel

Apart from being used as energy source, PKS can also be converted into carbon or activated carbon, to be used as water treatment material (Okoroigwe et al., 2013; Razi et al., 2018). It is also used as concrete aggregate to make roads (Ndoke, 2006). Although these are high profit and potential pathways for PKS, the high demand of domestic and foreign markets may refer to these other pathways for quite a foreseeable future.

4.4 *Palm Kernel Cake (PKC) and Its Opportunities*

Palm kernel cake (PKC) (also known as palm kernel expeller, Fig. 23 is a by-product of mechanical extraction of palm kernels during the manufacturing of palm kernel oil. On average, about 5.8% of kernels can be recouped from a ton of FFB being produced (Kamyab et al., 2018).

PKC is a dark brown protein meal with relatively high levels of oil, fiber, and medium-grade protein. This makes it a good source of protein and fat (for energy) and has been mainly used as one of the main components for feed meal formulation for the ruminant industry or rabbit feeding. Note however that it has relatively low adoption for the non-ruminant and poultry industry due to its high fiber content.

4.5 *Palm Oil Mill Effluent (POME) and Its Opportunities*

POME is the only liquid waste generated from the CPO manufacturing process.

It is originated from various milling processes such as sterilization and pressing operations (Fig. 24). Of every ton of FFB processed, 60–67% of POME is being emitted (Akhbari & Kutty, 2019). In the conventional practice, POM would channel this liquid effluent into waste treatment ponds for further settlement process. Several ponds are needed for the treatment process, which contributes to high land footprint (Fig. 25). Besides, these treatment processes release high amounts of methane and CO_2 which contribute to global warming. It has been estimated that about 60 million tons per year of POME (see Fig. 13) has been produced by 451 POMs throughout Malaysia in year 2020. In recent years, POME has been utilized for power generation via the biological pathway. According to Sustainable Energy Development Authority (SEDA) Malaysia, the biogas installed capacity in Malaysia was reported as 123 MW as of the year 2020 (Roadmap and pathway towards low carbon energy system, 2021). It has been targeted that a total of 410 MW of renewable energy is to be generated from biogas by year 2028 according to The National Renewable Energy Policy and Action Plan (Sustainable Energy Development Authority Malaysia, 2008).

Fig. 23 Palm kernel cake/expeller

Fig. 24 A close-up picture of POME originated from the sterilization and extracting processes in a POM

Fig. 25 A bird eye view of a POM with large area for POME treatment ponds

The palm industry is regulated by Malaysian Palm Oil Board (MPOB), which is imposing the mandatory installation of biogas trapping or methane avoidance facilities in palm oil mills as a condition for any new mill construction or existing mills applying for throughput expansion to achieve the targeted goal.

However, less than 100 recovery projects were reported among the 457 POMs in Malaysia that have registered with The Clean Development Mechanism (CDM).

As defined in Article 12 of the Protocol, CDM allows a country with an emission-reduction or emission-limitation commitment under the Kyoto Protocol to implement an emission-reduction project in developing countries (Sarwani et al., 2019).

There is generally a common consensus among the POM operators in their biogas investment for POME management to recover methane gas and generate sufficient power and steam to run their processing facilities. The scheme allows those POMs to be connected to the power grid, so that excess power may be supplied to the grid. Hence, POM operators only use the conventional open pond system to comply with government regulations without biogas capturing is considered as an outdated practice.

Note however that with the current comprehensive promotion of renewable energy policies and frameworks, the current development of biogas industry in Malaysia is still considered slow as compared to other countries. The main problem faced by the mill operators are mostly related to difficulty in financial acquisition. Factors such as high investment cost, new technology, skilled personnel, and termination of CDM program are forcing oil palm companies to adjust their adaption for the renewable energy strategy (Fig. 26).

Apart from biogas generation, POME sludge can also be used as fertilizers in the plantations (Khairuddin et al., 2016). However, the yield is not great. Furthermore, it will take a long time to dispose all the POME that is stored in the existing ponds, due to its daily generation. Since the time needed to produce compost/fertilizers is slower than the POME being generated on an hourly basis, this downstream option clearly is not an efficient way to solve the accumulating POME problem.

Fig. 26 A biogas system—tank digester system, covered lagoon digester system, and a gas engine

5 Strategies to Explore Full Potential of Biomass from Plantation

There are still many obstacles to be overcome to unlock the full potential of palm plantation biomass. In author's opinion, collaboration between industry players and academia is an obvious key for innovation and technological breakthroughs. More conversations are needed to allow both parties to understand challenges and objectives of each other. This will enable the academia to be more business conscious when formulating research problems. On the other hand, industrial players should understand the hurdles and limitations of academic researcher, which then allow them to find the right research team. Through collaboration, both industry players and academia can harness the uniqueness of each other to reach the win–win scenarios.

The author also feels that the industry players may want to consider approaching biomass technology innovation outside the conventional lenses of mechanical approaches. Other engineering disciplines or even the cross-disciplinary approaches may be adopted to unlock the true potential of the abundant biomass, to solve these old industrial problems.

6 Conclusion

Malaysia is blessed with sustainable and abundant of palm biomasses and their potential are yet to be fully unlocked. The challenges are the oldest tales in a modern industrial world. Hurdles faced by the industrial sectors include raw material availability and sustainability, maturity of technology, human capital development, absence of healthy investment and financial eco-system, stable market demand both domestically and foreign, clear guiding hands of policies, among others.

The prominent sustenance of future growth of biomass industry in Malaysia requires sound management of the nation's biomass feedstock. Positive interventions are needed to steer the industry toward high-value creation, strategically and ready position, and clear direction of making Malaysia a biomass regional hub, and hence a regional renewable energy reservoir.

Opportunities are aplenty in the biomass industry. Malaysia can and will enjoy the benefit economically and environmentally from her abundant biomass only when all stakeholders are moving toward a common direction. Other important factors are also necessary, such as the encouragement from government, firmly execute friendly policies, and coupled with innovative SME and local industry. With these joint effects, the potential of the biomass industry can be optimized and in turn, contributes to economic growth through sustainable production.

References

Agensi Inovasi Malaysia (AIM). (2013). National Biomass Strategy 2020: New wealth creation for Malaysia's biomass industry Version 2.0. Available from: https://www.nbs2020.gov.my/nbs2020-v20-2013.

Akhbari, A., Kutty, P. K., Chuen, O. C., & Ibrahim, S. (2020). A study of palm oil ill processing and environmental assessment of palm oil mill effluent treatment. *Environmental Engineering Research, 25*(2), 212–221.

Alaw, F. A., & Sulaiman, N. S. (2019). A review of boiler operational risks in empty fruit bunch fired biopower plant. *Journal of Chemical Engineering and Industrial Biotechnology (JCEIB), 05*(05), 29–35.

Bukhari, N. A., Loh, S. K., Baar, N. A., Luthfi, A. A. I., Adbul, P. M., & Jahim, J. M. (2019). Potential of oil palm trunk (OPT) sap as biofuels. Palm Oil Engineering Bulletin No. 133.

dan Komoditi, K. P. P. (2019). (MPIC) dateset: Monthly fresh fruit bunch (FFB) yield (Tonnes/Hectare), [Internet]. Available from: www.data.gov.my.

Department of Statistics Malaysia (2021b). Products: 1404909100 (Malaysia, I&E) Other than of a kind used primarily in tanning or dyeing and kapok palm kernel shells; Other than of a kind used primarily in tanning or dyeing and kapok palm kernel shells.

Department of Statistics Malaysia (2021a). Products: 4401310000 (Malaysia, I&E) Sawdust and wood waste and scrap, whether or not agglomerated in logs, briquettes, pellets or similar forms: Wood pellets; sawdust and wood waste and scrap, whether or not agglomerated in logs, briquettes, pellets or similar forms: Wood pellets.

Eco Synergy Solutions Sdn. Bhd. (2017). Proposed oil palm plantation development on Lot PT 4310 (649.4 ha) and Lot PT 4314 (1,685.51 ha) In Mukim Ulu Lepar, District of Kuantan, Pahang Darul Makmur. [Research]. Available from: https://enviro2.doe.gov.my.

EU-Malaysia Chamber of Commerce and Industry (EUMCCI). (2017). Oil palm biomass and biogas in Malaysia. [Internet]. Available from: www.eurocham.my.

Ibrahim, Z., Dr. Mokhtar, A., Dr. Aziz, A. A., Dr. Ramli, R., Datuk, D., Choo, Y. M., Dr. Lee, S., Omar, R., & Tabari, A. S. (2014b). Utilization of oil palm trunk (Opt) for the production of medium density fibreboard (MDF) (ITEX 2014b). MPOB International Invention, Innovation and Technology Exhibition 2014b. Available from: http://palmoilis.mpob.gov.my/V4/utilization-of-oil-palm-trunk-opt-for-the-production-of-medium-density-fibreboard-mdf-itex-2014/.

Ibrahim, Z., Aziz, A. A., Ramli, R., Mokhtar, A., Omar, R., & Lee, S. J. (2014). Production of medium density fibreboard (MDF) from oil palm trunk (OPT). *Journal of Applied Sciences, 14*, 1174–1179.

Ismail, A., & Mamat, M. N. (2002). The optimal age of oil palm replanting. *Oil Palm Industry Economic Journal* 2(1).

Jalil, R., & Hossain, N. (2015). Sugar and bioethanol production from oil palm trunk (OPT). *Asia Pacific Journal of Energy and Environment* 2(2)

Kamyab, H., Chelliapan, S.,Din, M. F. M., Rezania, S., Khademi, T., & Kumar, A. (2018). Palm oil mill effluent as an environmental pollutant. In V. Y. Waisundara (Ed.) *Chapter 2 of Palm Oil*, published by Intechopen. 78771, ISBN: 978–1–78923–427–5.

Khairuddin, M. N., Zakaria, A. J., Isa, I. M., & Jol, H. (2016). The potential of treated palm oil mill effluent (POME) sludge as an organic fertilizer. *AGRIVITA Journal of Agricultural Science, 38*(2), 142–154.

Kumneadklang, S., Larpkiattaworn, S., Niyasom, C., & Thong, S. O. (2015). Bioethanol production from oil palm frond by simultaneous saccharification and fermentation. *Energy Procedia, 79*, 784–790.

Levinson, R. (2020). The growing importance of PKS in the Japanese biomass market. Biomass Magazine. Available from: https://biomassmagazine.com.

Levinson, R. (2021). PKS market gears up for Japanese sustainability requirements. Hawkins Wright Ltd. Available from: www.hawkinswright.com.

Lim, S. H., Ibrahim, D., & Omar, I. C. (2012). Oil palm frond to produce bioethanol. *International Journal of Biochemistry and Biotechnology, 1*(1), 001–007.
Malaysia Industry-Government Group for High Technology (MIGHT). MIGHT Anual Report 2019 [Internet], Available from: https://www.might.org.my/download/might-annual-report-2019/.
Malaysia Palm Oil Board MPOB. (2020). Malaysia oil palm statistics 2020, 40th Edn, p. 23.
Malaysian Palm Oil Board (MPOB). (2011). The untapped potential of oil palm biomass and its potential applications, [Internet]. Available from: www.mpoc.org.myf.
Malaysian Palm Oil Board (MPOB). (2021). Oil Palm Planted Area 2020. [Internet]. Available from: https://bepi.mpob.gov.my (Access Jun 2021).
Malaysia Renewable Energy Roadmap. (2021). Pathway towards low carbon energy system, in accordance to Energy Supply Industry 2.0 (MESI 2.0) plan 2019, by Sustainable Energy Development Authority (SEDA) Malaysia, pp. 34, 58.
Ministry of International Trade and Industry. MITI report 2019 [Internet], 2019. Available from: https://www.miti.gov.my/miti/resources/MITI%20Report/MITI_REPORT_2019.pdf
Nambiappan, B., Ismail, A., Hashim, N., Ismail, N., Shahari, D. N., Idris, N. A. N., Omar, N., Salleh, K. M., Hassan, N. A. M., & Kushairi, A. (2018). Malaysia: 100 years of resilient palm oil economic performance. *Journal of Oil Palm Research, 30*(1), 13–25.
Ndoke, P. N. (2006). Performance of palm kernel shells as a partial replacement for coarse aggregate in asphalt concrete, published by Leonardo *Electronic Journal of Practices and Technologies 9*, 145–152
Northport (Malaysia) Bhd. (Apr 2021) South point biomass logistics hub, MBIC—Japan external trade development organization (JETRO) Meeting presentation.
Ofori-Boateng, C., & Lee, K. T. (2013). Sustainable utilization of oil palm wastes for bioactive phytochemicals for the benefit of the oil palm and nutraceutical industries. *Phytochemistry Reviews, 12*, 173–190.
Okoroigwe, E. C., Oformatach, A. C., Oparaku, N. F., & Unachukwu, G. O. (2013). Production and evaluation of activated carbon from palm kernel shell (PKS) for economic and environmental sustainability. *International Journal of Physical Sciences, 8*(19), 1036–1041.
Onoja, E., Chandren, S., Razak, F. I. A., Mahat, N. A., & Wahab, R. A. (2018). Oil palm (Elaeis guineensis) biomass in Malaysia: The present and future prospects. *Waste and Biomass Valorization 10*(28).
Parveez, G. K. A., Hishamuddin, E., Loh, S. K., Ong-Abullah, M., Salleh, K.M., Izuddin, Z. B. M. N., Sundram, S., Hasan, Z. A. A., & Idris, Z. (2020). Oil palm economic performance in Malaysia and R&D progress in 2019. *Journal of Oil Palm Research.* Available from: https://doi.org/10.21894/jopr.2020.0032.
Rasat, M. S. M., Wahab, R., Khalid, I. (2014). Bio-composite lumber from oil palm fronds: An alternative in wood-based industry from utilization of the oil palm agricultural residues. LAP Lambert Academic Publishing, ISBN 978–3–659–63359–1.
Razi, M. A. M., Al-Gheethi, A., Al-Qaini, M., & Yousef, A. (2018). Efficiency of activated carbon from palm kernel shell for treatment of lake and greywater. *Arab Journal of Basic and Applied Sciences 25*(4).
Rosli, F., Ghazali, C. M. R., Abdullah, M. M. A. B., & Hussin, K. (2016). A review: Characteristics of oil palm trunk (OPT) and quality improvement of palm trunk plywood by resin impregnation. *BioResources, 11*(2), 5565–5580.
Roundtable on Sustainable Palm Oil. (2021). https://rspo.org.
Sarwani, M. K. I., Fawzi, M., Osman, S. A., & Nasrin, A. B. (2019). Bio-methane from palm oil mill effluent (POME): Transportation fuel potential in Malaysia. *Journal of Advanced Research in Fluid Mechanics and Thermal Sciences, 63*(1), 1–11.
Singh, R., Lee, K. T., Ooi, L. C. L., Low, E. T. L., Abdullah, M. O., Sambanthamurthi, R., & Azman, I. (2021). An overview of the development of the oil palm industry and impact of the shell gene innovation as a quality control tool to improve productivity. [Internet] *Journal of Oil Palm Research.* Available from: https://doi.org/10.21894/jopr.2021.0001.

Subramaniam, V., Chin, C. M., & Ngan, M. A. (2004). Energy database of the oil palm. Palm Oil Engineering Bulletin No. 70.

Suhaily, S. S., Jawaid, M., Abdul Khalil, H. P. S., Mohamed, A. R., & Ibrahim, F. (2012). A review of oil palm biocomposites for furniture design and applications: Potential and challenges. *BioRes., 7*(3), 4400–4423.

Sustainable Energy Development Authority Malaysia. (2008). National renewable energy policy and action plan, by ministry of energy, green technology and water, Malaysia. Editor., p. 38

Teoh, C.H. (2002). The palm oil industry in Malaysia—From seed to frying pan. WWF Switzerland Study Report, [Internet]. Available from: www.rspo.org

World Bank Group. (2019). Global knowledge and research hub in malaysia, agricultural transformation and inclusive growth—The Malaysian experience, November 2019. The Malaysia Development Experience Series, The World Bank Group 2019 publication. [Internet]. Available at: https://documents1.worldbank.org/curated/en/617611574179512389/pdf/Agricultural-Transformation-and-Inclusive-Growth-The-Malaysian-Experience.pdf.

Zahari, M. A. K. M., Ariffin, H., Mokhtar, M. N., & Salihon, J. (2015). Case study for a palm biomass biorefinery utilizing renewable non-food sugars from oil palm frond to produce poly3-hydroxybutyrate bioplastic. *Journal of Cleaner Production, 87*, 284–290.

Zahari, M. W., Hassan, O. A., Wong, H. K., & Liang, J. B. (2003). Utilization of oil palm frond—Based diets for beef and dairy production in Malaysia. *Asian-Australasian Journal of Animal Sciences., 16*(4), 625–634.

Zulkifli, H., Halimah, M., Chan, K. W., Choo, Y. M., & Basri, M. W. (2010). Life cycle assessment for oil palm fresh fruit bunch production from continued land use for oil palm planted on mineral soil (Part 2). *Journal of Oil Palm Research, 22*, 887–894.

Aviation Biofuels: Conversion Routes and Challenges

Jia Wen Chong, Nishanth G. Chemmangattuvalappil, and Suchithra Thangalazhy-Gopakumar

Abstract According to Energy outlook 2020, the aviation fuel consumption is expected to increase 13% by year 2050, as compared to year 2019. Increased cost of petroleum fuels and growing environmental concerns have made the aviation industry to include biofuel as a possible fuel source. Non-edible oils, energy crops and algae are the common feedstock for jet fuel production. There are several biomasses-to-jet fuel conversion pathways such as lipid hydro-processing, Fischer–Tropsch synthesis, alcohol-to-jet fuel, pyrolysis process, hydrothermal liquefaction and blending of fatty acid methyl ester. However, the sustainable production of jet biofuel is still under research. This chapter discusses the opportunities and challenges of these technologies from sustainable development perspective.

Keywords Sustainable aviation fuel (SAF) · Lipid hydroprocessing (HRJ) · Fischer–Tropsch synthesis (F-T) · Alcohol-to-jet fuel (ATJ) · Hydrothermal liquefaction (HTL) · Standards and specifications

1 Introduction

The aviation industry has grown tremendously over the last few decades, evolving from a luxury to an integral part of commuting. Nowadays, air transport has become a common means of transportation, carrying 4.5 billion passengers around the world (ATAG, 2020). By 2021, the global demand of passengers and cargo flight were forecasted to increase by 50.4% and 13.1%, respectively, as compared to 2020, reaching 4.39 trillion revenue passenger kilometres and 254 billion cargo and mail tonne-kilometres (IATA, 2020a). There was also an increase of 44% in the number of city pairs connections compared to 1998, reaching 14,765 city pairs connections in year 2020.

The global economic impact of the aviation industry is estimated to be $3.5 trillion, which is equivalent to 4.1% of world gross domestic product in year 2019 (ATAG,

J. W. Chong · N. G. Chemmangattuvalappil · S. Thangalazhy-Gopakumar (✉)
Department of Chemical and Environmental Engineering, University of Nottingham Malaysia, Broga Road, 43500 Semenyih, Selangor, Malaysia
e-mail: Suchithra.Thangalazhy@nottingham.edu.my

D. C. Y. Foo et al. (eds.), *Sustainable Technologies for the Oil Palm Industry*,
https://doi.org/10.1007/978-981-19-4847-3_2

2020). In 2019, the aviation industry provided a total of 87.7 million of direct and indirect jobs worldwide. The aviation industry is one of the key players in the global economy, where 35% of all international trades and 58% of international tourist travel are by air. In 2019, there was a total fleet of 33,299 in-service commercial aircrafts, operated by 1,478 commercial airlines globally. According to the statistics, 46.8 million flights were scheduled in 2019, consuming a total of 363 billion litres of jet fuels, which is roughly 8% of the global fuel used (ATAG, 2020). As a result, 914 million tonnes of carbon dioxide (CO_2) were produced in year 2019, representing 2.1% of the global manmade CO_2 emissions. Although the greenhouse gas (GHG) emission from the aviation industry is less significant as compared to other sectors, it is expected to increase dramatically unless further actions are taken (EIA, 2019). For instance, the International Civil Aviation Organisation (ICAO) (ICAO, 2019) forecasted an increment of 2.4–3.8 times the 2015 value, in CO_2 emission from aviation industry by the year of 2050, if no additional measures are taken. In that case, it is predicted that the aviation industry could own up to 25% of the global carbon budget by 2050 (Tabuchi, 2019).

In 2008, the aviation industry came to a mutual agreement on a global, sector-wide climate framework by announcing Carbon Offsetting Reduction Scheme for International Aviation (CORSIA). CORSIA aims to cap the net emissions through the introduction of a carbon–neutral growth target at the beginning of 2020. As a result, there will be a net 50% cut in CO_2 emission of the 2005 level by year 2050. Approximately 2.5 billion metric tonnes of CO_2 is predicted to be mitigated between 2021 and 2035 under the enforcement of CORSIA (IATA, 2019). In December 2015, the Paris Agreement of the United Nations Framework Convention on Climate Change was joined by 197 countries. The goal of the Paris Agreement, is to limit the increases in the global average temperature to well below 2 ºC, preferably to 1.5 °C to that of the pre-industrial levels (UNFCCC, 2015). Thus, the demand to develop alternatives to conventional fossil-derived jet fuel for commercial aircraft has intensified.

Development of Sustainable Aviation Fuel (SAF) has emerged as the only promising solution to reduce GHG emission due to its economic and environmental benefits (Köhler et al., 2014). SAF is well known as the clean substitute for conventional fossil-derived jet fuels. Unlike conventional fossil-derived jet fuels, SAF is generated from sustainable resources such as agricultural residues, waste oils or non-fossil CO_2 through various conversion pathways. The blend of SAF with conventional jet fuel can be applied as a drop-in fuel where no additional modifications are required on the aircraft's infrastructure or engine (SkyNRG, 2021). The key difference of SAF and fossil-derived jet fuels lies in the source of carbon. Carbon that was previously stored in carbon reservoirs was released back into the atmosphere through the burning of fossil fuels. However, almost the same amount of CO_2 was captured from the atmosphere via photosynthesis by the plants that are source of biomass, which make biomass a carbon–neutral energy source (EIA, 2020a). It was reported that the blend of conventional aviation kerosene with 1% synthetic paraffinic biokerosene (SPK-like) fuel was capable to mitigate approximate 0.76–0.81% of the CO_2 emission (Silva et al., 2020). In addition, SAF can also reduce direct emissions of particulate matter (PM) and sulphur oxides while improving the fuel efficiency.

Many international organisations such as Air Transport Action Group (ATAG), International Air Transport Association (IATA) and the European Union (EU) initiated the development of alternative aviation fuel, by funding projects to bring SAF in commercial flights through strategic applications. In addition to the cut down of carbon footprint in aviation industry, SAF also mitigates the risks associated with jet fuel price volatility while reducing the dependency on the petroleum industry (Brooks et al., 2016). According to the report by Midwest Aviation Sustainable Biofuels Initiative (MASBI), jet fuel cost is a significant proportion of an airline's total operating costs, roughly 40% (MASBI, 2013). Hence, airlines are particularly vulnerable to the high volatility of jet fuel price. For instance, the jet fuel price quadrupled from \$0.72 to \$2.98 per gallon between 2002 and 2013, resulted in lower profitability for airlines and have contributed to airline bankruptcies and failures (Dillingham, 2014).

There is various ongoing research on converting renewable biomass feedstocks into liquid fuels which are applicable in the aviation industry. However, strict international standards and regulations for jet fuels narrowed down the possible range of molecules that can be used in jet fuels. The only unimolecular species that are suitable for jet fuel applications are unsaturated, branched or cyclic hydrocarbons ranging from C_{10} to C_{16} region (kerosene) (Jenkins et al., 2016). Although various certifications for blends of SAF up to 50% by volume were granted since 2009, many challenges to widespread commercialisation remain (Gegg et al., 2014). The present review provides an overview of the standards and specification of aviation fuel, classification of biomass feedstock, conversion routes and challenges faced in commercialising the bio-aviation jet fuel.

2 Aviation Jet Fuel's Standards and Specifications

To be used in commercial aviation applications, SAF must first fulfil the requirements described in the relevant specifications published by various bodies that specialise in fuel certification. These certification bodies include international standards associations, national governments and corporations. Typically, these specifications are established to control the chemical and physical properties of aviation jet fuel and allow product to be periodically checked for compliance (IATA, 2012). Commonly used specification and grades for conventional and military aviation jet fuel are shown in Table 1.

It is important for SAF to possess similar fuel characteristics and specifications as conventional aviation jet fuel, requiring no modification or redesign on aircraft's infrastructure and engine, airport infrastructure and supply method. In September 2009, American Society for Testing and Materials (ASTM) International issued the specification ASTM D7566—"Standard Specification for Aviation Turbine Fuel Containing Synthesized Hydrocarbons" for derivation of Synthetic Paraffinic Kerosene (SPK) from coal, gas, and biomass through Fischer–Tropsch (F-T) process. The phrase "Synthetic Paraffinic Kerosene" (SPK) was used to describe the synthetic blending components in aviation jet fuel. This specification covers the

Table 1 Common jet fuel grades and respective standards used around the world

Standards	Jet fuel grade	Locations
Conventional aviation jet fuels		
ASTM D1655—"Standard specification for aviation Turbine fuel" (ASTM 2019a)	Jet A and Jet A-1	US and International
UK Defence Standard (DEF STAN) 91-91 "Turbine fuel, Kerosene type, Jet A-1" (UK Defence Standardization, 2019)	Jet A-1	UK and International
GOST 10227 (ICAO, 2012)	TS-1	Russia, Commonwealth of Independent States (CIS) and parts of Eastern Europe
GB 6537 (ICAO, 2012)	Jet Fuel No. 3	China
Military aviation jet fuels		
US Military Specification MIL-T-83188D (Shell, 2021)	JP-8	US and International
DEF STAN 91–87 AVTUR/FSII (formerly DERD 2453) (Shell, 2021)	JP-8	UK and International
US Military Specification MIL-PRF-5624S (Shell, 2021)	JP-5	US and International
DEF STAN 91–86 AVCAT/FSII (formerly DERD 2452) (Shell, 2021)	JP-5	UK and International

manufacture of aviation jet fuel that consists of conventional and synthetic blending components (ASTM, 2020). In June 2011, SPK from Hydroprocessed Esters and Fatty Acids (HEFA) was approved by ASTM for incorporation into ASTM D7566 as Annex 2 (IATA, 2020b). For every newly approved process, a new annex is incorporated into the amended standard ASTM D7566. As of April 2021, ASTM D7566-20d has a total of seven annexes consisting seven approved alternative conversion pathways for the use on commercial airliners (Green Car Congress, 2020; ICAO, 2020):

(i) Fischer–Tropsch hydroprocessed synthesised paraffinic kerosene (FT-SPK)
(ii) Hydroprocessed esters and fatty acids synthesised paraffinic kerosene (HEFA-SPK)
(iii) Hydroprocessed fermented sugars synthesised iso-paraffins (HFS-SIP)
(iv) Synthesised kerosene with aromatics derived by alkylation of light aromatics from non-petroleum sources (FT-SPK/A)
(v) Alcohol-to-jet synthetic paraffinic kerosene (ATJ-SPK)
(vi) Catalytic hydrothermolysis synthesised kerosene (CH-SK, or CHJ)
(vii) Synthesised paraffinic kerosene from hydrocarbon-hydroprocessed esters and fatty acid (HC-HEFA-SPK).

In accordance with ASTM D7566, the maximum permissible blend fractions of FT-SPK (Annex 1), HEFA-SPK (Annex 2), FT-SPK/A (Annex 4), ATJ-SPK (Annex 5) and CHJ (Annex 6) are restricted to 50% by volume. Meanwhile, HFS-SIP (Annex 3) and HC-HEFA-SPK (Annex 7) are restricted to have 10% maximum blend fractions. The summary of the approved annexes to ASTM D7566 is shown in Table 2 (IATA, 2020b; UK Defence Standardization, 2019).

ASTM D7566 certified SAF can be regarded as specification D1655 turbine fuel upon release from blending with conventional jet fuel, up to 50% SAF (ASTM, 2020). Once the final fuel blend has demonstrated compliance with the relevant product specification such as ASTM D1655 and/or DEF STAN 91-91 (Table 3), it may be regarded as on par with conventional jet fuel for most applications (JIG, 2019).

Table 2 Summary of approved annexes to ASTM D7566

Annexes	Conversion process	Feedstocks	Blending ratio by volume (%)	Timeline
Annex 1	FT-SPK	Biomass, coal and natural gas used for syngas production	50	Sept 2009
Annex 2	HEFA-SPK	Bio-oils, animal fats and recycled oils	50	Jun 2011
Annex 3	HFS-SIP	Biomass (sugarcane and sugar beet) used for sugar production	10	Jul 2014
Annex 4	FT-SPK/A	Biomass, coal and natural gas used for syngas production	50	Nov 2015
Annex 5	ATJ-SPK	Starch/sugar-rich feedstocks or cellulosic biomass (sugarcane, sugar beet, saw dust and straw) used for ethanol and iso-butanol production	50	Apr 2016
Annex 6	CHJ	Fatty acids and fatty acid esters, or various lipids from plant and animal fats, oils and greases such as soybean oil, jatropha oil, camelina oil, carinata oil and tung oil	50	Feb 2020
Annex 7	HC-HEFA-SPK	Oil produced from algae	10	May 2020

Table 3 Parts of Jet A-1 requirements specified by ASTM D1655 and DEF STAN 91–91

Fuel property requirement	Limits	Test Methods	
		IP	ASTM
Visual appearance	Clear, bright and visually free from solid matter and un-dissolved water at ambient fuel temperature	N/A	N/A
Maximum particulate contamination	1 mg/L	423	D5452
Maximum total acidity	0.015 mg KOH/g	354	D3242
Maximum total aromatics	26.5% v/v	436	D6379
Maximum total sulphur	0.3% m/m	336	D1266, D2622, D4294, D5453
Minimal flash point	38 °C	170, 523	D56, D3828
Density at 15 °C	775–840 kg/m^3	160, 365	D1298, D4052
Maximum freezing point	−47 °C	16, 435, 528, 529	D2386, D5972, D7153, D7154
Maximum viscosity at – 20 °C	8 mm^2/s	71	D445, D7945, D7042
Minimum specific energy, net	42.8 MJ/kg	12, 355	D3338, D4809
Minimum smoke point	18 mm	598	D1322
Maximum naphthalene content	3% vol	N/A	D1840
Copper strip corrosion, 2 h at 100 °C	No. 1	154	D130
Thermal stability maximum filter pressure drops	25 mm Hg	323	D3241
Maximum existent gum	7 mg/100 ml	540	D381
Maximum BOCLE wear scar diameter	0.85 mm	N/A	D5001
Maximum antioxidant additives	24 mg/l	N/A	N/A

3 Classification of Feedstock for Bio-Jet Fuel Production

Biomass feedstock is referred to the organic material, waste or residues that can be used to produce energy. These biological materials originate from agriculture, including plants and animal wastes as well as aquatic and industrial waste (OJEU, 2009). Almost all biomass feedstocks can be converted to energy fuels by biochemical or thermochemical conversion routes.

3.1 *Wastes/Low-Valued Co-Products*

A large number of agriculture residues are generated annually, especially in agriculture-based countries like China and India (Ross, 2019). However, only 27% of the agricultural residues generated were further processed as fuel and wood products (DOSM, 2019). On the other hand, biomass wastes generated from wood processing industry come mainly from sawmilling and logging activities such as sawdust, wood chips, damaged or unwanted stem wood and logs. Animal fats such as beef tallow, chicken fats and pork lard are collected as by-products from meat processing facilities and rendering processes. Usage of animal fats as biodiesel feedstock offers advantages in economic, environmental and food security aspects over the used edible vegetable oils. However, the high concentration of saturated fatty acid and free fatty acid (FFA) in animal fats require production techniques with higher complexity, resulting in biodiesel with lower physical and chemical quality. High FFA content also leads to soap formation during the base-catalysed transesterification process due to the reaction between FFAs and base catalyst. This results in the loss of catalyst and ester products which increases the production costs (Encinar et al., 2011). On the other hand, low unsaturation of FFAs offers several advantages which include high calorific value, high cetane number and high oxidation stability (Adewale et al., 2015).

Waste cooking oil (WCO) is easily available from restaurants, cafeterias and household kitchens. It was estimated that approximately 25 million tonnes of WCO are generated globally, where the United States alone generates 10 million tonnes of WCO annually (Yaakob et al., 2013). WCO collected from kitchen can be categorised as either yellow grease or brown grease. Yellow grease means used cooking oil, spent shortenings or any other vegetable oil, with FFAs content of less than 15% (Goering et al., 1982). It is considered as a "clean" type of grease with little to no contamination and can be recycled. On the other hand, brown grease includes the fat, oil and grease materials collected in grease traps or waste traps, with FFAs content of more than 15% (Goering et al., 1982). Brown grease is often associated with issues such as food and trash contamination, heavy emulsification, foul odour and cold flow. WCO is reported to be a cheaper alternative of biodiesel feedstock compared to vegetable oils, lowering the biodiesel production cost by 60–90% (Talebian-Kiakalaieh et al., 2013). Figure 1 illustrated the examples of feedstocks categorised under wastes and low-valued co-products.

3.2 *Dedicated Energy Crops*

Dedicated energy crops are crops grown specifically for energy production which require low cost and maintenance. Energy crops can be generally categorised as grassy, woody and oil crops (Fig. 2). Both fast-growing woody and grassy plant species like poplar, willow, miscan thus and switchgrass are cultivated in perennial

Fig. 1 Type of wastes/low-valued co-products

plantation. Typical rotation periods for woody plants species are between three to seven years and one year for grassy plant species (Rosillo-Calle et al., 2008). Grassy plant species can be classified as high yield and low-energy input crops (McCalmont et al., 2017). Switchgrass is a promising biofuel feedstock as it can be cultivated on land with minimal agricultural value while adaptable to a wide range of climates. On the other hand, miscan thus is a perennial, warm-season Asian grass with rapid growth rate and low fertiliser and pesticide inputs (McCalmont et al., 2017).

Oil energy crops are the base feedstocks for biodiesel production. Grown mainly in Malaysia and Indonesia, oil palm is reported to be the most efficient oil seed crop in the world. Approximately 3.95 tonnes of palm oil and 0.47 tonnes of palm kernel oil can be produced with 1 hectare of land (MPOC, 2020). Based on the total yield of about 4.5 tonnes per hectare, oil palm is 10 times more productive than soybean (MPOB, 2020). Unlike soybean and rapeseed, palm oil is a perennial crop which starts bearing palm fruits for oil about three years after planting. Besides, palm oil has a relatively long productive lifespan of 25–30 years (Kurnia et al., 2016). Crude palm oil refers to palm fruit oil extracted from the fruit's flesh. Crude palm oil will be sent to palm oil refineries to be refined, followed by conversion to methyl esters. Methyl esters from palm oil can be directly used as biodiesel or by blending with petroleum diesel. They also exhibit low engine emission and high oxidation stability (Mekhilef et al., 2011).

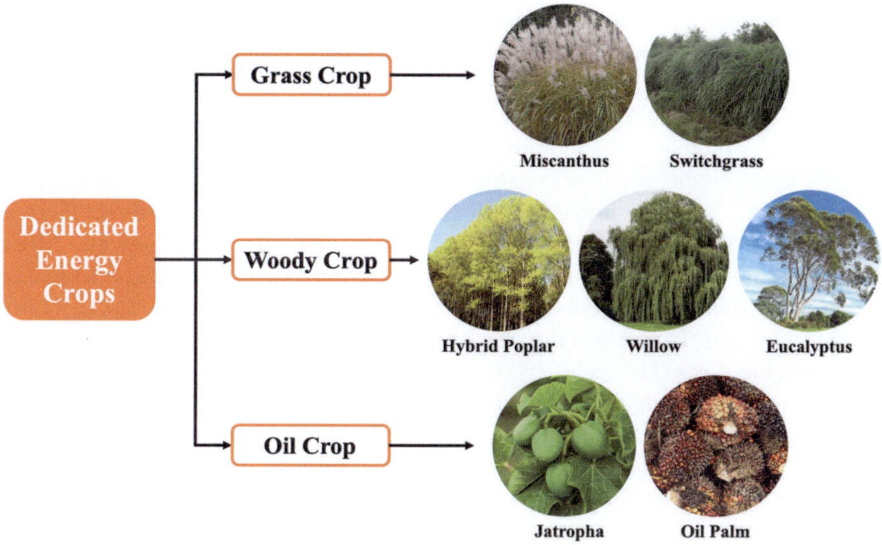

Fig. 2 Type of dedicated energy crops

3.3 Algae

Algae are very diverse, ranging from microscopic (microalgae) to large seaweed (macroalgae), as shown in Fig. 3. They can be found almost everywhere on earth. In recent years, algae have emerged as one of the most promising alternative sources for biodiesel production for their high lipid content, high rate of CO_2 absorption, higher growth rate and productivity (Kandaramath Hari et al., 2015).

Microalgae are capable to reproduce themselves through photosynthesis, converting sunlight, CO_2 and nutrients such as nitrogen and phosphorus into biomass feedstock. A complete growth cycle of microalgae takes only a few days, resulting in higher biomass productivity and oil yield compared to other crops (Ahmad et al.,

Fig. 3 Type of algae biomass feedstock

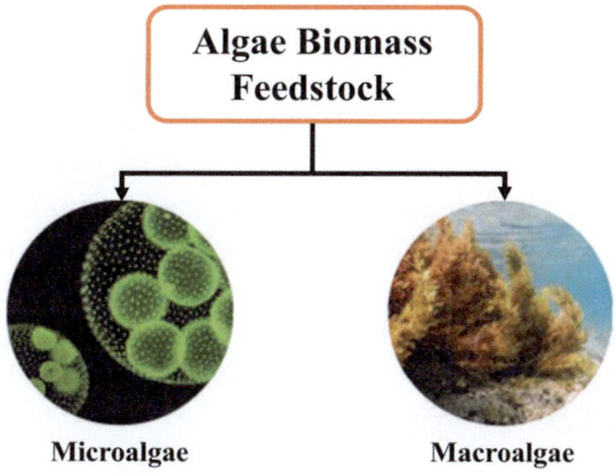

2011). Moreover, they can be grown almost everywhere and can be harvested all year. Algae can be grown using water resources such as seawater, brackish water and wastewater which are unsuitable for human consumption and thus reduce the food–fuel conflict (Mata et al., 2010). Microalgae with an oil production of at least 70 wt% of dry biomass only requires 0.1m^2 year per kg biodiesel of land to produce 121 104 kg of biodiesel per year (Ahmad et al., 2011). In recent years, the focus of research in this area has shifted from direct extraction of lipids to the direct thermal processing and/or fermentation of algae biomass into SAF (Chuck et al., 2016). Algae biodiesels are reported to be zero sulphur while having the same performance as petroleum diesel with reduced emissions of PM, carbon monoxide (CO), hydrocarbons and SO_x. However, emission of nitrogen oxides (NO_x) may be higher in certain engine types (Delucchi, 2003).

Similar to microalgae, seaweed (macroalgae)-derived SAF is also gaining increasing attention as a potential feedstock. Seaweed biomass can be produced either by cultivation and harvesting or by collecting wild drift seaweed, followed by dewatering process (Singh et al., 2017). In 2017, 30 million fresh weight tonnes of seaweed were produced globally as biomass supply (Buschmann et al., 2017). Chuck et al. (2016) estimated that production from offshore farms could achieve 110 EJ, coastal farms 35 EJ and open sea colonies could even reach 6000 EJ. However, the carbohydrates content of seaweed is fairly low while the ash content is significant, which leads to lower efficiency in conversion through thermal processing and fermentation.

4 Conversion Routes of Bio-aviation Fuel

4.1 Hydroprocessed Renewable Jet Fuel

Hydroprocessed renewable jet fuel (HRJ) is a high energy SAF which can be produced by hydro-processing of vegetable oils, animal fats, waste grease, algal oil, and bio-oil. HRJ fuel is free of aromatics and sulphur while exhibit high cetane number, high thermal stability, low tailpipe emission and reducing greenhouse gas emission (Gong et al., 2010). These fuels are also reported to be stable and resistant to microbial growth (Aatola et al., 2008). However, the absence of oxygen and sulphur decreases its lubricity, which can be improved by blending with conventional jet fuel or with the addition of additives (Kandaramath Hari et al., 2015). In the past, HRJ fuel has been used for commercial and military flight. For instance, Air France, KLM and Lufthansa have performed commercial passenger flight with blends of HEFA fuel (Winchester et al., 2013). In June 2011, the conversion route for HEFA-SPK has been approved for SAF production. According to ASTM D7566 Annex 2, the blending ratio of HEFA-SPK in jet fuel can reach up to 50% by volume (ICAO, 2020).

Other than commercial application, various studies were conducted to produce HRJ fuel from palm oil feedstock. A pilot plant scale of hydro-processing conversion technology was proposed to produce renewable diesel from crude palm oil (Guzman et al., 2010). High paraffinic renewable diesel with promising cetane index was obtained from the hydro-processing of palm oil using conventional hydrotreating catalyst such as $NiMo/\gamma Al_2O_3$. Other than crude palm oil biomass, refined palm oil such as degummed palm oil, and palm fatty acid distillate were also used to produce diesel fuel through hydro-processing conversion process (Kiatkittipong et al., 2013). The catalytic hydro-processing reaction was performed in a small shaking batch reactor using Pd/C and sulphided $NiMo/\gamma Al_2O_3$. Pd/C showed better catalytic activity for fatty acid feedstocks and NiMo/γ-Al2O3 better catalytic activity for triglyceride feedstocks as compared to their counterparts. Similar approach was also applied in the work of Wang et al. (Wang & Hsieh, 2020). In this study, catalytic hydro-processing of palm oil over Pd/C and $NiMo/\gamma Al_2O_3$ catalysts for producing straight chain C_{15}–C_{18} alkanes was carried out in a fixed bed reactor. In another study, several Ni-loaded zeolite catalysts such as Ni/SAPO-34, Ni/MCM-41, Ni/HY, Ni/SAPO-11 and Ni/Hbeta were used to produce HRJ fuel via hydro-processing of palm oil (Li et al., 2016). Among the catalysts studied, the Ni/SAPO-34 catalyst demonstrated the highest alkane selectivity and lowest arene selectivity.

One advantages of HRJ fuel is the less vulnerability of fuel to oxidation, making it to stand out as a suitable SAF (Doliente et al., 2020). This is a crucial factor for jet fuels as it helps to prevent the contamination of the fuel supply due to oxidation. Unlike fatty acid methyl ester (FAME) fuel in which the fuel quality is heavily dependent on the choice of feedstock, quality of HRJ fuel is independent on the feedstock used. Furthermore, less reactive soot was emitted when blend of Jet A-1 fuel with 35% of HRJ fuel was used at ground idle and climb-up engine thrust (Liati et al., 2019). As aircraft soot has a significant impact on air pollution, usage of SAF blend with HRJ fuel can prove to be beneficial in improving the air quality at airport areas.

4.1.1 Hydro-Processing

Today, hydro-processing conversion technologies are widely used in oil refineries due to its high maturity level and commercial availability (Sotelo-Boyas et al., 2012). Hydro-processing conversion is a catalytic process that converts liquid-phase unsaturated fatty acids into saturated fatty acids with the addition of hydrogen (Kalnes et al., 2010). Hydro-processing conversion involves reactions such as hydrogenation, dehydrogenation, cracking, isomerisation, cyclisation, and aromatisation occur during lipid hydro-processing. These reactions can be categorised into two main reactions: hydrotreating and hydrocracking (Sinha et al., 2016), as shown in Fig. 4 (Wang, 2016).

(a) *Hydrotreating Reaction*

Hydrotreating reaction, also known as the hydrofining reaction, involves non-destructive hydrogenation. By undergoing catalytic hydrogenation, the carbon double

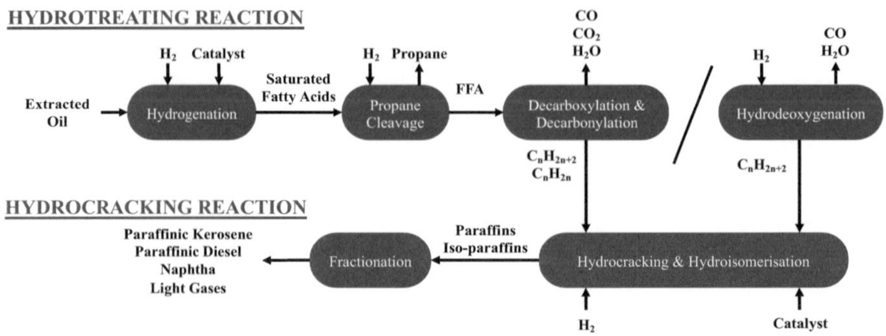

Fig. 4 General process flow of hydro-processing conversion process

bond, C = C in the fatty acids is converted into carbon single bond, C–C. This is then followed by cleaving of the propane backbone which produces three moles of FFAs. At this stage, the glycerol part of triglyceride molecule is converted into propane with the addition of hydrogen (Wang, 2016). The alternate option for conversion of glycerides to FFAs is the thermal hydrolysis (Wang & Tao, 2016; Wang et al., 2012a, 2012b). This approach converts crude lipids into FFAs and glycerol with the addition of water. One mole of triglyceride reacts with three moles of subcritical water, producing three moles of FFAs with one mole of glycerol (Wang et al., 2012c).

The process continues with alkane production through direct decarboxylation and decarbonylation or hydrodeoxygenation (Chiaramonti et al., 2017). Long hydrocarbon chain with carbon number between 8 and 22 is generated depending on the feedstock used (Chiaramonti et al., 2017). In direct decarboxylation, carboxyl group is removed from the saturated fatty acids, producing alkane and releasing CO_2 (Eq. 1) (Snåre et al., 2006). Meanwhile, direct decarbonylation produces an alkene via removal of carboxyl group, by forming CO and water (Eq. 2) (Stepacheva et al., 2016):

$$C_nH_{2n+1}COOH \leftrightarrow C_nH_{2n+2} + CO_2 \tag{1}$$

$$C_nH_{2n+1}COOH \leftrightarrow C_nH_{2n} + CO + H_2O \tag{2}$$

Additionally, fatty acids can be deoxygenated via hydrodeoxygenation by adding hydrogen to produce linear hydrocarbons. The reactions occur during hydrodeoxygenation are shown in Eqs. 3 and 4 (Stepacheva et al., 2016):

$$C_nH_{2n+1}COOH + H_2 \leftrightarrow C_nH_{2n+2} + H_2O \tag{3}$$

$$C_nH_{2n+1}COOH + H_2 \leftrightarrow C_nH_{2n+2} + CO + H \tag{4}$$

Hetero-elements in hydrocarbons such as oxygen, nitrogen and sulphur were removed as water, ammonia and hydrogen sulphide, respectively (Kröger & Müller-Langer, 2012). Mild operating conditions with a temperature of ranging between 350 °C and 450 °C and elevated partial hydrogen pressure are employed (Farhat & El Bassam, 2005). Hydrotreating reaction is carried out with catalyst with hydrogenation functionality, such as Pd, NiMo and CoMo in sulphide state supported over nonacidic support like γ–Al_2O_3 or activated carbon (Farhat & El Bassam, 2005; Krár et al., 2010; Liu et al., 2009; Morgan et al., 2010; Murata et al., 2010; Rozmysłowicz et al., 2010).

(b) *Hydrocracking Reaction*

The final bio-jet fuel produced should possess high flash point and good cold flow properties as set in the jet fuel specification. Thus, isomerisation and cracking of alkanes produced from decarboxylation and hydrodeoxygenation is required to produce a SPK product with carbon length ranging from C_9 to C_{15} (Pearlson, 2011). Hydrocracking reaction involves destructive hydrogenation where triglyceride components with higher molecular weight are cracked into smaller and lighter hydrocarbon distillates. The cracking and isomerisation reactions can be performed as concurrent or sequential (Dunn, 2010). However, studies have shown that isomerisation of straight chain alkanes takes place first, followed by the cracking process (Wang & Tao, 2016). During the isomerisation and cracking process, straight chain hydrocarbons are converted into branched structure through cracking and saturation of paraffins. However, over cracking often results in lower yields of jet fuel range alkanes and higher yields of product with short carbon length, ranging from C_1 to C_8 (Wang, 2016). Hydrocracking reaction is catalysed by bi-functional catalyst supported by crystalline supports (zeolites), amorphous supports (silica-alumina), silico-aluminophosphates and titonasilicates with both hydrogenation and acidic functionality incorporated in it (Kumar et al., 2010; Tiwari et al., 2011; Verma et al., 2011). High operating temperature and usage of high hydrogen pressure are required in order to minimise the condensed chain polymerisation reactions which lead to coke formation and catalyst deactivation (M. Farhat & El Bassam, 2005). Finally, the hydro-processing conversion technologies end with a fractionation process, which separates the mixtures to paraffinic kerosene (HRJ-SPK), paraffinic diesel, naphtha and light gases.

One of the major challenges faced in the hydro-processing technology is the high hydrogen requirement. In hydro-processing of biomass feedstock, the requirement of hydrogen is approximately two to three times higher than that in typical hydrocracker due to the high concentration of unsaturated components originated from feedstock and formed during cracking reactions (Sinha et al., 2016). In addition, other reactions such as hydrodeoxygenation and high concentration of unsaturated components in feedstock also contributed to the high hydrogen consumption. Besides, highly exothermic nature of hydro-processing process often results in excess cracking and coke formation reaction in the catalyst pores. This phenomenon can further lead to problems such as high-pressure drops, low catalyst life and high deactivation rate.

Thus, process design including process parameters, reactor type and catalyst properties need to be tailored to increase the product yield. Furthermore, feedstock with high viscosities, densities and bulky structure often leads to low selectivity in hydroprocessing reaction, causing low yield of the desired product. Impurities such as metals and large amount of phosphorus can be found in most of the vegetable oil feedstock may cause catalyst poisoning. Thus, pre-treatment like water degumming process is required prior to hydro-processing process to completely remove metals and impurities from the feed (Sinha et al., 2016).

4.2 Fischer–Tropsch Synthesis

Fischer–Tropsch (F-T) synthesis is a conversion technology which converts synthesis gas (syngas) over a heterogenous catalyst, producing oil product that can be further refined to conventional transportation fuels and petrochemical products. Syngas that is required as feed material for the F-T synthesis process can be produced by gasification of coal, biomass, organic waste or any other carbon-based raw material or by the reforming of natural gas. The products produced by the F-T synthesis commonly referred to as synthetic crude oil (syncrude) (de Klerk, 2013). Similar to conventional crude oil, further refining is required for the F-T syncrude to produce final products (de Klerk, 2011). Conversion of biomass feedstock to SAF via F-T synthesis consist of three main process steps, which are the conversion of biomass to syngas, conversion of syngas to syncrude and syncrude refining to aviation fuels (Fig. 5) (de Klerk, 2016).

F-T fuels are typically ultra-clean fuels, free of sulphur and aromatic compounds (Wang & Tao, 2016). F-T fuels also display low emission of CO, nitrogen oxides, hydrocarbons and PM (Kim et al., 2009), thus making it one of the most promising alternative green fuel. Various studies have shown that applications of F-T fuels in diesel engine results in significant reduction in CO with no loss of efficiency, total hydrocarbon and PM emissions than those of an engine fuelled with conventional petroleum diesel fuel (Armas et al., 2010; Czerwinski et al., 2007; Krahl et al., 2009; Lapuerta et al., 2010; Schaberg et al., 2000). Under the same operating conditions,

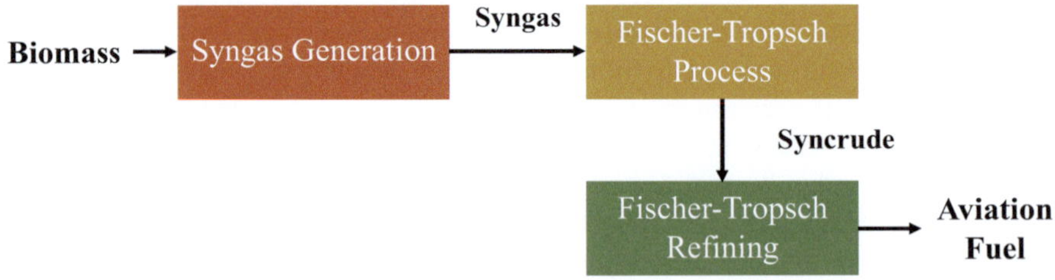

Fig. 5 General biomass-to-liquid process based on Fischer–Tropsch synthesis to produce aviation fuels

Table 4 General information of high-temperature and low-temperature FT synthesis

F-T synthesis process	High-temperature F-T	Low-temperature F-T
Operating temperature (Saeidi et al., 2014)	300–350 °C	200–240 °C
Type of catalyst used (Saeidi et al., 2014)	Fe	Fe/Co
Reactor types (de Klerk, 2020)	Fluidised bed	Fixed bed, slurry bubble column
Synthetic oil composition (de Klerk, 2016)		
Alkanes (paraffin)	20–30%	> 70%
Cycloalkanes (naphthene)	< 1%	< 1%
Alkenes (olefin)	> 50%	15–20%
Aromatics	1–5%	< 1%
Oxygenated	10–15%	~ 5%

lower measurement of exhaust particle number concentration and size distribution from an engine fuelled with F-T fuels can be obtained, with respect to those obtained with engine fuelled with conventional diesel (Li et al., 2007).

F-T fuels of different carbon number and product composition can be synthesised with different catalyst type, operating temperature and reactor type. Of these three differentiating features, operating temperature has the most influence on the final F-T product composition. Despite the diversity of F-T process condition, industrial F-T process can be classified into two main types based on their operating conditions, as shown in Table 4 (de Klerk, 2013, 2016).

Up to date, limited studies were conducted for producing biofuels from palm oil-based biomass via the F-T synthesis process (Laohalidanond et al., 2006). However, the F-T synthesis is one of the most promising conversion routes in converting syngas to biofuels from oil palm biomass feedstock (Chew & Bhatia, 2008). In addition, the production of liquid fuels from Tunisian palm residue via F-T process was discussed and compared by Hnich et al. (Ben Hnich et al., 2020). Computational modelling tools such as process design simulation, chemical reactions investigation, along with energy and exergy analyses were applied. Among the case studies conducted, the F-T process added with auto-thermal-reforming section with oxygen to carbon ratio of 1, and steam to carbon ratio of 2.25 was found to be the most efficient in terms of F-T liquid fuels productivity.

4.2.1 Syngas Generation

In the first step, syngas is produced from selected biomass feedstock through gasification process. Prior to this, pre-treatment step such as drying, size reduction, screening and slurry may be performed depending on the properties of raw biomass feed and the requirements of the selected gasification technology. As gasification is a partial oxidation process, oxygen is co-fed into the gasifier as oxidant feed for gasification process. Additionally, steam is co-fed into the gasifier as a source of hydrogen. The final syngas composition can be controlled by the ratio of steam to raw material feed (de Klerk, 2013). High operating temperature, ranging from 900 °C to 1600 °C, is required due to the endothermic nature of gasification reaction. At the gasification temperature, all volatile elements will be converted to raw syngas, while non-volatile elements will be ended up either as soot, ash or slag. Depending on the raw biomass source feed, raw syngas produced from gasification contains many other compounds. For instance, if the raw biomass fed contain sulphur or nitrogen, the raw syngas would contain hydrogen sulphide and ammonia. However, pure synthesis gas should only consist of hydrogen and CO. Furthermore, some of the contaminants contained in raw syngas are poisonous to F-T catalyst. Thus, syngas cleaning is required to remove these contaminants, such as hydrogen sulphide, halogenated compounds, CO_2 and water, before it can be used in the next stage, the F-T synthesis.

4.2.2 F-T Synthesis

The F-T process is an indirect liquefaction process, and it is less dependent on the nature of raw biomass feedstock, where the difference in fuel properties is mainly due to the operating conditions (de Klerk et al., 2013). The cleaned syngas from any source can be used in the F-T synthesis process. The catalytic conversion of syngas through F-T synthesis is a stepwise reaction. At the first stage, CH_2 intermediates (Eq. 5) representing a methylene group of a saturated hydrocarbon were formed. These intermediates are then converted into a larger chain by the surface polymerisation (Douvartzides et al., 2019). Equations 6 and 7 represent the catalytic conversions of syngas to straight saturated hydrocarbons and unsaturated hydrocarbons (Gruber et al., 2019). However, the exact composition of the product varies depending on the H_2/CO molar ratio, thermodynamic conditions and catalyst used.

$$2H_2 + CO \rightarrow CH_2 + H_2O \tag{5}$$

$$(2n + 1)H_2 + nCO \rightarrow C_nH_{2n+2} + nH_2O \tag{6}$$

$$(2n)H_2 + nCO \rightarrow C_nH_{2n} + nH_2O \tag{7}$$

Application of iron catalyst is favoured in the F-T process due to its availability and low cost as compared to any other active materials. Iron catalyst functions well in both low-temperature (~230 °C) and high-temperature (~340 °C) operations, resulting in products with a lower alkane-to-alkane ratio (Müller-Langer et al., 2017). On the other hand, cobalt catalyst was also used in the F-T process, along with noble metals and metal oxides as catalyst promoters. Cobalt catalyst generally shows a longer

Table 5 Summary of product phases produced by F-T synthesis

Product phases (Mass basis)	High-temperature F-T (%)	Low-temperature F-T (%)
Gases (C_1–C_4)	20–25	5–10
Oil	20–25	15–20
Wax	0	20–25
Aqueous Organics	~ 5	1–2
Water[a]	45–50	50–55

[a]Closed gas loop—no net water gas shift conversion

lifetime as compared to other catalysts. However, cobalt catalyst costs more than iron catalysts and is only suitable for low-temperature (200 ºC–240 ºC) operations due to excessive methane formation at temperature above 250 ºC (Müller-Langer et al., 2017).

Commercially established F-T reactors can be categorised into three main groups: (i) fixed bed, (ii) fluidised (circulating) bed and (iii) slurry bubble column reactors (Hu et al., 2012). Fluidised-bed reactor exhibits a higher conversion rate per amount of catalyst and thus is preferred in commercial F-T synthesis. Other than that isothermal operation of the fluidised bed allows the process to be carried out at a higher temperature in comparison with fixed bed and slurry reactors (Müller-Langer et al., 2017).

The final composition of the produced syncrude depends only on the operating condition of the F-T synthesis process. The syncrude is a multiphase mixture as shown in Table 5 (de Klerk, 2013). Depending on the F-T technology employed, the syncrude product is typically distributed among gaseous phase, liquid oil phase, liquid aqueous phase and solid waste phase.

The length of hydrocarbon chains in the syncrude is influenced by the catalyst used, pressure and temperature conditions of the process (Kandaramath Hari et al., 2015). The probability of chain growth is high at a lower temperature range of 200 ºC to 240 ºC, resulting in high yields of saturated hydrocarbons and high molecular weight linear waxes. On the other hand, at a higher temperature range of 300 ºC to 350 °C, gasoline (C_5–C_{10}), diesel (C_{11}–C_{19}) together with linear low-molecular-mass olefins are produced (Wang & Tao, 2016). Other than alkanes and alkenes, oxygenated compounds such as alcohols, aldehydes and carboxylic acids along with aromatics and ketones are formed as well (Gruber et al., 2019). F-T synthesis process is a highly exothermic process, and thus, the removal of heat from the reaction is crucial to prevent issues such as overheating, catalyst deactivation and production of undesired methane. However, the F-T synthesis is rather costly due to lower process efficiency of 25–50%, offering lower power and fuel economy (Thomas et al., 2008).

4.2.3 F-T Refining

The third step involves syncrude refining to produce Jet A-1 aviation fuel. Syncrude from F-T synthesis is further refined through refinery processes such as alkene oligomerisation, hydrocracking, hydro-isomerisation, aromatisation/naphtha reforming, aromatic alkylation and hydrogenation (de Klerk, 2016). In alkene oligomerisation, light F-T alkenes are oligomerised to produce highly branched products. On the other hand, hydrocracking and hydro-isomerisation processes convert the high molecular weight waxes into lighter products with shorter chain length and lower boiling points. Products from hydrocracking and isomerisation are then heated and distilled to produce jet fuel, diesel fuel and lubricant (Wang et al., 2016). As the minimum aromatics content for jet fuel is 8 vol%, thus aromatisation/naphtha reforming and aromatic alkylation is required to produce aromatics. Aromatic alkylation converts alkanes in the syncrude into aromatics mainly ranging from C_6–C_{10}. Lastly, hydrotreating is necessary to hydrogenate alkenes from the alkene oligomerisation unit. Refined F-T fuel exhibits clean-burning, low-aromatic and almost zero-sulphur content. F-T fuels are characterised as non-toxic, zero nitrogen oxides emission, high cetane number and reduced particulate emission (Kandaramath Hari et al., 2015). Absence of sulphur in F-T fuels leads to low lubricity, which can be improved by blending with conventional fuel (Kandaramath Hari et al., 2015).

4.3 Alcohol-To-Jet Fuel Process

Alcohol-to-jet fuel (ATJ), also known as the alcohol oligomerisation, is fuel converted from bio-alcohols (Wang & Tao, 2016). Among these alcohols, usage of ethanol in ATJ process is preferred in view of its current production, consumption and global application (Bo Han et al., 2020). Alcohol used in ATJ process can be obtained via various pathways depending on the biomass feedstock. Furthermore, process criteria of biomass to alcohol such as pre-treatment method, microorganism's choice, alcohol yield and process economics are highly dependent on the selection of biomass feedstock (Wang et al., 2016). The biomass feedstock used to produce bio-alcohol may include sugarcane bagasse, corn stover, wheat straw, switchgrass, wood lignocellulosic, microbes, etc. (Canilha et al., 2012; Chung et al., 2014; Sánchez & Montoya, 2013). The produced bio-alcohol can then be further processed into aviation fuel via conventional fermentation of carbohydrates or advanced fermentation using highly modified microbes (Geleynse et al., 2018). Product from ATJ pathway is commonly known as ATJ-SPK. However, only ATJ-SPK produced from either ethanol or bioethanol intermediate is approved by ASTM D7566 as part of Annex A5, up to a 50 vol% maximum blending (ICAO, 2020).

A major benefit of bio-jet fuel via ATJ pathway is that the aromatic content contained is within the permitted range. Although the specification on Jet A fuel limits the maximum total aromatic content at 25%, aromatic content in the fuel is essential for the seal of aircraft and engine, preventing leakage problem (Liu et al.,

2013). SAF with zero or low-aromatic content is not suitable to be applied in aircraft's engine without blending with Jet A-1 fuel. However, SAF with high aromatics content could lead to engine failure and erosion on turbine blades by forming a significant amount of carbonaceous particles (Hemighaus et al., 2006). Hence, it is expected that SAF via ATJ pathway could obtain approval for use without blending (Doliente et al., 2020).

In recent years, bioethanol derived from lignocellulosic biomass gained a huge amount of attention in the petrochemical industry as a component of automobile fuel. Approximate 75% of the bioethanol produced worldwide is being used to power automobiles. Furthermore, the usage of bioethanol-based diesel fuel for automotive has increased since 2001. For instance, the biodiesel consumption in the United States grew from 10 million gallons in 2011 to 1.81 billion gallons in 2019 (EIA, 2020b). Most motor gasoline sold in the United States consist of 10% ethanol by volume (EIA, 2020c). Thus, the commercialisation of SAF via ATJ route may induce the competition on feedstock availability between aviation and automotive industry (Doliente et al., 2020). Besides, one of the technical downside of ATJ conversion process is the lower yield of bio-alcohol production (Wei et al., 2019).

Various studies were conducted to explore the application of oil palm biomass in the production of bioethanol, where bioethanol can be further used to produce ATJ fuel. For instance, a novel approach was proposed to convert oil palm trunk biomass (starch-rich biomass) into bioethanol. This method extracted most of the starch from biomass through autoclaving and subsequent amylolytic hydrolysis prior to hydrothermal pre-treatment. Therefore, this approach would be capable to reduce the operational and capital cost while maximising the ethanol production (Eom et al., 2015). Besides, bioethanol was produced from oil palm frond via simultaneous saccharification and fermentation (Kumneadklang et al., 2015). In another study, the feasibility of ethanol production from oil palm empty fruit bunch via acid impregnation-steam explosion as a pre-treatment technique was conducted. The study concluded that higher ethanol yield can be obtained via separate hydrolysis and fermentation process of pre-treated empty fruit bunch fibre (Siramon et al., 2018). An overview on the applications of oil palm empty fruit bunches as a feedstock for bioethanol production in Malaysia was presented by Derman et al. (2018). Various bioconversion process of empty fruit bunches to bioethanol along with the process analysis were presented in the review paper.

The physicochemical properties of the bioethanol should be comparable with those of existing petroleum-based aviation fuel to ensure the smooth conversion of bioethanol to bio-jet fuel. The ATJ process can be categorised into four main pathways, as shown in Fig. 6 (Morgan et al., 2019).

4.3.1 Alcohol Dehydration

Back in 1960s, catalytic dehydration of alcohol has been applied as a conversion pathway of renewable ethanol to produce ethylene as an alternative source to petrochemically derived ethylene (Geleynse et al., 2018). During the catalytic dehydration

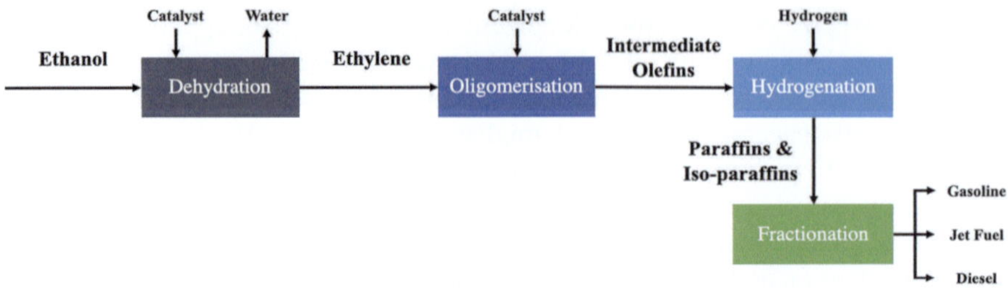

Fig. 6 Overview of ATJ conversion process. Initially, alcohol dehydration is performed to generate olefins, followed by oligomerisation to generate middle distillates in the presence of catalyst. Then, hydrogenation of the oligomerised olefins, also known as the middle distillates, to produce jet fuel ranged hydrocarbons. Finally, the hydrocarbons undergo fractionation to purify the bio-jet fuel product. Short chain alcohol, such as ethanol, butanol and iso-butanol, was commonly used as intermediate feedstock for commercial production of ATJ fuels from biomass feedstock (García-Sánchez et al., 2019). Due to the underlying reaction mechanisms of the dehydration and oligomerisation processes, alcohol with higher carbon number results in a higher theoretical yield through ATJ upgrading although the maximum theoretical carbon yield is identical. However, lower degree of oligomerisation is required for alcohols with higher carbon number, thus lower oligomerisation costs (Geleynse et al., 2018). On the other hand, final product with even distribution of carbon number can be obtained with alcohol of lower carbon numbers. This enables the final product from alcohol of lower carbon numbers to show smoother distillation curve which resembles more to the conventional petroleum-based jet fuel as compared to the product from high alcohol (Geleynse et al., 2018)

of ethanol, hydroxy group (-OH) was removed, generating water as product. On the other hand, the methyl group is deprotonated by the conjugate base of the catalyst, which then rearrange the hydrocarbon into ethylene (Fan et al., 2012). Two reactions occur simultaneously during the dehydration reaction of ethanol (Eq. 8 and 9).

$$C_2H_5OH \rightarrow C_2H_4 + H_2O \quad (+44.9\ \text{kJ/mol}) \tag{8}$$

$$2C_2H_5OH \rightarrow C_2H_5OC_2H_5 + H_2O \quad (-25.1\ \text{kJ/mol}) \tag{9}$$

The operating temperature of this process is fairly high, ranging from 180–500 °C due to the endothermic nature of the main dehydration reaction. Lower temperature favours the side reaction. A variety of catalysts have been explored extensively, such as pure or doped alumina, zeolites, silicoaluminumphosphates and heteropolyacid catalyst for the application in alcohol dehydration pathway in order to increase ethylene yield while lowering the reaction temperature (Fan et al., 2012; Golay et al., 1999; Wu & Wu, 2017). Various studies have shown that complete conversion with ethylene selectivity up to 99.7% can be achieved over these catalysts at moderate operating temperature of 250 °C (Fan et al., 2012; Hu et al., 2010; Silvester et al., 2015; Zhan et al., 2010). Syndol catalyst, which is the specialised ethanol-to-ethylene heterogenous catalyst, has been developed by Halcon SD and applied commercially (Chen et al., 2007).

4.3.2 Olefin Oligomerisation

In the oligomerisation of alkene, the formed ethylene from previous step is converted into higher olefins by using homogenous or heterogeneous catalysts in single or multiple reactor configurations. The oligomerisation of alkene is mainly performed through carbenium ion mechanisms (Carlson et al., 2009). Currently, there are a few commercially developed ethylene oligomerisation processes such as the one-step Zeigler–Natta process (Chevron–Phillips's chemicals), two-step Zeigler–Natta process (INEOS) and the shell higher olefins process (SHOP) (Díaz-Pérez & Serrano-Ruiz, 2020; Geleynse et al., 2018; Lutz, 1986). Direct oligomerisation of ethylene is the simplest available route for ethylene conversion to jet and diesel fuels. Heterogeneous catalysts such as zeolites and acidic mesoporous catalyst are generally preferred in the direct oligomerisation of ethylene (Brooks et al., 2016). The conversion of ethylene can be further improved by using nickel incorporated zeolites, amorphous and mesoporous silica-alumina catalyst (Heveling et al., 1998; Heydenrych et al., 2001; Liu, 2018). However, the selectivity of this process is not high, while significant recycle is required to obtain the C_{10}–C_{16} range olefins. Furthermore, the economics of these reaction have not been favourable, and thus, minimal progression was observed on this process (Brooks et al., 2016). As the direct oligomerisation of ethylene is rather challenging, a two-step oligomerisation of ethylene has been developed as an alternative pathway. In this process, ethylene is converted into a mixture of intermediate olefin ranging from C_4–C_{10} prior to oligomerisation to jet fuel-length olefins. The two-step oligomerisation shows a greater flexibility in product distribution, however, at the cost of increased process complexity and increased branching of heavier fractions (Singh & Naveen, 2017).

4.4 Pyrolysis Process

Pyrolysis involves thermal cracking of biomass feedstock in an inert atmosphere, producing biochar, bio-oil and syngas. Pyrolysis process can be compared to charcoal production that was used for centuries (Jenkins et al., 2016). Final product distribution of pyrolysis is dependent on the operating conditions, such as temperature, vapour residence time, heating rate and feedstock size. However, temperature has the most influence on the final pyrolysis product composition. At lower temperature (< 300 °C) and heating rate, the conversion of biomass to biochar and gaseous products such as CO_2 and water dominates. On the other hand, higher temperature (400 °C – 500 °C) and heating rate promote the formation of levoglucosan, a substance that kick-starts the production of smaller molecules, thus producing liquid products. At temperature above 650 °C, the decomposition of the biomass and liquid products dominates, generating gaseous products such as CO, hydrogen and methane (Müller-Langer et al., 2017). In fast and flash pyrolysis, the high operating temperature and heating rate enable the process to go beyond the temperature range which the first reaction dominates, straight to the second reaction pathway, producing products with mainly

Table 6 Summary of different pyrolysis technologies

Conversion technology	Slow pyrolysis	Fast pyrolysis	Flash pyrolysis
Operating conditions			
Temperature (°C)	300–700	400–800	800–1000
Vapour residence time (min)	10–100	0.5–5	< 0.5
Heating rate (°C/s)	0.1–1	10–200	> 1000
Feedstock size (mm)	5–50	< 3	< 0.2
Product yield			
Bio-oil (wt%)	~ 30	~ 50	~ 75
Biochar (wt%)	~ 35	~ 20	~ 12
Gases (wt%)	~ 35	~ 30	~ 13

liquid fraction. To maximise the bio-oil production, a finely ground feedstock (< 3 mm) is preferred for their low thermal conductivity, which improves the heating rates and heat transfer rates. Besides that, maximum pyrolysis bio-oil yield can be achieved under the operating temperature of around 500 °C and short vapour residence time (< 2 s) while limiting the secondary reactions. A rapid removal of product char and cooling of pyrolysis vapours results in minimal vapour cracking and higher bio-oil yield (Bridgwater, 2012).

Pyrolysis technologies can be categorised into three reaction pathways: (i) slow pyrolysis, (ii) fast pyrolysis and (iii) flash pyrolysis, depending on the operating conditions. Table 6 summarised the operating conditions and approximate product yields for these processes (Nanda et al., 2014). Fast pyrolysis method is commonly preferred in industrial bio-oil production due to its scalability and economics of particle grinding (Jenkins et al., 2016).

4.4.1 Slow Pyrolysis

In slow pyrolysis, biomass feedstock is pyrolysed at lower heating rates (0.1–1 °C/s). Conversion of biomass to char product dominates at slow pyrolysis and thus leads to less liquid and gaseous product. A considerable amount of study has been done on this process. Slow pyrolysis of palm kernel cake was carried out in a fixed bed reactor at 700 °C, heating rate of 20 °C/min, nitrogen gas flow rate of 200 cm^3 per minute to obtained bio-oil yield of 54.3 wt% (Weerachanchai et al., 2011). Ronsse et al. (2013) conducted slow pyrolysis of pine wood, wheat straw, green waste and dried algae and obtained up to 98.4 wt% yield of biochar. Hernandez-Mena et al. (2014) performed slow pyrolysis of woody bamboo in a fixed bed reactor at temperature ranging from 300–600 °C, at a 10 °C/min heating rate. A maximum biochar yield of 80% was attained at 300 °C. However, the yield of biochar decreases as the pyrolysis temperature increases, due to the increased in thermal degradation rate. Similar trend was also obtained for slow pyrolysis of coconut fibre and willow (Cai et al., 2020; Dhar et al., 2020). On the other hand, slow pyrolysis of palm empty fruit bunch was conducted to product bio-oil in a bench scale tubular furnace reactor, with temperature ranging from 400 °C to 600 °C at heating rate of 10 °C/min (Sutrisno & Hidayat, 2018). Maximum bio-oil yield of 44.5 wt% was obtained at 450 °C. The most commonly used reactors in this process are fixed bed and tubular reactor (Canabarro et al., 2013).

4.4.2 Fast Pyrolysis

In fast pyrolysis, biomass is rapidly heated to a high temperature in the absence of oxygen. However, the heating rate of fast pyrolysis is not as fast as it is in flash pyrolysis. In general, product from fast pyrolysis can be categorised as high-grade bio-oil (Canabarro et al., 2013). Gupta et al. (2019) performed fast pyrolysis of teak sawdust in the temperature range of 400 °C to 700 °C. Maximum bio-oil yield of 48.8% was observed at 600 °C. In Varma et al.'s (2019) work, maximum yield of wood sawdust pyrolysis bio-oil (44.16 wt.%) was observed at temperature of 500 °C and heating rate of 50 °C/min. In another study, date palm tree mixture wastes, and date seed biomass were used to product bio-oil by fast pyrolysis. Maximum yield of 68 wt.% date seed bio-oil was obtained at a temperature of 500 °C (Bharath et al., 2020). Fast pyrolysis of oil palm empty fruit bunch for bio-oil production was conducted in a bubbling fluidised-bed reactor (Park et al., 2019). Improved liquid yield was observed by pre-treating the empty fruit bunch via acid washing with dilute nitric acid. Besides, oil palm frond and palm oil empty fruit bunch were also used to produce pyrolysis bio-oil through fast pyrolysis process (Solikhah et al., 2018). The higher heating value of oil palm frond and empty fruit bunch bio-oils were reported to be 12.19 and 26.49 MJ/kg, respectively. Various reactor configurations such as fluidised bed, entrained flow reactor, wire mesh reactor, vacuum furnace reactor, vortex reactor, rotating reactor and circulating fluidised-bed reactor are suitable to perform fast pyrolysis of biomass for bio-oil (Goyal et al., 2008). Among the abovementioned reactors, fluidised bed is the most promising technology in fast pyrolysis as it allows high heating rate, rapid de-volatilisation, easy control, simple char collection and low cost (Luo et al., 2004).

4.4.3 Flash Pyrolysis

Flash pyrolysis is a process in which the reaction time is only a few seconds or even less, which require high heating rate. This involves special reactor configurations such as entrained flow reactor and fluidised-bed reactor, where biomass residence times are only a few seconds (Luo et al., 2004). Biomass particle size for flash pyrolysis should be fairly small as rapid heating is required. The effect of temperature and pressure in flash pyrolysis of palm kernel shell is investigated in Matamba et al.'s work (2020). The study concluded that higher operating temperatures and pressures favoured the generation of polycyclic aromatic hydrocarbons and hydrogen gas. In another work, Maliutina et al. (2017) concluded that maximum bio-oil yield of 73.74 wt. % can be obtained with flash pyrolysis of palm kernel shell at 600 °C. Flash pyrolysis of biomass can achieve up to 75% of bio-oil yield (Jahirul et al., 2012). However, this process has some technological challenges such as corrosiveness of the bio-oil and solids in the bio-oil (Jahirul et al., 2012). Besides, ash content in the bio-oil has negative impact on the quality and stability of the produced bio-oil. Furthermore, occurrence of catalytic repolymerisation reactions inside the bio-oil due to char fines often leads to a higher viscosity (Canabarro et al., 2013).

4.4.4 Pyrolysis Bio-Oil Upgrading

Bio-oil obtained from the pyrolysis of biomass contains several organic compounds such as hydrocarbons, acids, alcohols, ketones, aldehydes, phenols and large molecular oligomers, as well as inorganic species. However, these compounds often lead to poor fuel properties and performance of pyrolysis bio-oil. Problems like high oxygen content, high solids content, high viscosity, chemical instability, acidic and corrosive inhibit its application as jet fuel. Depending on the operating condition and type of feedstock used, the pyrolysis products are typically distributed among gaseous phase, liquid oil phase, liquid aqueous phase and solid phase (Müller-Langer et al., 2017). Pyrolysis oil produced will be in stable mixtures of each state, which is impossible to separate them mechanically. Further suspension or thermal drying process is necessary to remove water from the pyrolysis oil. Upgradation of bio-oil can take place in three different pathways:

(a) *Physical Upgrading of Pyrolysis Bio-Oil*

Addition of polar solvents has been proved to homogenise and reduce the viscosity of bio-oil. Usage of common polar solvents such as ethyl acetate, acetone, methanol and ethanol is extensively studied in the past few decades (Zhang et al., 2019). On the other hand, addition of polar solvents also showed a significant improvement in oil stability, pH value and high heating value. Blending of bio-oil and solvent under supercritical reactions further enhanced the bio-oil properties by promoting reactions such as esterification and hydrogenation, resulting improvement in physicochemical properties of bio-oil (Omar et al., 2019).

In addition, pyrolysis bio-oils are not miscible with hydrocarbon fuels like diesel. However, the bio-oil can be emulsified with diesel or biodiesel with the aid of surfactants. In work done by Liu et al. (2019), emulsified oil with fuel properties such as density, viscosity, corrosivity and heating value is close to those of diesel oil can be produced under optimal emulsification conditions, with a mixture of Span 80 and Tween 80 emulsifiers. Emulsification is a comparatively straightforward technique to upgrade bio-oil with diesel or biodiesel; however, the downside of this technique includes the high cost of surfactants and high energy requirement for this process.

(b) *Chemical Upgrading of Pyrolysis Bio-Oil*

Pyrolysis can act as thermal pre-treatment before gasification via synthesis gas (Müller-Langer et al., 2017). A small amount of energy penalty can be observed from the lower pyrolysis energy efficiency, transportation energy and additional bio-oil gasification stand (Bridgwater, 2009). As reported by Dahmen et al. (2012), gasification of solid and liquid suspensions of pyrolysis product is possible to conduct in a large experimental scale. Synthetic hydrocarbons from gasification include diesel, gasoline, kerosene, liquefied petroleum gas and synthetic natural gas and are suitable for conventional fuel application, but much cleaner (Bridgwater, 2012).

(c) *Catalytic Upgrading of Pyrolysis Bio-Oil*

In conventional refinery fluid catalytic cracking, petroleum fractions with high molecular weight and high boiling points were converted into lighter products like gasoline or gases. The cracking of these heavy distillates was originally carried out by the thermal cracking process, which requires high temperature with high-pressure hydrogen flow (Zhang et al., 2019). However, this process is then replaced by catalytic cracking as higher production yield of gasoline with a high-octane rating can be achieved. Catalytic vapour cracking, also known as zeolite cracking, has reactions like those of conventional refinery fluid catalytic cracking. Zeolite cracking typically takes place between 300 °C and 600 °C at atmospheric pressure over a zeolite catalyst. HZSM-5 is the common catalyst used in this reaction, which is also applicable in fluid catalytic cracking industry (Chen & Yoshikawa, 2018). An increase in temperature results in a decrease in oil production and an increase in gas production (Mortensen et al., 2011). Higher temperature promotes cracking reactions, resulting in the production of smaller volatile compounds. To remove oxygen elements from bio-oil, a certain amount of cracking at high temperature is required. However, if the rate of cracking is too high, at increased temperature, degradation of the bio-oil to light gases and carbon will occur instead.

4.5 Hydrothermal Liquefaction

Hydrothermal liquefaction (HTL), also known as the hydrous pyrolysis or direct liquefaction, converts biomass feedstock into liquid biocrude in the presence of a solvent, with or without the catalyst. In other word, HTL is a thermochemical conversion which converts biomass into liquid fuels by breaking down the solid bio-polymeric structure to liquid components in a hot and pressurised water environment (Gollakota et al., 2018). Operating condition of HTL can be compared to that of the geological formation of fossil fuels. Fossil fuels are formed when biomass that are buried deep within the earth are subject to high temperature and pressures over millions of years, whereas in HTL, liquid biofuel can be obtained within a short time period (Patil et al., 2008). The HTL process operates at a temperature of 250 °C–370 °C under elevated pressures between 10 and 25 MPa and residence times of 10 to 60 min (Müller-Langer et al., 2017). High process pressure sustains the produced solvents in the liquid state and increases the yield of liquid biocrude. However, high process pressure increases the cost of HTL technology. Compared to pyrolysis, HTL process was conducted at a lower operating temperature and heating rate (Jindal & Jha, 2016). HTL is attractive compared to other thermochemical processes where utilisation of wet biomass is possible, with no requirement for biomass drying, thus having higher energy efficiency. Water is the most common solvent used in HTL process but results in a lower yield of biocrude with high oxygen content and low heating values. Higher biocrude yield with low oxygen content can be produced with other organic solvents such as alcohols, phenol, esters, ketones, tetralin and various alcohol–water mixtures (Douvartzides et al., 2019). Homogenous catalysts such as sulphuric acid, oxalic acid, hydrochloric acid sodium hydroxide and metallic salt are commonly used in the process to hinder the formation of char and tar (Douvartzides et al., 2019).

HTL biocrudes are high viscosity dark-coloured semi-fluids with a smoke-like smell (Jena & Das, 2011). Chemical composition of HTL biocrudes relies on the HTL operation conditions such as type of biomass feedstock, temperature, solvent, solvent density, reaction time and gas used as reaction atmosphere (Ramirez et al., 2015). However, the composition of biomass fed into the HTL process has the greatest influence on the final biocrude properties. HTL biocrude is a complex mixture, mainly comprising oxygenated organic chemicals, aliphatic, sugars, oligomers, nitrogenous aliphatic and nitrogenous aromatic. Table 7 shows the main chemical groups for HTL biocrude (Ramirez et al., 2015).

Generally, HTL biocrude exhibits higher viscosity than pyrolysis bio-oil, although HTL biocrude is less dense than pyrolysis bio-oil (Ramirez et al., 2015). In addition, HTL biocrude is deoxygenated by undergoing condensation reaction of light fragment produced from biomass, resulting in a more hydrophobic phase with less dissolved water. SAF produced via HTL process possesses high combustion quality,

Table 7 Chemical groups of HTL biocrude

Main components	Area (%)[a]
Phenolics	6–65
Esters	2–44
Aromatics and heterocyclics	6–35
Aldehydes	0–18
Carboxylic acids	2–40
Ketones	0–38
Alkanes	9–13
Nitrogenates	12–23

[a]Area% from gas chromatography–mass spectrometry results

cold flow properties and stability, which fulfils the ASTM and military jet fuel specifications (Wang & Tao, 2016). Research from Li et al. (2010) and McAfee (2012) shows that production of SAF from various triglyceride-based feedstocks such as camelina, jatropha, soybean and tung oil via HTL process is feasible. Besides, the production of HTL fuel from palm kernel shell and oil palm fronds were investigated as well (Chan et al., 2018; Jadhav et al., 2019). The work done by Chan et al. (2015) involved the subcritical and supercritical HTL of oil palm biomass such as oil palm empty fruit bunch, palm mesocarp fibre and palm kernel shell to produce bio-oil. In another study, the effect of different organic solvent on the HTL of palm oil empty fruit bunch was investigated and found that water as solvent provided the highest conversion rate of 49.14% (Sarwono et al., 2016). In addition, the effect of metal oxide catalyst on the product yields and characteristics of HTL bio-oil from oil palm empty fruit bunch was investigated (Yim et al., 2017).

4.5.1 Reaction Mechanisms

The reaction chemistry and pathways of biomass liquefaction mechanism can be complicated due to the nature of biomass as a complex mixture of carbohydrates, lignin, proteins and lipids (Jindal & Jha, 2016). However, the key pathway of HTL comprises three major steps (Toor et al., 2011), which are the.

(i) Depolymerisation of biomass
(ii) Decomposition of biomass monomers
(iii) Recombination and repolymerisation of reactive fragments.

In general, HTL process decomposed and depolymerised the biomass feedstock into compounds of smaller size, which are usually highly reactive. Thus, polymerising and forming biocrude, gas and solids compounds is essential. Critical HTL process parameters such as temperature, residence time, the process of repolymerisation, condensation and decomposition of the compounds from the different phases may differ accordingly (Gollakota et al., 2018).

In depolymerisation, macromolecules in the structure of biomass feedstock are dissolved and depolymerise into smaller compounds. The intrinsic hemicellulose and cellulose biopolymers further strengthen the thermal stability of the biofuel (Jae et al., 2010). The obstinate properties of lignocellulose biomass that resembles the natural geological processes of forming fossil fuels could poses challenges to the process. However, this phenomenon can be easily overcome by carrying out depolymerisation process (Toor et al., 2011). Structure of long chain polymers containing hydrogen, oxygen and carbon is altered into shorter chain hydrocarbon (Zein & Winter, 2000).

In the decomposition of biomass monomers, dehydration, decarboxylation and deamination are carried out (Jena et al., 2015). Oxygen is removed from the biomass feedstock in the form of water and CO_2 through dehydration and decarboxylation, respectively. This is then followed by hydrolysis, where macromolecules in biomass are converted into polar oligomers and monomers (Reaney & Ratanapariyanuch, 2010). High temperature and pressurised water solvent break down the hydrogen bonded structure of cellulose, forming glucose monomers. High reactivity of fructose accelerates the decomposition of glucose monomers via various reaction pathways, including isomerisation, hydrolysis, dehydration, reverse-aldol fragmentation, rearrangement and recombination reactions (Zhang et al., 2016). The degradation product mainly consisted of polar organic molecules, furfurals, glycolaldehyde, phenol and organic acids which are highly soluble in water (Toor et al., 2011).

In the third step, recombination and repolymerisation occurred due to deficiency of hydrogen compound (Toor et al., 2011). With sufficient hydrogen supply in the organic matrix, free radical activity will be capped yielding the stable molecular weight species. Under condition of high concentration of free radicals or hydrogen deficit, reactive fragments are most likely to recombine or repolymerised, leading to undesirable coke formation (Yatish, 2015).

4.5.2 Upgrading of HTL Biocrude

HTL biocrude is highly oxygenated and possess low heating value as compared to that of conventional petroleum. This often hinders the direct application of biocrudes with conventional petroleum feedstocks. Further upgrading of HTL biocrude to liquid transportation fuels can be done through separation, hydrodeoxygenation, catalytic cracking and esterification. The main objectives of HTL upgrading process are to remove heteroatoms (nitrogen, sulphur and oxygen), reduce viscosity, reduce the acid number and increase heating value using heterogenous catalysts (Jindal & Jha, 2016; Sun et al., 2010).

Outcome of HTL process usually consists of a gas phase portion, a liquid oily fraction, a liquid aqueous fraction and solid residue. In this approach, high-value products such as biocrude are separated out from the mixture product for further processing to produce aviation jet fuel. Besides, existence of water may cause catalyst inactivity, and thus, removal of water content from oil fraction is especially important (Venderbosch et al., 2010). Separation of HTL product can be done by means of solvent extraction and distillation. In solvent extraction approach, selected solvent is added to the two-phase HTL product to enhance the separation and extraction. This is followed by decanting of the liquid product to separate aqueous and oil portions (Peterson et al., 2008). Polar solvents are commonly applied in this approach, such as acetone, ethanol and dichloromethane (Ramirez et al., 2015). On the other hand, HTL product can also be separated via various methods of distillation. The selection of distillation process was based on the physical and chemical characteristic of the product fed and the range of its components (Ramirez et al., 2015).

The heating value of biocrude is highly influenced by the hydrogen and oxygen content. Thus, it is crucial to increase the hydrogen content and reduce oxygen content in order to improve the biocrude's heating value (Huber et al., 2006). Besides, the occurrence of heteroatoms such as sulphur, nitrogen and oxygen increases the chance of catalyst deactivation and coke formation in further processing. By removing these heteroatoms via hydrodeoxygenation process, the saturation of hydrocarbons in biocrude can be increased as well thereby improving the fuel characteristics (Elvers &

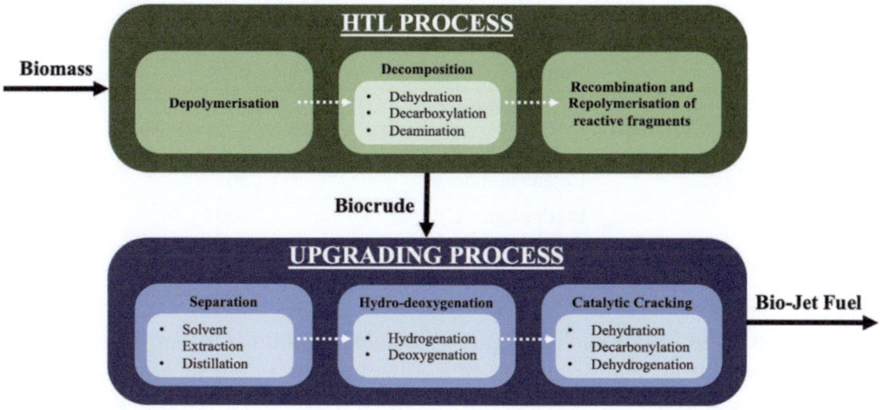

Fig. 7 General HTL conversion and upgrading process to product bio-jet fuel

Schutze, 2007). Hydrodeoxygenation of HTL biocrude involves hydrogenation and deoxygenation processes simultaneously. In this approach, high temperature, high hydrogen pressure and metal catalyst are required to remove oxygenated components from the HTL biocrude. The usage of CoMo and nickel catalyst is common for hydro-processing in oil refineries. However, the selection of suitable catalysts is determined by the properties of crude oil, including metal nitrogen and sulphur content.

The cracking of biocrude results in lighter products with better properties. Cracking can be conducted via thermal cracking and catalytic cracking. However, thermal cracking is not a feasible option as the highly oxygenated compound in biocrudes promotes coke formation (Melero et al., 2011). On the other hand, catalytic cracking is preferable for its enhanced selectivity, moderate operating conditions and lower production of undesirable by-products such as gases and coke (Melero et al., 2011). Furthermore, higher liquid product yield can be observed in catalytic cracking to that of thermal cracking at the same temperature (Ramirez et al., 2015). The commonly used catalysts in the catalytic cracking process include the natural clay materials, synthetic amorphous silica-alumina and synthetic crystalline zeolites (Elvers & Schutze, 2007). Catalytic cracking approach consists of reaction pathways such as dehydration, decarbonylation, dehydrogenation, hydrogenation and hydrogen-transfer reactions (Corma et al., 2007). Figure 7 summarised the pathways of HTL conversion and upgrading process for bio-jet fuel production.

4.6 Blending of Fatty Acid Methyl Ester (FAME) in Jet Fuel

Fatty acid methyl esters (FAME) is a non-toxic, biodegradable biodiesel with chemical formula of $CH_3(CH_2)nCOOCH_3$. Feedstock for FAME's production includes a

Table 8 Comparison of fuel properties

Properties	FAME	Diesel
Density at 20 °C (kg/l)	0.88	0.83
Lower heating value (MJ/kg)	37.1	43.1
Viscosity at 20 °C (mm^2/s)	7.5	5.0
Cetane number	56	50
Flash point (°C)	120	80
Fuel equivalence	0.91	1

variety of fatty acid source such as algae, vegetable oils, animal fats or waste cooking oils (F3centre, 2017). The properties of the final product depend on the composition of the fatty acid feedstock. Generally, both saturated and unsaturated fatty acids exhibit low melting points. However, high concentration of polyunsaturated fatty acids often leads to high oxidation tendency and results in short storage time of fuel.

FAME can be used as a pure fuel or a drop-in fuel. Pure FAME fuel possesses similar physical properties to those of conventional fuel as shown in Table 8 (FNR, 2012). However, short storage time of pure FAME fuel was reported due to problems like oxidation and polymerisation. Besides, issues such as deterioration of oil quality, clogging of particulate filters, poor injection performance and cold-start properties are commonly faced by pure FAME fuel (IEA AMF, 2021). Thus, it is more common for FAME to be used as a blend-in component in fossil diesel to increase the renewable content of the fuel. The blend of FAME and petroleum diesel are commonly referred to as BX, where X refers to the volume per cent of FAME in the biodiesel blend. For example, B5 consists of 5% FAME, and B10 consists of 10% FAME (Baljet, 2009). In Europe, up to 7% v/v of FAME is allowed in diesel fuel without any modification in vehicles or the distribution system. Meanwhile in the U.S., limit of FAME allowed in diesel fuel is 5% v/v (McGill et al., 2009).

Palm kernel oil (PKO) has been considered as a promising feedstock for transesterification process. A maximum biodiesel yield of 96% was achieved during the transesterification of PKO with ethanol in presence of 1% of KOH catalyst (Alamu et al., 2007). Another study on transesterification of from palm oil and PKO using lipase enzyme demonstrated that palm oil-biodiesel has a better FAME and fuel properties as compared to PKO-biodiesel (Kareem et al., 2017). The blending of PKO FAME with petroleum diesel had similar performance as petroleum diesel in a direct injection diesel engine (Lin et al., 2008). However, more studies on PKO transesterification are required as this oil can provide methyl esters of shorter chain fatty acids. Fatty acid profile of PKO showed that around 48% lauric acid (C_{12}) as the major fatty acid followed by 16% miristic acid (C_{14}) and 15% oleic acid (C_{18}) (Mayorga et al., 2020). The methyl esters of short fatty acids can improve cold properties, such as freezing and cloud point.

However, FAME is not approved as a jet fuel additive as of May 2021. In Europe, up to 50 mg/kg of FAME as an "Identified Incidental Contaminant" in jet fuel was approved by the Joint Inspection Group (JIG) in May 2015 (ASTM, 2020). The aviation industry intended to further increase the maximum permissible level of

FAME to 100 mg/kg, which is not yet approved by the turbine engine manufacturers (JIG, 2018). Hence, there remains a need to support the technical case by collecting data from the field to monitor FAME levels and correlate them with any impact on aircraft or engine maintenance interval.

There is a significant number of literatures on the production and blending of FAMEs with diesel for automotive applications. However, research work on the application of FAMEs in aviation industry is still scarce. Mayorga et al. (2019) found that FAME from palm kernel oil mixed with jet fuel possessed better cold properties due to the shorter chains of methyl esters obtained from palm oil. Improvement on the cold flow properties of jet fuel blend can be achieved with the addition of additives (Mayorga et al., 2020). Other than the abovementioned contributions, bio-jet fuel blends of commercial Jet-A1 fuel with FAMEs derived from macaúba (Silva et al., 2020), jatropha (Ranucci et al., 2018), babassu (Llamas et al., 2012a; Oliveira et al., 2018; Ranucci et al., 2018), palm kernel (Llamas et al., 2012b; Ranucci et al., 2018), coconut (Llamas et al., 2012b) and camelina (Llamas et al., 2012a) were reported as well. Most of the FAME-kerosene blends are reported to comply with most of the specification listed in ASTM1655 standard, such as density, flash point, oxidation stability, viscosity at −20 °C and more. However, the blends do not meet the specification for lower heating value. Results from Llamas et al. (2012a, b) show that it would be feasible to blend babassu, camelina, coconut and palm kernel FAMEs with Jet-A1, up to 10% of the former, as a partial substitution of fossil jet fuels. Work from Baroutian et al. (2013) also concluded that blend of jet biofuel with 10 to 20% waste vegetable and jatropha FAMEs have comparable properties with that of commercially available aviation fuel. Kilaz et al. (2014) observed that 2 vol% of FAME in Jet A-1 fuel does not have a significantly negative effect on main jet fuel ASTM D1655 specifications. Ranucci et al.'s study (2018) further supported on the feasibility of jet fuel blend with FAME content up to 10% for aviation applications.

4.6.1 Transesterification Process

FAME can be obtained from different routes such as transesterification and esterification. Transesterification process is less expensive as compared to the other bio-jet conversion route while reducing the food–fuel crisis as non-edible oils are utilised as feedstock (Kandaramath Hari et al., 2015). Transesterification is a process where organic alkyl groups of ester and alcohol exchange in the presence of strong acid or strong base catalyst, with mono- and diacylglycerols are formed as intermediates of the reaction (Muthukumaran et al., 2020) (Fig. 8).

In this reaction, methanol and ethanol are the most used alcohol. Sodium and potassium methanolate are commonly used as catalysts in the process (Patel & Shah, 2015). However, for transesterification process on the basis of homogenous system, process with alkali catalyst such as sodium hydroxide (NaOH) or potassium hydroxide (KOH) is a much more rapid process than acid catalyst (Dunn, 2010). To achieve the maximum yield of transesterification process, the alcohol used should be free of moisture, and the FFA content of the oil should be less than 0.5%.

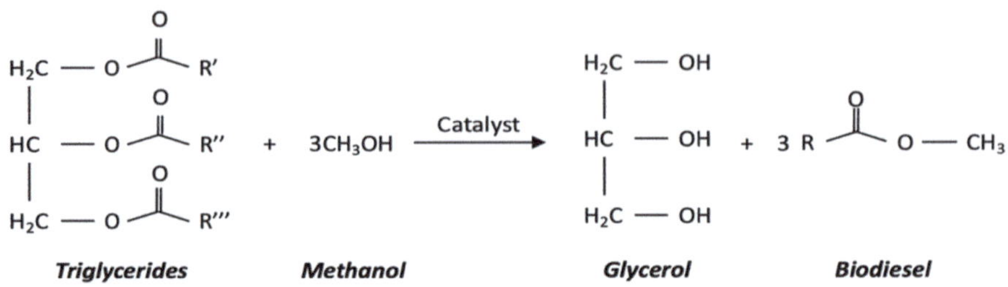

Fig. 8 Chemical structure of the transesterification process

Existing moisture (water molecule) in the transesterification process will react with the triglyceride to produce FFA (Eq. 10).

$$R - COOCH_3 + H_2O \leftrightarrow R - COOH + CH_3OH \quad (10)$$

In addition, the water molecule dissociates the sodium (Na) or potassium (K) atom from the base catalyst, forming ions Na^+ or K^+ along with the hydroxide ions. This will increase the possibility of unwanted side reaction, the saponification reaction, which produces soap and water as the final product. It also complicates the separation and recovery the catalyst. Generally, for a biomass source that contains a considerable amount of FFA, esterification process can be performed as a pre-treatment step to convert the FFA to methyl esters, thereby reducing the FFA level. Figure 9 illustrated the conversion process flow of crude palm oil to bio-jet fuel using the transesterification process (How, 2019).

Several reviews covering on the production of FAME biodiesel by transesterification have been published (Dash & Lingfa, 2017; Ishak & Kamari, 2019; Leung et al., 2010; Nasreen et al., 2018). In addition to that, a comprehensive review on the catalysts used in transesterification reaction of oils in biodiesel generation was provided (Thangaraj et al., 2019). The applications of heterogenous catalyst such as heteropolyacids, zeolites and polymeric catalysts in the transesterification process

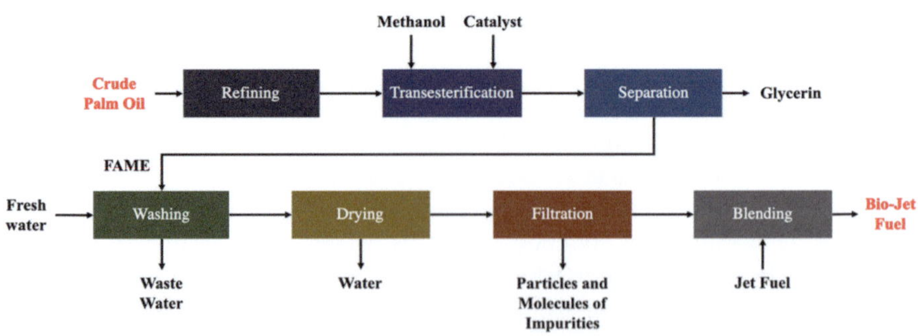

Fig. 9 Process flow for conversion of crude palm oil to bio-jet fuel via transesterification process

were discussed by Ramos et al. (Ramos et al., 2014). Work from Norshahidatul and Fatin (Transesterification of waste cooking oil, 2018) concluded that two-step (esterification–transesterification) transesterification reaction of waste cooking oil results in higher FAME yield as compared to one-step (esterification) reaction. Other than the work discussed above, efforts have also been made to search for a more energy-efficient and cost-effective transesterification process. An efficacious and heterogenous approach to produce FAME from jatropha oil catalysed by sulphated metal oxide catalysts has been developed by Chen et al. (Chen & Yoshikawa, 2018). Besides, work had been done in exploring the catalyst-free technologies for transesterification process (Okoro et al., 2018). The study shows that direct supercritical transesterification process is capable to reduce the heating and cooling utilities up to 95% as compared to the integrated subcritical lipid hydrolysis and supercritical esterification process. Findings from Encinar et al.'s work showed that the application of ultrasound irradiation in transesterification process could benefit the biodiesel manufacture in energy savings as it does not require additional heating (Encinar et al., 2018).

5 SAF in Aircraft Engines

In the past, various researches on the effect of SAF application in engine operation were conducted by evaluating the performance and emission of the jet engines. The effect of rapeseed SAF and its blends with diesel on the thermal performance and emissions of a diesel micro-turbine engine was reported (Nascimento et al., 2008). Engine burning with SAF and its blend show no obvious difference in engine performance and behaviour compared to the diesel fuel. In addition, slightly higher CO, lower NO_x and zero SO_2 can be achieved with the usage of SAF. However, significant fouling and damage can be observed at the turbine nozzle and rotor after the application of SAF. On the other hand, a 30 kW gas turbine runs on Jet A fuel, Soy Methyl Ester, Canola Methyl Ester, recycled rapeseed methyl ester, hot-fag SAF and their 50% blends by volume with Jet A fuel (Habib et al., 2010). The engine was operated over a range of throttle settings. Engine burning on SAF resulted in reduced static thrust and thrust-specific fuel consumption, lower CO and NO emissions and improved the thermal efficiency. Similar work is carried out by utilising jatropha SAF in IS/60 Rovers gas turbine (Rehman et al., 2011).

Experiments and simulations were conducted to investigate the performance of Armfield CM4 turbojet engine with palm oil methyl ester biodiesel and its blends with Jet-A1 fuel (Abu Talib et al., 2014). SAF blend with 20% palm oil methyl ester shows comparable level of performance as compared to that of Jet-A1 fuel, especially in terms of thrust and thermal efficiency. With the usage of SAF, efficiency of combustor improved while efficiencies for other components remained collectively consistent. Besides, the PM emissions from CFM56-7B commercial jet engine fuelled with blends of Jet-A1 fuel and alternative biomass-based and F-T-based fuel were measured (Lobo et al., 2011). The blends show reduction in PM emissions when compared to the emissions from pure Jet-A1 fuel. Investigation on the emissions characteristic of two combustion platforms, a T63 turboshaft engine and an atmospheric swirl-stabilised research combustor was conducted (Corporan et al., 2007). The combustion platforms are fuelled with conventional military jet fuel (JP-8), F-T-synthetic jet fuel and the blends of these two. Huge reduction in the particle concentration and mean size on both combustion platforms can be observed with the neat F-T and blends relatively to operation with JP-8. Furthermore, reductions of over 90% in particle number were achieved for both platforms with the usage of F-T fuel under different operating conductions. Reduction in smoke number, sulphur oxide emissions and slight increases in water vapour can also be observed (Corporan et al., 2007).

The combustion emissions data of F-T synthetic fuel, FAME fuel and their blends with Jet-A1 fuel for gas turbine engine, CFM56-7 was studied (Timko et al., 2011). Reduced or negligible aromatic content, increased oxygen content and reduction in NO_x and CO emissions are reported with the usage of F-T synthetic fuel and FAME blends. In addition, particle emissions of aromatic hydrocarbons decreased in particle size, number density and total mass when F-T synthetic fuel and FAME blends are used. Other than the engine performance and emissions, safety risks and reliability of SAF utilisation in the operation of aircraft engines are relatively important. Čerňan et al. (2017) ran a small jet engine with blends of FAME biofuels and Jet-A1 fuel at different concentration. The rubber seals used in the fuel systems of the aviation turbo-compressor engines show degradation as the effect of FAME biofuel. However, by complying with restrictions, blends of FAME biofuel are reported to be usable for propulsion of aviation turbo-compressor engines. Figure 10 summarised the performance and emission of SAF application in aircraft engine operations.

Apart from this, test-flights using SAF from various sources at different mixing ratio are conducted by major airlines. In the 2 h test-flight by Air New Zealand, jatropha SAF was blended with Jet-A1 fuel in one of the four tanks in Rolls Royce RB-211 Engines (Kanter, 2008). In January 2009, Continental Airlines was the first airline to utilise algae as a SAF feedstock in a twin-engine commercial aircraft (BBC, 2009). Demonstration flights by Japan Airlines used a combination of three sustainable SAF feedstock on Pratt and Whitney JT9D engines (JAL, 2009). On the other hand, all four of the Boeing's General Electric GEnx-2B engines were fuelled with a blend of camelina SAF and Jet-A1 fuel (Coxworth, 2011). National Research Council of Canada (NRC) flew the first civil jet powered by 100% unblended SAF in October 2012 (ANN, 2012). In June 2011, KLM flew the world's first commercial SAF flight, carrying 171 passengers from Amsterdam to Paris (Paur, 2011). Tables 9 and 10 listed the data for demonstration flights and commercial flights using SAF.

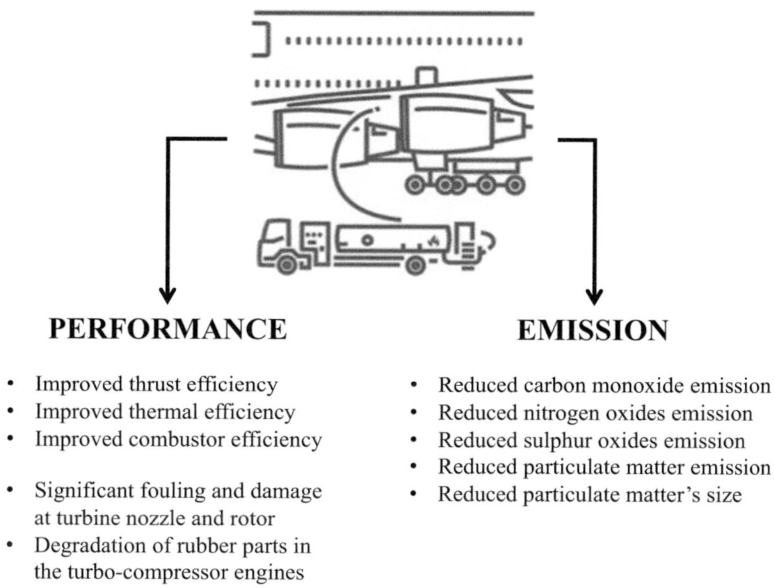

Fig. 10 Effect of SAF applications in aircraft engine operation

Table 9 Demonstration flights using bio-jet fuels

Year	Airlines	Aircraft Model	Raw Materials	Mixing Ratio (%)
2008	Virgin Atlantic (Dunn, 2008)	Boeing 747–400	Coconut and Babassu	20
2008	Air New Zealand (Kanter, 2008)	Boeing 747–400	Jatropha	50
2009	Continental Airlines (BBC, 2009)	Boeing 747–800	Algae and Jatropha	50
2009	Japan Airlines (JAL, 2009)	Boeing 747–300	Camelina, Jatropha and Algae	50
2011	Boeing (Coxworth, 2011)	Boeing 747-8F	Camelina	15
2011	Air China (DingDing & Wen, 2011)	Boeing 747–400	Jatropha	50
2012	Qantas (Aviation, 2012)	Airbus A330	Camelina and Brassica carinata	50
2012	Canada's National Research Council (ANN, 2012)	Dassault Falcon 20	Brassica carinata	100
2013	Paramus Flying Club (AOPA, 2013)	Cessna 182 Skylane	Used vegetable oil	50
2018	Spicejet Airlines (PTI, 2018)	Bombardier Q400	Jatropha	25

Table 10 Commercial flight using bio-jet fuels

Year	Airlines	Aircraft Model	Raw Materials	Mixing Ratio
2011	KLM (Paur, 2011)	Boeing 737–800	Used cooking oil	50%
2011	Lufthansa (Airbus, 2011)	Airbus A321	Jatropha, Camelina and Animal fats	50%
2011	Continental Airlines (Hilkevitch, 2011)	Boeing 737–800	Algae	20%
2014	Scandinavian Airlines (SAS, 2014)	Boeing 737–700	Used cooking oil	48%
2016	Alaska Airlines (Airlines et al., 2016)	Boeing 737–800	ATJ-SPK	20%
2017	Singapore Airlines (Airbus, 2017)	Airbus A350-900	HEFA	N/A
2018	Qantas (Demaio, 2018)	Boeing 787–9	Carinata	10%
2019	Etihad Airways (UAE, 2019)	Boeing 787	Salicronia	50%
2019	China Southern Airlines (ATAG, 2019)	A320NEO	Sugarcane	10%
2019	Egypt Airlines (AirlineGeeks, 2019)	Boeing 787–9 Dreamliner	Agricultural waste	30%

6 Challenges and Opportunities

Although there is an increasing need for alternative aviation biofuel, there are still many obstacles to be tackled to achieve the goal. Sustainability and financial problems in commercialisation are two major challenges for SAF. Another major concern of alternative jet fuel development is the price difference between SAF and conventional jet fuels. Increases in the production cost have significant impacts on the price of jet fuel, thus increasing the operating costs for the aviation business. Hence, it is crucial to reduce the overall cost of production and develop conversion pathways with comparable costs of crude derived jet fuel. The issue of sustainability is highly dependent on the availability of feedstock. It affects not only environmental issues such as land–water usage, GHG and particulate emissions but also social and economic issues such as usage of crop lands for the usage of non-edible biomass, coordination of policies.

6.1 Challenges in Biomass Feedstock

Feedstock is one of the major factors that contributes the production cost of bio-jet fuel. The availability and price of feedstock act as an indicator for the commercialisation feasibility of each bio-jet fuel technologies (Zhang et al., 2020). To ensure that the process has a continuous flow of feedstock to be converted into fuel, the establishment of a long-term feedstock supply that guarantees at least 10 to 15 years is crucial. If there is no continuity in feedstock supply, food-versus-fuel issues and deforestation might arise (Sinha et al., 2016). The feedstock used in SAF production should be cost-effective. Besides, feedstock should be cultivated with minimal impact on land use, water intake and low greenhouse gas emission (Kandaramath Hari et al., 2015). Palm-based biomass has high potential as a feedstock for SAF through different routes. However, the potential of palm kernel oil with HRJ and ATJ conversion pathway must be explored.

In the 2016 Billion-Ton Report, U.S. Department of Energy estimated the availability of some biomass feedstock at a range of price $40–60 per dry tonnes from 2017 to 2040 (Langholtz et al., 2016). Table 11 shows the prices and availabilities of biomass material from different categories, which will be fed into various conversion pathway to product SAF (Langholtz et al., 2016).

As the demand for SAF increases, issues like deforestation and degradation of low fertility of soil arise. Burn-offs, net reduction in global vegetation and power supply for fuel processing further increase CO_2 output to the atmosphere (Charles et al., 2007). Other than that, afforestation issues will also affect the fertility of soil and biodiversity, further add on the CO_2 output to atmosphere (Kandaramath Hari et al., 2015). Trees in native forests are cleared to make way for oilseed plantation. Besides, deforestation can directly lead to the loss of biodiversity. Usage of pesticides and fertilisers further induce soil erosion and water pollution.

Table 11 Feedstock availability and pricing

Biomass feedstock	Range of minimum prices per dry tonnes ($)	Availability, million dry tonnes per year	Effective year of availability
Animal fats and vegetable oils	550–1200	5.00–5.30	2012–2014
Agricultural wastes	40–60	27–38	2017–2040
Municipal Solid Wastes	40–60	51–55	2017–2040
Forestry and wood wastes	40–60	36–53	2017–2040
Other wastes[a]	40–60	8.7–9.4	2017–2040
Energy crops	40–80	78–411	2017–2040
Algae	490–2900	47–132	–

[a]Consisting of biosolids; trap grease; food processing wastes from industrial, institutional, and commercial sources; utility tree trimmings

6.2 Challenges in Production Process

The efficiency of the production cost and flexibility of feedstock are two of the biggest challenges faced in the production process. As algae emerges as the new biodiesel feedstock, problems and difficulties such as the expensive start-up phase and maintenance fees arise. Besides, challenges like the uncertainties in cultivation strategies and the complexity of lipid extraction and biodiesel conversion technology are often encountered as well. Lipid feedstock, specifically vegetable oil, contains a substantial amount of impurities such as metals, sodium, potassium, calcium and phosphorus. Presence of these metal impurities in the feedstock can lead to catalyst poisoning, reduced catalyst life and deactivation of the catalyst. Thus, pre-treatment of feedstock is crucial for the complete removal of the impurities. One of the suggested pre-treatment methods is "water degumming" process with guard bed catalyst where metals and phospholipids present in the feedstock can be removed completely (Sinha et al., 2016).

As the production process must be consistent, competent and highly efficient, it is of utmost importance that a suitable catalyst is selected with the right designing process (Kandaramath Hari et al., 2015). Further improvements and optimisations in SAF production technology are a must to reduce the current production cost, producing SAF which is economically viable to compete with petroleum-based jet fuels (ATAG. Begineer's, 2009). Based on the availability of feedstock and conversion yield illustrated in U.S. Department of Energy and National Renewable Energy Laboratory's report, it is estimated that 2–65.5 billion gallons of SAF can be produced annually (Langholtz et al., 2016; Wang et al., 2016). Table 12 shows the potential conversion pathway for each category of biomass feedstock and the estimated conversion yield and SAF production capacity, respectively (Wang et al., 2016).

According to the U.S. Energy Information Administration, the price of crude oil is forecasted to average $64 per barrel for the year 2021, up from an average of $43 per barrel in 2020 (EIA, 2021). As of April 2021, the jet fuel is priced at $66.58 per barrel, equivalent to $489 per tonnes jet fuel (IATA, 2021). The estimated price of animal fats and vegetable oils as well as algae feedstock (Table 12) is much higher when compared to the price of crude oil and the final produce crude derived jet fuel. Hence, it is economically impractical to perform the conversion of these biomass materials into bio-jet fuel.

6.3 Challenges in Commercialisation of Bio-Jet Fuels

Bio-aviation fuels with zero sulphur and aromatics, low freezing points and high auto-ignition temperature are of foremost importance for compatibility of renewable jet fuels with the conventional jet fuels. Thermal and storage stability of bio-aviation fuel are important during storage and transportation. In addition to this, the process for testing new bio-aviation fuel is particularly rigorous, which must be carried out

Table 12 Feedstock potential conversion pathway and the respective conversion yield and SAF production capacity

Biomass feedstock	Conversion pathway	Conversion yield, gallons per dry tonnes	SAF production capacity, billion gallons per year
Animal fats and vegetable oils	HRJ	27.8–86.7	0.14–0.46
	HTL	7.6–122	0.04–0.65
Agricultural wastes	ATJ	10.5–79.2	0.28–3.0
	Pyrolysis	18.5	0.50–0.70
	F-T	8.7–87.5	0.23–3.33
MSW	F-T	8.7–87.5	0.44–4.8
Forestry and wood wastes	ATJ	10.5–79.2	0.38–4.2
	Pyrolysis	18.5	0.67–0.98
	F-T	8.7–87.5	0.31–4.6
Energy crops	ATJ	10.5–79.2	0.82–32.5
	Pyrolysis	18.5	1.4–7.6
	F-T	8.7–87.5	0.68–36.0
Algae	HRJ	27.8–86.7	1.3–11.4
	HTL	7.6–122	0.36–16.1
Total			2.0–65.5

in the laboratory and on the ground. This is to ensure the fuel developed shows good performance and is safe to be used under a wide range of operational circumstances (ATAG, 2009).

High operational cost and technical challenges faced in bio-jet fuel conversion process as well as the scaling up of bio-jet fuel technologies contributed to the high selling price of bio-aviation fuels, which often hinders the commercialisation of bio-jet fuel (Zhang et al., 2020). It has been reported that the minimum selling price of SAF can range from \$2.2 to \$34.7 per gallon (\$92 to \$1 457 per barrel), which are approximate two to twenty times higher than conventional jet fuel price (Wang et al., 2016). To achieve cost-competitiveness with crude oil-based jet fuel, more efforts are required to be invest in the research and development for a more efficient SAF conversion process.

In addition, the expenses and time required to obtain the certification and approval SAF are one of the biggest challenges for the commercialisation of SAF. This may be due to the engagement of several functional departments such as the ASTM and the Federal Aviation Administration (FAA), which specialised in the certification of SAF (Zhang et al., 2020). ASTM D4054—"Standard Practice for Qualification and Approval of New Aviation Turbine Fuels and Fuels Additives" was developed as a guide by the engine and aircrafts original equipment manufacturers with ASTM International member supports, to facilitate the certification process. Generally, the

Table 13 Fuel volume required in each tier for D4054 testing

Tiers	Fuel volume, gallons (litres)
Tier 1—Fuel specification properties	Up to 10 (38)
Tier 2—Fit-For purpose (FPP) properties	10–100 (38–380)
Tier 3—Component/Rig tests	250–10,000 (950–38,000)
Tier 4—Engine/APU tests	Up to 225,000

ASTM D4054 standard comprised of a four-tiered process for the testing and evaluation process of new bio-aviation fuels and fuels additives (Rumizen, 2013). The D4054 standard is an iterative and rigorous evaluation process that requires candidate fuel producers to generate a significant volume of fuel sample for testing of fuel properties, composition and performance (EERE, 2016). Table 13 listed out the tiers comprised in the D4054 testing and the fuel volume required for each test. While Tier 1 and Tier 2 of D4054 testing may be provided at no cost to the fuel producer, a significant amount of sample fuel are required for Tier 3 and Tier 4 testing, posing a challenge to the fuel producers. The total volume of sample fuel required is roughly estimated to be 200 000 gallons and have run up costs nearly $10–15 million (Colket et al., 2017; Zhang et al., 2020). Also, few weeks or months are usually required just to just to generate 5–10 gallons of SAF fuel in a laboratory setting. This is in addition to the capital and operating costs for pilot-scale production required to generate significant volume of SAF for the fuel certification process (EERE, 2016). Typically, the certification process takes about 3 to 7 years (Hileman, 2019).

7 Conclusion

In this chapter, non-edible crops and algae biomasses derived from agricultural, industrial waste and aquatic are reviewed. Different conversion routes of bio-aviation fuels are considered, such as blending of fatty acid methyl ester (FAME), hydroprocessed renewable jet (HRJ), Fischer–Tropsch (F-T) synthesis, pyrolysis with bio-oil upgrading, hydrothermal liquefaction (HTL) process and alcohol-to-jet (ATJ) route. For the blending of FAME, blends of FAME and Jet-A1 fuel show promising properties such as density, flash point and oxidation stability except for the lower heating value. However, more research work and data collections are required to support the approval process for blending of FAME in jet fuel application. Hydro-processing is a well-established technology. However, further blending with conventional jet fuel is required for HRJ fuel to improve its lubricity. As for F-T fuels, they are characterised as high quality with zero-toxicity and free of contaminants. Furthermore, F-T synthesis can produce hydrocarbons of different lengths from various biomass. However, the operating cost of F-T synthesis is rather high. Besides, the overall efficiency of the F-T synthesis is relatively low. Other than that, the pyrolysis process has the advantage of being a relatively simple process and inexpensive technology. However, pyrolysis bio-oil produced will be in stable mixtures of solid, liquid and gaseous state, which is mechanically impossible to separate them. Further suspension or thermal drying process is necessary to remove water from the pyrolysis bio-oil. With regards to HTL process, the capability to process wet biomass makes HTL process stand out from the other conversion technologies. However, the high operating pressure in HTL process leads to high operating cost of the technology. Finally, the key advantage of ATJ fuels has over other SAF is the wide availability of feedstock, hence having a better chance of being scaled up commercially viable quantities. Nonetheless, this may raise the competition on feedstock availability between aviation and automotive industries. Along with the development of bio-aviation fuel, challenges from various aspects like social, environmental, economic and production arise as well. Hence, efforts and coordination of various parties are important in the scaling up and commercialisation of bio-aviation fuel production.

Acknowledgements The authors would like to express sincere gratitude to the Ministry of Higher Education of Malaysia for the realisation of this research project under the Grant FRGS/1/2019/TK02/UNIM/02/1.

References

Aatola, H., Larmi, M., Sarjovaara, T., & Mikkonen, S. (2008). Hydrotreated vegetable oil (HVO) as a renewable diesel fuel: Trade-off between NOx, particulate emission, and fuel consumption of a heavy duty engine. *SAE International Journal of Engines, 1*(1), 2008-01-2500. https://doi.org/10.4271/2008-01-2500

Abu Talib, A. R., Gires, E., & Ahmad, M. T. (2014). Performance evaluation of a small-scale turbojet engine running on palm oil biodiesel blends. *Journal of Fuels, 2014*, 1–9. https://doi.org/10.1155/2014/946485

Adewale, P., Dumont, M.-J., & Ngadi, M. (2015). Recent trends of biodiesel production from animal fat wastes and associated production techniques. *Renewable and Sustainable Energy Reviews, 45*, 574–588. https://doi.org/10.1016/j.rser.2015.02.039

Ahmad, A. L., Yasin, N. H. M., Derek, C. J. C., & Lim, J. K. (2011). Microalgae as a sustainable energy source for biodiesel production: A review. *Renewable and Sustainable Energy Reviews, 15*(1), 584–593. https://doi.org/10.1016/j.rser.2010.09.018

Airbus. (2011). *World's first scheduled passenger biofuel flights commence.* Airbus S.A.S. https://www.airbus.com/newsroom/press-releases/en/2011/07/world-s-first-scheduled-passenger-biofuel-flights-commence.html. Published 2011.

Airbus. (2017). *Singapore Airlines launches trans-Pacific biofuel flights with the A350 XWB.* Airbus S.A.S. https://www.airbus.com/newsroom/news/en/2017/05/singapore-airlines-launches-trans-pacific-biofuel-flights-with-the-a350-xwb.html. Published 2017.

AirlineGeeks. (2009). *Egypt air flies newest dreamliner home using biofuel, sets distane record.* AirlineGeeks. https://airlinegeeks.com/2019/07/26/egyptair-flies-newest-dreamliner-home-using-biofuel-sets-distance-record/. Published 2019.

Alamu, O., Waheed, M. A., & Jekayinfa, S. (2007). Biodiesel production from Nigerian palm kernel oil: Effect of KOH concentration on yield. *Energy for Sustainable Development, 11*, 77–82.

Alaska Airlines. (2016). *Alaska Airlines flies first commercial flight with new biofuel made from forest residuels*. Alaska Air. https://blog.alaskaair.com/alaska-airlines/company-news/nara-flight/. Published 2016.

ANN. (2012). NRC flies world's first civil jet powered by 100 percent biofuel. *Aero News Network*.

AOPA. *Flying club's Cessna 182 makes biofuel flight*. Aircraft Owners and Pilot Association.

Armas, O., Yehliu, K., & Boehman, A. L. (2010). Effect of alternative fuels on exhaust emissions during diesel engine operation with matched combustion phasing. *Fuel, 89*(2), 438–456. https://doi.org/10.1016/j.fuel.2009.09.022

ASTM. (2020a). *Standard specification for aviation turbine fuels*. ASTM Int. 2020a; ASTM D1655. https://doi.org/10.1520/D1655-20D

ASTM. (2020b). *Standard specification for aviation turbine fuel containing synthesized hydrocarbons*. ASTM International. https://doi.org/10.1520/D7566-20C

ATAG. (2019). *China southern airlines takes delivery of A320NEO powered with biofuel mix*. ATAG.

ATAG. (2020). Aviation benefits beyond borders.

ATAG. *Begineer's guide to aviation biofuels*.

Australian Aviation. (2012). Qantas spruiks biofuels with A330 flight. *Australian Aviation*.

Baljet, M. (2009). FAME/airport aviation fueling. *International Airport Review*.

Baroutian, S., Aroua, M. K., Raman, A. A. A., Shafie, A., Ismail, R. A., & Hamdan, H. (2013). Blended aviation biofuel from esterified Jatropha curcas and waste vegetable oils. *Journal of the Taiwan Institute of Chemical Engineers, 44*(6), 911–916. https://doi.org/10.1016/j.jtice.2013.02.007

BBC. (2009, January 8). First flight of algae-fuelled jet. *BBC News*.

Ben Hnich, K., Khila, Z., & Hajjaji, N. (2020). Comprehensive study of three configurations coproducing synthetic fuels and electricity from palm residue via Fischer-Tropsch process. *Energy, 205*, 118027. https://doi.org/10.1016/j.energy.2020.118027

Bharath, G., Hai, A., Rambabu, K., et al. (2020). Systematic production and characterization of pyrolysis-oil from date tree wastes for bio-fuel applications. *Biomass and Bioenergy, 135*, 105523. https://doi.org/10.1016/j.biombioe.2020.105523

Bo Han, G., Hee Jang, J., Hwei Ahn, M., & Hun Jung, B. (2020). Recent application of bio-alcohol: Bio-jet fuel. In *Alcohol fuels—current technologies and future prospect*. IntechOpen. https://doi.org/10.5772/intechopen.89719

Bridgwater, A. V. (2009). *Technical and economic assessment of thermal processes for biofuels*.

Bridgwater, A. V. (2012). Review of fast pyrolysis of biomass and product upgrading. *Biomass and Bioenergy., 38*, 68–94. https://doi.org/10.1016/j.biombioe.2011.01.048

Brooks, K. P., Snowden-Swan, L. J., & Jones, S. B., et al. (2016). Low-carbon aviation fuel through the alcohol to jet pathway. In *Biofuels for aviation* (pp. 109–150). Elsevier. https://doi.org/10.1016/B978-0-12-804568-8.00006-8

Buschmann, A. H., Camus, C., Infante, J., et al. (2017). Seaweed production: Overview of the global state of exploitation, farming and emerging research activity. *European Journal of Phycology, 52*(4), 391–406. https://doi.org/10.1080/09670262.2017.1365175

Cai, N., Zhang, H., Nie, J., Deng, Y., & Baeyens, J. (2020). Biochar from biomass slow pyrolysis. *IOP Conference Series: Earth Environmental Science, 586*, 012001. https://doi.org/10.1088/1755-1315/586/1/012001

Canabarro, N., Soares, J. F., Anchieta, C. G., Kelling, C. S., & Mazutti, M. A. (2013). Thermochemical processes for biofuels production from biomass. *Sustainable Chemical Processes, 1*(1), 22. https://doi.org/10.1186/2043-7129-1-22

Canilha, L., Chandel, A. K., & Suzane dos Santos Milessi T, et al. (2012). Bioconversion of sugarcane biomass into ethanol: An overview about composition, pretreatment methods, detoxification of hydrolysates, enzymatic saccharification, and ethanol fermentation. *Journal of Biomedicine and Biotechnology, 2012*, 1–15. https://doi.org/10.1155/2012/989572

Carlson, T. R., Tompsett, G. A., Conner, W. C., & Huber, G. W. (2009). Aromatic production from catalytic fast pyrolysis of biomass-derived feedstocks. *Topics in Catalysis, 52*(3), 241–252. https://doi.org/10.1007/s11244-008-9160-6

Čerňan, J., Hocko, M., & Cúttová, M. (2017). Safety risks of biofuel utilization in aircraft operations. *Transportation Research Procedia, 28*, 141–148. https://doi.org/10.1016/j.trpro.2017.12.179

Chan, Y. H., Yusup, S., Quitain, A. T., et al. (2015). Effect of process parameters on hydrothermal liquefaction of oil palm biomass for bio-oil production and its life cycle assessment. *Energy Conversion and Management, 104*, 180–188. https://doi.org/10.1016/j.enconman.2015.03.075

Chan, Y. H., Quitain, A. T., Yusup, S., Uemura, Y., Sasaki, M., & Kida, T. (2018). Optimization of hydrothermal liquefaction of palm kernel shell and consideration of supercritical carbon dioxide mediation effect. *Journal of Supercritical Fluids, 133*, 640–646. https://doi.org/10.1016/j.supflu.2017.06.007

Charles, M. B., Barnes, P., Ryan, N., & Clayton, J. (2007). Airport futures: Towards a critique of the aerotropolis model. *Futures, 39*(9), 1009–1028. https://doi.org/10.1016/j.futures.2007.03.017

Chen, L., & Yoshikawa, K. (2018). Bio-oil upgrading by cracking in two-stage heated reactors. *AIMS Energy, 6*(1), 203–315. https://doi.org/10.3934/energy.2018.1.203

Chen, G., Li, S., Jiao, F., & Yuan, Q. (2007). Catalytic dehydration of bioethanol to ethylene over $TiO_2/\gamma\text{-}Al_2O_3$ catalysts in microchannel reactors. *Catalysis Today, 125*(1–2), 111–119. https://doi.org/10.1016/j.cattod.2007.01.071

Chew, T. L., & Bhatia, S. (2008). Catalytic processes towards the production of biofuels in a palm oil and oil palm biomass-based biorefinery. *Bioresource Technology, 99*(17), 7911–7922. https://doi.org/10.1016/j.biortech.2008.03.009

Chiaramonti, D., & Nogueira, L. A. H. (2017). Aviation biofuels: Processes, opportunities, constraints, perspectives. In M. R. Riazi & D. Chiaramonti (Eds.), *Biofuels production and processing technology* (1st ed., pp. 295–318). CRC Press.

Chuck, C. J., McManus, M., Allen, M. J., & Singh, S. (2016). Feedstocks for aviation biofuels. In *Biofuels for aviation* (pp. 17–34). Elsevier. https://doi.org/10.1016/B978-0-12-804568-8.00002-0

Chung, D., Cha, M., Guss, A. M., & Westpheling, J. (2014). Direct conversion of plant biomass to ethanol by engineered Caldicellulosiruptor bescii. *Proceedings of the National Academy of Sciences, 111*(24), 8931–8936. https://doi.org/10.1073/pnas.1402210111

Colket, M., Heyne, J., Rumizen, M., et al. (2017). Overview of the National Jet Fuels Combustion Program. *The American Institute of Aeronautics and Astronautics, 55*(4), 1087–1104. https://doi.org/10.2514/1.J055361

Corma, A., Huber, G., Sauvanaud, L., & OConnor, P. (2007). Processing biomass-derived oxygenates in the oil refinery: Catalytic cracking (FCC) reaction pathways and role of catalyst. *Journal of Catalysis, 247*(2), 307–327. https://doi.org/10.1016/j.jcat.2007.01.023

Corporan, E., DeWitt, M. J., Belovich, V., et al. (2007). Emissions characteristics of a turbine engine and research combustor burning a Fischer–Tropsch jet fuel. *Energy & Fuels, 21*(5), 2615–2626. https://doi.org/10.1021/ef070015j

Coxworth, B. (2011). Boeing 747–8 Freighter to make first biofuel-powered transatlantic flight. *New Atlas.*

Czerwinski, J., Zimmerli, Y., Neubert, T., Heitzer, A., & Kasper M. (2007). *Injection, combustion and (nano) particle emissions of a modern HD-diesel engine with GTL, RME & ROR.* https://doi.org/10.4271/2007-01-2015

da Silva, J. Q., Santos, D. Q., Fabris, J. D., Harter, L. V. L., & Chagas, S. P. (2020). Light biodiesel from macaúba and palm kernel: Properties of their blends with fossil kerosene in the perspective of an alternative aviation fuel. *Renewable Energy, 151*, 426–433. https://doi.org/10.1016/j.renene.2019.11.035

Dahmen, N., Dinjus, E., Kolb, T., Arnold, U., Leibold, H., & Stahl, R. (2012). State of the art of the bioliq® process for synthetic biofuels production. *Environmental Progress & Sustainable Energy, 31*(2), 176–181. https://doi.org/10.1002/ep.10624

Dash, S. K., & Lingfa, P. (2017). *A review on production of biodiesel using catalyzed transesterification* (pp. 020100). https://doi.org/10.1063/1.4990253
de Klerk, A. (2011). *Fischer-Tropsch refining*. Wiley-VCH Verlag GmbH & Co. KGaA. https://doi.org/10.1002/9783527635603
de Klerk, A. (2013). Fischer-Tropsch process. In: *Kirk-Othmer encyclopedia of chemical technology*. Wiley. https://doi.org/10.1002/0471238961.fiscdekl.a01
de Klerk, A. (2016). Aviation turbine fuels through the Fischer–Tropsch process. In *Biofuels for aviation* (pp. 241–259). Elsevier. https://doi.org/10.1016/B978-0-12-804568-8.00010-X
de Klerk, A. (2020). Transport fuel. In *Future energy* (pp. 199–226). Elsevier. https://doi.org/10.1016/B978-0-08-102886-5.00010-4
de Klerk, A., Li, Y.-W., & Zennaro, R. (2013). Fischer-Tropsch technology. In: P. M. Maitlis, & A. de Klerk (Eds.), *Greener Fischer-Tropsch processes for fuels and feedstocks* (pp. 53–79). Wiley-VCH Verlag GmbH & Co. KGaA. https://doi.org/10.1002/9783527656837.ch3
de Oliveira, V. F., Parente, E. J. S., Cavalcante, C. L., & Luna, F. M. T. (2018). Short-chain esters enriched biofuel obtained from vegetable oil using molecular distillation. *Canadian Journal of Chemical Engineering, 96*(5), 1071–1078. https://doi.org/10.1002/cjce.23044
Delucchi, M. A. (2003) *A lifecycle emissions model (LEM): Lifecycle emissions from transportation fuels, motor vehicles, transportation modes, electricity use, heating and cooking fuels, and materials.*
Demaio, D. (2018). *Qantas 787 Dreamliner takes off fuelled by mustard seed biofuel on Los Angeles-Melbourne*. Traveller. https://www.traveller.com.au/qantas-787-dreamliner-takes-off-fuelled-by-mustard-seed-biofuel-on-los-angelesmelbourne-flight-h0py1r. Published 2018.
Derman, E., Abdulla, R., Marbawi, H., & Sabullah, M. K. (2018). Oil palm empty fruit bunches as a promising feedstock for bioethanol production in Malaysia. *Renewable Energy, 129*, 285–298. https://doi.org/10.1016/j.renene.2018.06.003
Dhar, S. A., Sakib, T. U., & Hilary, L. N. (2020). Effects of pyrolysis temperature on production and physicochemical characterization of biochar derived from coconut fiber biomass through slow pyrolysis process. *Biomass Conversion and Biorefinery*. https://doi.org/10.1007/s13399-020-01116-y
Díaz-Pérez, M. A., & Serrano-Ruiz, J. C. (2020). Catalytic production of jet fuels from biomass. *Molecules, 25*(4), 802. https://doi.org/10.3390/molecules25040802
Dillingham, G. L. (2014). *Impact of fuel price increases on the aviation industry.*
DingDing, X., & Wen, W. (2011, October 29). China conducts its first jet biofuel trial. *China Daily*.
Doliente, S. S., Narayan, A., Tapia, J. F. D., Samsatli, N. J., Zhao, Y., & Samsatli, S. (2020). Bio-aviation fuel: A comprehensive review and analysis of the supply chain components. *Front Energy Research, 8*(10), 1–38. https://doi.org/10.3389/fenrg.2020.00110
DOSM. (2019). *Selected agricultural indicators, Malaysia, 2019.*
Douvartzides, S. L., Charisiou, N. D., Papageridis, K. N., & Goula, M. A. (2019). Green diesel: Biomass feedstocks, production technologies, catalytic research, fuel properties and performance in compression ignition internal combustion engines. *Energies, 12*(5), 809. https://doi.org/10.3390/en12050809
Dunn, G. (2008). Partners carry out first biofuel flight using Virgin 747. *Flight Global*.
Dunn, R. O. (2010). Other alternative diesel fuels from vegetable oils and animal fats. In: *The biodiesel handbook*. (pp. 405–437). Elsevier. https://doi.org/10.1016/B978-1-893997-62-2.50015-2
EERE. (2016). *Alternative aviation fuels: Overview of challenges, opportunities, and next steps.* United States; 2016.
EIA. (2020c). *Ethanol*. U.S Energy Information Administration. https://www.eia.gov/energyexplained/biofuels/ethanol.php. Published 2020c.
EIA. (2020a). *Biomass explained: Biomass and the environment.* U.S. Energy Information Administration. https://www.eia.gov/energyexplained/biomass/biomass-and-the-environment.php. Published 2020a.

EIA. (2020b). *Use of biomass-based diesel fuel*. U.S Energy Information Administration. https://www.eia.gov/energyexplained/biofuels/use-of-biodiesel.php. Published 2020b.

EIA. (2019). *International energy outlook 2019*.

EIA. *Short-term energy outlook April 2021*.

Elvers, B., & Schutze A. (Eds.). (2007). *Handbook of fuels: Energy sources for transportation* (2nd ed.). Wiley.

Encinar, J. M., Sánchez, N., Martínez, G., & García, L. (2011). Study of biodiesel production from animal fats with high free fatty acid content. *Bioresource Technology, 102*(23), 10907–10914. https://doi.org/10.1016/j.biortech.2011.09.068

Encinar, J., Pardal, A., Sánchez, N., & Nogales, S. (2018). Biodiesel by transesterification of rapeseed oil using ultrasound: A kinetic study of base-catalysed reactions. *Energies, 11*(9), 2229. https://doi.org/10.3390/en11092229

Eom, I.-Y., Yu, J.-H., Jung, C.-D., & Hong, K.-S. (2015). Efficient ethanol production from dried oil palm trunk treated by hydrothermolysis and subsequent enzymatic hydrolysis. *Biotechnology for Biofuels, 8*(1), 83. https://doi.org/10.1186/s13068-015-0263-6

F3centre. (2017). FAME, Fatty Acid Methyl Esters. Swedish Knowledge Center Renewable Transport Fuels (pp. 1–2).

Fan, D., Dai, D.-J., & Wu, H.-S. (2012). Ethylene formation by catalytic dehydration of ethanol with industrial considerations. *Materials (basel), 6*(1), 101–115. https://doi.org/10.3390/ma6010101

Farhat, M. A., & El Bassam, A. (2005). Handbook of industrial chemistry: Organic chemicals. *Choice Review Online, 43*(01), 43–0322–43–0322. https://doi.org/10.5860/choice.43-0322

FNR. (2012). *Basisdaten Bioenergie Deutschland*.

GAR. *How it works: Biodiesel production*. Golden Agri-Resources Ltd. Accessed September 6, 2021. http://www.goldenagri.com.sg/how-it-works-biodiesel-production/#. Published 2019.

García-Sánchez, M., Sales-Cruz, M., Lopez-Arenas, T., Viveros-García, T., & Pérez-Cisneros, E. S. (2019). An intensified reactive separation process for bio-jet diesel production. *Processes., 7*(10), 655. https://doi.org/10.3390/pr7100655

Gegg, P., Budd, L., & Ison, S. (2014). The market development of aviation biofuel: Drivers and constraints. *Journal of Air Transport Management, 39*, 34–40. https://doi.org/10.1016/j.jairtraman.2014.03.003

Geleynse, S., Brandt, K., Garcia-Perez, M., Wolcott, M., & Zhang, X. (2018). The alcohol-to-jet conversion pathway for drop-in biofuels: Techno-economic evaluation. *Chemsuschem, 11*(21), 3728–3741. https://doi.org/10.1002/cssc.201801690

Goering, C. E., Schwab, A. W., Daugherty, M. J., Pryde, E. H., & Heakin, A. J. (1982). Fuel properties of eleven vegetable oils. *Transactions of ASAE, 25*(6), 1472–1477. https://doi.org/10.13031/2013.33748

Golay, S., Doepper, R., & Renken, A. (1999). Reactor performance enhancement under periodic operation for the ethanol dehydration over γ-alumina, a reaction with a stop-effect. *Chemical Engineering Science, 54*(20), 4469–4474. https://doi.org/10.1016/S0009-2509(99)00105-0

Gollakota, A. R. K., Kishore, N., & Gu, S. (2018). A review on hydrothermal liquefaction of biomass. *Renewable and Sustainable Energy Reviews, 81*, 1378–1392. https://doi.org/10.1016/j.rser.2017.05.178

Gong, Y., Kaario, O., Tilli, A., Larmi, M., Tanner, F. X. (2010). *A computational investigation of hydrotreated vegetable oil sprays using RANS and a modified version of the RNG k–ε model in OpenFOAM*. https://doi.org/10.4271/2010-01-0739

Goyal, H. B., Seal, D., & Saxena, R. C. (2008). Bio-fuels from thermochemical conversion of renewable resources: A review. *Renewable and Sustainable Energy Reviews, 12*(2), 504–517. https://doi.org/10.1016/j.rser.2006.07.014

Green Car Congress. (2020). *ASTM approves 7th annex to D7566 sustainable jet fuel specification: HC-HEFA*. BioAge Group. https://www.greencarcongress.com/2020/05/20200514-ihi.html. Published 2020.

Gruber, H., Groß, P., Rauch, R., et al. (2019). Fischer-Tropsch products from biomass-derived syngas and renewable hydrogen. *Biomass Conversion Biorefinery*. https://doi.org/10.1007/s13399-019-00459-5

Gupta, G. K., Gupta, P. K., & Mondal, M. K. (2019). Experimental process parameters optimization and in-depth product characterizations for teak sawdust pyrolysis. *Waste Management, 87*, 499–511. https://doi.org/10.1016/j.wasman.2019.02.035

Guzman, A., Torres, J. E., Prada, L. P., & Nuñez, M. L. (2010). Hydroprocessing of crude palm oil at pilot plant scale. *Catalysis Today, 156*(1–2), 38–43. https://doi.org/10.1016/j.cattod.2009.11.015

Habib, Z., Parthasarathy, R., & Gollahalli, S. (2010). Performance and emission characteristics of biofuel in a small-scale gas turbine engine. *Applied Energy, 87*(5), 1701–1709. https://doi.org/10.1016/j.apenergy.2009.10.024

Hemighaus, G., Boval, T., & Bacha, J., et al. (2006). *Aviation fuels technical review*.

Hernandez-Mena, L. E., Pecora, A. A. B., & Beraldo, A. L. (2014). Slow pyrolysis of bamboo biomass: Analysis of biochar properties. *Chemical Engineering Transactions, 37*, 115–120. https://doi.org/10.3303/CET1437020

Heveling, J., Nicolaides, C. P., & Scurrell, M. S. (1998). Catalysts and conditions for the highly efficient, selective and stable heterogeneous oligomerisation of ethylene. *Applied Catalysis, a: General, 173*(1), 1–9. https://doi.org/10.1016/S0926-860X(98)00147-1

Heydenrych, M. D., Nicolaides, C. P., & Scurrell, M. S. (2001). Oligomerization of ethene in a slurry reactor using a Nickel(II)-exchanged silica-alumina catalyst. *Journal of Catalysis, 197*(1), 49–57. https://doi.org/10.1006/jcat.2000.3035

Hileman, J. (2019). *Fuel approval process & status*.

Hilkevitch, J. (2011, November 11). Continental airlines flight is first in U.S. to use biofuel. *Los Angeles Times*.

Hu, Y., Zhan, N., Dou, C., et al. (2010). Selective dehydration of bio-ethanol to ethylene catalyzed by lanthanum-phosphorous-modified HZSM-5: Influence of the fusel. *Biotechnology Journal, 5*(11), 1186–1191. https://doi.org/10.1002/biot.201000139

Hu, J., Yu, F., & Lu, Y. (2012). Application of Fischer-Tropsch synthesis in biomass to liquid conversion. *Catalysts, 2*(2), 303–326. https://doi.org/10.3390/catal2020303

Huber, G. W., Iborra, S., & Corma, A. (2006). Synthesis of transportation fuels from biomass: Chemistry, catalysts, and engineering. *Chemical Reviews, 106*(9), 4044–4098. https://doi.org/10.1021/cr068360d

IATA. (2012). *IATA guidance material for Biojet fuel management*.

IATA. 2019. *Annual review 2019*.

IATA. (2020a). *Economic performance of the airline industry*.

IATA. (2020b). *Sustainable aviation fuel: Technical certification*.

IATA. *Jet fuel price monitor*. IATA Economics. https://www.iata.org/en/publications/economics/fuel-monitor/). Published 2021.

ICAO. (2012). *Manual on civil aviation jet fuel supply*. International Civil Aviation Organisation.

ICAO. (2019). *2019 Environmental report: Aviation and environment*.

ICAO. (2020). *Conversion processes*. International Civil Aviation Organisation. https://www.icao.int/environmental-protection/GFAAF/Pages/Conversion-processes.aspx. Published 2020.

IEA AMF (2021). How much FAME biodiesel can be blended in diesel fuel? *Technology collaboration programme on advanced motor fuels*. https://www.iea-amf.org/app/webroot/files/file/AnnexReports/AMF_Annex_34-1.pdf. Published 2021.

Ishak, S., & Kamari, A. (2019). A review of optimum conditions of transesterification process for biodiesel production from various feedstocks. *International Journal of Environmental Science and Technology, 16*(5), 2481–2502. https://doi.org/10.1007/s13762-019-02279-6

Jadhav, A., Ahmed, I., Baloch, A. G., et al. (2019). Utilization of oil palm fronds for bio-oil and bio-char production using hydrothermal liquefaction technology. *Biomass Conversion and Biorefinery*. https://doi.org/10.1007/s13399-019-00517-y

Jae, J., Tompsett, G. A., Lin, Y.-C., et al. (2010). Depolymerization of lignocellulosic biomass to fuel precursors: Maximizing carbon efficiency by combining hydrolysis with pyrolysis. *Energy & Environmental Science, 3*(3), 358. https://doi.org/10.1039/b924621p

Jahirul, M., Rasul, M., Chowdhury, A., & Ashwath, N. (2012). Biofuels production through biomass pyrolysis—A technological review. *Energies, 5*(12), 4952–5001. https://doi.org/10.3390/en5124952

JAL. (2009). *JAL flight brings aviation one step closer to using biofuel*. Japan Airlines. Accessed August 21, 2020. https://press.jal.co.jp/en/release/200901/003159.html. Published 2009.

Jena, U., & Das, K. C. (2011). Comparative evaluation of thermochemical liquefaction and pyrolysis for bio-oil production from microalgae. *Energy & Fuels, 25*(11), 5472–5482. https://doi.org/10.1021/ef201373m

Jena, U., McCurdy, A. T., Warren, A., et al. (2015). Oleaginous yeast platform for producing biofuels via co-solvent hydrothermal liquefaction. *Biotechnology for Biofuels, 8*(1), 167. https://doi.org/10.1186/s13068-015-0345-5

Jenkins, R. W., Sutton, A. D., & Robichaud, D. J. (2016). Pyrolysis of biomass for aviation fuel. In *Biofuels for aviation* (pp. 191–215). Elsevier. https://doi.org/10.1016/B978-0-12-804568-8.00008-1

JIG. (2018). Guidance on managing FAME in jet fuel. *Production Quality Bulletin, 106*, 1–10.

JIG. (2019). Aviation fuel quality requirements for jointly operated systems. *Production Specificity Bulletin,* 31, 1–9.

Jindal, M. K., & Jha, M. K. (2016). Hydrothermal liquefaction of wood: a critical review. *Reviews in Chemical Engineering, 32*(4). https://doi.org/10.1515/revce-2015-0055

Kalnes, T. N., McCall, M. M., & Shonnard, D. R. (2010). Chapter 18. Renewable diesel and jet-fuel production from fats and oils (pp. 468–495). https://doi.org/10.1039/9781849732260-00468

Kandaramath Hari, T., Yaakob, Z., & Binitha, N. N. (2015). Aviation biofuel from renewable resources: Routes, opportunities and challenges. *Renewable and Sustainable Energy Reviews, 42*, 1234–1244. https://doi.org/10.1016/j.rser.2014.10.095

Kanter, J. (2008, December 30). Air New Zealand flies on engine with jatropha biofuel blend. *The New York Times.*

Kareem, S. O., Falokun, E. I., Balogun, S. A., Akinloye, O. A., & Omeike, S. O. (2017). Enzymatic biodiesel production from palm oil and palm kernel oil using free lipase. *Egyptian Journal of Petroleum, 26*(3), 635–642. https://doi.org/10.1016/j.ejpe.2016.09.002

Kiatkittipong, W., Phimsen, S., Kiatkittipong, K., Wongsakulphasatch, S., Laosiripojana, N., & Assabumrungrat, S. (2013). Diesel-like hydrocarbon production from hydroprocessing of relevant refining palm oil. *Fuel Processing Technology, 116*, 16–26. https://doi.org/10.1016/j.fuproc.2013.04.018

Kilaz, G., Bist, S., Lopp, D. W., Stanley, D. L., & Tao, B. Y. (2014). Investigation of fatty acid methyl esters in jet fuel. *International Journal of Sustainable Aviation, 1*(1), 103. https://doi.org/10.1504/IJSA.2014.062872

Kim, Y. H., Jun, K.-W., Joo, H., Han, C., & Song, I. K. (2009). A simulation study on gas-to-liquid (natural gas to Fischer-Tropsch synthetic fuel) process optimization. *Chemical Engineering Journal, 155*(1–2), 427–432. https://doi.org/10.1016/j.cej.2009.08.018

Köhler, J., Walz, R., Marscheder-Weidemann, F., & Thedieck, B. (2014). Lead markets in 2nd generation biofuels for aviation: A comparison of Germany, Brazil and the USA. *Environmental Innovation and Societal Transitions, 10*, 59–76. https://doi.org/10.1016/j.eist.2013.10.003

Krahl, J., Knothe, G., Munack, A., et al. (2009). Comparison of exhaust emissions and their mutagenicity from the combustion of biodiesel, vegetable oil, gas-to-liquid and petrodiesel fuels. *Fuel, 88*(6), 1064–1069. https://doi.org/10.1016/j.fuel.2008.11.015

Krár, M., Kovács, S., Kalló, D., & Hancsók, J. (2010). Fuel purpose hydrotreating of sunflower oil on CoMo/Al_2O_3 catalyst. *Bioresource Technology, 101*(23), 9287–9293. https://doi.org/10.1016/j.biortech.2010.06.107

Kröger, M., & Müller-Langer, F. (2012). Review on possible algal-biofuel production processes. *Biofuels, 3*(3), 333–349. https://doi.org/10.4155/bfs.12.14

Kumar, R., Rana, B. S., Tiwari, R., et al. (2010). Hydroprocessing of jatropha oil and its mixtures with gas oil. *Green Chemistry, 12*(12), 2232. https://doi.org/10.1039/c0gc00204f

Kumneadklang, S., Larpkiattaworn, S., Niyasom, C., & O-Thong, S. (2015). Bioethanol production from oil palm frond by simultaneous saccharification and fermentation. *Energy Procedia, 79*, 784–790. https://doi.org/10.1016/j.egypro.2015.11.567

Kurnia, J. C., Jangam, S. V., Akhtar, S., Sasmito, A. P., & Mujumdar, A. S. (2016). Advances in biofuel production from oil palm and palm oil processing wastes: A review. *Biofuel Research Journal, 3*(1), 332–346. https://doi.org/10.18331/BRJ2016.3.1.3

Langholtz, M., Stokes, B., & Eaton, L. (2016). *2016 Billion-ton report: Advancing domestic resources for a thriving bioeconomy.*

Laohalidanond, K., Heil, J., & Wirtgen, C. (2006). The Production of synthetic diesel from biomass. *Current Applications and Science Technology, 6*(1), 35–45.

Lapuerta, M., Armas, O., Hernández, J. J., & Tsolakis, A. (2010). Potential for reducing emissions in a diesel engine by fuelling with conventional biodiesel and Fischer-Tropsch diesel. *Fuel, 89*(10), 3106–3113. https://doi.org/10.1016/j.fuel.2010.05.013

Leung, D. Y. C., Wu, X., & Leung, M. K. H. (2010). A review on biodiesel production using catalyzed transesterification. *Applied Energy, 87*(4), 1083–1095. https://doi.org/10.1016/j.apenergy.2009.10.006

Li, X., Huang, Z., Wang, J., & Zhang, W. (2007). Particle size distribution from a GTL engine. *Science of the Total Environment, 382*(2–3), 295–303. https://doi.org/10.1016/j.scitotenv.2007.04.032

Li, L., Coppola, E., Rine, J., Miller, J. L., & Walker, D. (2010). Catalytic hydrothermal conversion of triglycerides to non-ester biofuels. *Energy & Fuels, 24*(2), 1305–1315. https://doi.org/10.1021/ef901163a

Li, T., Cheng, J., Huang, R., Yang, W., Zhou, J., & Cen, K. (2016). Hydrocracking of palm oil to jet biofuel over different zeolites. *International Journal of Hydrogen Energy, 41*(47), 21883–21887. https://doi.org/10.1016/j.ijhydene.2016.09.013

Liati, A., Schreiber, D., Alpert, P. A., et al. (2019). Aircraft soot from conventional fuels and biofuels during ground idle and climb-out conditions: Electron microscopy and X-ray micro-spectroscopy. *Environmental Pollution, 247*, 658–667. https://doi.org/10.1016/j.envpol.2019.01.078

Lin, B.-F., Huang, J.-H., & Huang, D.-Y. (2008). Effects of biodiesel from palm kernel oil on the engine performance, exhaust emissions, and combustion characteristics of a direct injection diesel engine. *Energy & Fuels., 22*(6), 4229–4234. https://doi.org/10.1021/ef800338j

Liu, Y. (2018). Catalytic ethylene oligomerization over Ni/Al-HMS: A key step in conversion of bio-ethanol to higher olefins. *Catalysts, 8*(11), 537. https://doi.org/10.3390/catal8110537

Liu, Y., Sotelo-Boyás, R., Murata, K., Minowa, T., & Sakanishi, K. (2009). Hydrotreatment of Jatropha oil to produce green diesel over trifunctional Ni–Mo/SiO_2–Al_2O_3 catalyst. *Chemistry Letters, 38*(6), 552–553. https://doi.org/10.1246/cl.2009.552

Liu, G., Yan, B., & Chen, G. (2013). Technical review on jet fuel production. *Renewable and Sustainable Energy Reviews, 25*, 59–70. https://doi.org/10.1016/j.rser.2013.03.025

Liu, K., Zhao, W., Guo, T., et al. (2019). Emulsification and performance measurement of bio-oil with diesel. *Waste and Biomass Valorization.* https://doi.org/10.1007/s12649-019-00917-1

Llamas, A., Al-Lal, A.-M., Hernandez, M., Lapuerta, M., & Canoira, L. (2012a). Biokerosene from Babassu and Camelina oils: Production and properties of their blends with fossil kerosene. *Energy & Fuels, 26*(9), 5968–5976. https://doi.org/10.1021/ef300927q

Llamas, A., García-Martínez, M., Al-Lal, A.-M., Canoira, L., & Lapuerta, M. (2012b). Biokerosene from coconut and palm kernel oils: Production and properties of their blends with fossil kerosene. *Fuel, 102*, 483–490. https://doi.org/10.1016/j.fuel.2012.06.108

Lobo, P., Hagen, D. E., & Whitefield, P. D. (2011). Comparison of PM emissions from a commercial jet engine burning conventional, biomass, and Fischer-Tropsch fuels. *Environmental Science and Technology, 45*(24), 10744–10749. https://doi.org/10.1021/es201902e

Luo, Z., Wang, S., Liao, Y., Zhou, J., Gu, Y., & Cen, K. (2004). Research on biomass fast pyrolysis for liquid fuel. *Biomass and Bioenergy., 26*(5), 455–462. https://doi.org/10.1016/j.biombioe.2003.04.001

Lutz, E. F. (1986). Shell higher olefins process. *Journal of Chemical Education, 63*(3), 202. https://doi.org/10.1021/ed063p202

Maliutina, K., Tahmasebi, A., Yu, J., & Saltykov, S. N. (2017). Comparative study on flash pyrolysis characteristics of microalgal and lignocellulosic biomass in entrained-flow reactor. *Energy Conversion and Management, 151*, 426–438. https://doi.org/10.1016/j.enconman.2017.09.013

MASBI. (2013). *Fueling a sustainable future for aviation.*

Mata, T. M., Martins, A. A., & Caetano, N. S. (2010). Microalgae for biodiesel production and other applications: A review. *Renewable and Sustainable Energy Reviews, 14*(1), 217–232. https://doi.org/10.1016/j.rser.2009.07.020

Matamba, T., Tahmasebi, A., Khoshk Rish, S., & Yu, J. (2020). Promotion effects of pressure on polycyclic aromatic hydrocarbons and H_2 formation during flash pyrolysis of palm kernel shell. *Energy & Fuels, 34*(3), 3346–3356. https://doi.org/10.1021/acs.energyfuels.9b04409

Mayorga, M. A., Lopez, M., Lopez, C. A., et al. (2020). Production of aviation biofuel from palm kernel oil. *Chemical Engineering Transactions, 80*, 319–324. https://doi.org/10.3303/CET2080054

Mayorga Betancout, M. A., Cadavid Estrada, J. G., Bonilla Paez, J. A., Lopez Santamaria, C. A., Gomez, M. L., & Galindo, J. M. (2019). Use of biofuels in the aeronautical industry: Colombian air force case. *Tecciencia, 14*(26), 53–63. https://doi.org/10.18180/tecciencia.2019.14.7

McAfee, E. A. (2012). *The aemetis biorefinery: 100% replacement, renewable jet and diesel fuels by conversion of existing biofuels refinery facilities.* Advanced Biofuels Marketing Conference. https://www.aemetis.com/wp-content/uploads/2012/11/Aemetis-Jet-Fuel-Website.pdf. Published 2012.

McCalmont, J. P., Hastings, A., McNamara, N. P., et al. (2017). Environmental costs and benefits of growing Miscanthus for bioenergy in the UK. *GCB Bioenergy, 9*(3), 489–507. https://doi.org/10.1111/gcbb.12294

McGill, R., Aakko-Saksa, P., & Nylund, N.-O. (2009). *Final report—Analysis of biodiesel options.*

Mekhilef, S., Siga, S., & Saidur, R. (2011). A review on palm oil biodiesel as a source of renewable fuel. *Renewable and Sustainable Energy Reviews, 15*(4), 1937–1949. https://doi.org/10.1016/j.rser.2010.12.012

Melero, J. A., García, A., & Iglesias, J. (2011). Biomass catalysis in conventional refineries. In *Advances in clean hydrocarbon fuel processing* (pp. 199–240). Elsevier. https://doi.org/10.1533/9780857093783.2.199

Morgan, T., Grubb, D., Santillan-Jimenez, E., & Crocker, M. (2010). Conversion of triglycerides to hydrocarbons over supported metal catalysts. *Topics in Catalysis, 53*(11–12), 820–829. https://doi.org/10.1007/s11244-010-9456-1

Morgan, T. J., Youkhana, A., Turn, S. Q., Ogoshi, R., & Garcia-Pérez, M. (2019). Review of biomass resources and conversion technologies for alternative jet fuel production in Hawai'i and tropical regions. *Energy & Fuels., 33*(4), 2699–2762. https://doi.org/10.1021/acs.energyfuels.8b03001

Mortensen, P. M., Grunwaldt, J.-D., Jensen, P. A., Knudsen, K. G., & Jensen, A. D. (2011). A review of catalytic upgrading of bio-oil to engine fuels. *Applied Catalysis, a: General, 407*(1–2), 1–19. https://doi.org/10.1016/j.apcata.2011.08.046

MPOB. (2020). *Environmental impact*. Official Palm Oil Information Source. Accessed June 6, 2020. http://www.palmoilworld.org/environment.html. Published 2020.

MPOC. (2020). *The Oil palm tree.* Malaysian Palm Oil Council (MPOC). Accessed June 6, 2020. http://mpoc.org.my/the-oil-palm-tree/. Published 2020.

Müller-Langer, F., Klemm, M., & Schneider, J. (2017). Biofuels production processes and technologies. In M. R. Riazi & D. Chiaramonti (Eds.), *Biofuels production and processing technology* (1st ed., pp. 153–182). CRC Press.

Murata, K., Liu, Y., Inaba, M., & Takahara, I. (2010). Production of synthetic diesel by hydrotreatment of Jatropha oils using Pt–Re/H-ZSM-5 catalyst. *Energy & Fuels, 24*(4), 2404–2409. https://doi.org/10.1021/ef901607t

Muthukumaran, C., Sharmila, G., Manojkumar, N., Gnanaprakasam, A., & Sivakumar, V. M. (2020). Optimization and kinetic modeling of biodiesel production. In *Encyclopedia of renewable and sustainable materials* (pp. 193–201). Elsevier. https://doi.org/10.1016/B978-0-12-803581-8.10578-8

Nanda, S., Mohammad, J., Reddy, S. N., Kozinski, J. A., & Dalai, A. K. (2014). Pathways of lignocellulosic biomass conversion to renewable fuels. *Biomass Conversion and Biorefinery, 4*(2), 157–191. https://doi.org/10.1007/s13399-013-0097-z

Nascimento, M. A. R., Lora, E. S., Corrêa, P. S. P., et al. (2008). Biodiesel fuel in diesel microturbine engines: Modelling and experimental evaluation. *Energy, 33*(2), 233–240. https://doi.org/10.1016/j.energy.2007.07.014

Nasreen, S., Nafees, M., Qureshi, L. A., Asad, M. S., Sadiq, A., & Ali, S. D. (2018). Review of catalytic transesterification methods for biodiesel production. In: *Biofuels—State of development*. InTech. https://doi.org/10.5772/intechopen.75534

OJEU. (2009). *Renewable energy directive (2009/28/EC).*

Okoro, O., Sun, Z., & Birch, J. (2018). Catalyst-free biodiesel production methods: A comparative technical and environmental evaluation. *Sustainability, 10*(2), 127. https://doi.org/10.3390/su10010127

Omar, S., Alsamaq, S., Yang, Y., & Wang, J. (2019). Production of renewable fuels by blending bio-oil with alcohols and upgrading under supercritical conditions. *Frontiers of Chemical Science and Engineering, 13*(4), 702–717. https://doi.org/10.1007/s11705-019-1861-9

Park, J.-W., Heo, J., Ly, H. V., Kim, J., Lim, H., & Kim, S.-S. (2019). Fast pyrolysis of acid-washed oil palm empty fruit bunch for bio-oil production in a bubbling fluidized-bed reactor. *Energy, 179*, 517–527. https://doi.org/10.1016/j.energy.2019.04.211

Patel, N. K., & Shah, S. N. (2015) Biodiesel from plant oils. In: *Food, energy, and water* (pp. 277–307). Elsevier. https://doi.org/10.1016/B978-0-12-800211-7.00011-9

Patil, V., Tran, K.-Q., & Giselrød, H. R. (2008). Towards sustainable production of biofuels from microalgae. *International Journal of Molecular Sciences, 9*(7), 1188–1195. https://doi.org/10.3390/ijms9071188

Paur, J. (2011). *KLM completes first schedules service flight using biofuel*. Wired. https://www.wired.com/2011/07/klm-completes-first-scheduled-service-flight-using-biofuel/. Published 2011.

Pearlson, M. N. (2011). *A techno-economic and environment assessment of hydroprocessed renewable distillate fuels.*

Peterson, A. A., Vogel, F., Lachance, R. P., Fröling, M., Antal, M. J., Jr., & Tester, J. W. (2008). Thermochemical biofuel production in hydrothermal media: A review of sub- and supercritical water technologies. *Energy & Environmental Science, 1*(1), 32. https://doi.org/10.1039/b810100k

PTI. (2018, August 27). SpiceJet operated India's first biofuel-powered flight from Dehradun to Delhi. *The Economic Times.*

Ramirez, J., Brown, R., & Rainey, T. (2015). A review of hydrothermal liquefaction bio-crude properties and prospects for upgrading to transportation fuels. *Energies, 8*(7), 6765–6794. https://doi.org/10.3390/en8076765

Ramos, L. P., Cordeiro, C. S., Cesar-Oliveira, M. A. F., Wypych, F., & Nakagaki, S. (2014). Applications of heterogeneous catalysts in the production of biodiesel by esterification and transesterification. In *Bioenergy research: Advances and applications* (pp. 255–276). Elsevier. https://doi.org/10.1016/B978-0-444-59561-4.00016-4

Ranucci, C. R., Alves, H. J., & Monteiro, M. R., et al. (2018). Potential alternative aviation fuel from jatropha (Jatropha curcas L.), babassu (Orbignya phalerata) and palm kernel (Elaeis guineensis) as blends with Jet-A1 kerosene. *Journal of Clean Production, 185*, 860–869. https://doi.org/10.1016/j.jclepro.2018.03.084

Reaney, M. J., & Ratanapariyanuch, K. (2010). *Process for the extraction of macromolecules from a biomass using thin stillage* (pp. 1–32).

Rehman, A., Phalke, D. R., & Pandey, R. (2011). Alternative fuel for gas turbine: Esterified jatropha oil–diesel blend. *Renewable Energy, 36*(10), 2635–2640. https://doi.org/10.1016/j.renene.2010.06.013

Ronsse, F., van Hecke, S., Dickinson, D., & Prins, W. (2013). Production and characterization of slow pyrolysis biochar: Influence of feedstock type and pyrolysis conditions. *GCB Bioenergy, 5*(2), 104–115. https://doi.org/10.1111/gcbb.12018

Rosillo-Calle, F., de Groot, P., Hemstock, S. L., & Woods, J. (2008). *The biomass assessment handbook: Bioenergy for a sustainable environment*. Earthscan.

Ross, S. (2019). 4 countries that product the most food. *Investopedia.*

Rozmysłowicz, B., Mäki-Arvela, P., Lestari, S., et al. (2010). Catalytic deoxygenation of tall oil fatty acids over a palladium-mesoporous carbon catalyst: A new source of biofuels. *Topics in Catalysis, 53*(15–18), 1274–1277. https://doi.org/10.1007/s11244-010-9581-x

Rumizen, M. (2013). *ASTM D4054 users' guide.*

Saeidi, S., Talebi Amiri, M., Saidina Amin, N. A., & Rahimpour, M. R. (2014). Progress in reactors for high-temperature Fischer-Tropsch process: Determination place of intensifier reactor perspective. *International Journal of Chemical Reactor Engineering, 12*(1), 639–664. https://doi.org/10.1515/ijcre-2014-0045

Sánchez, Ó. J., Montoya, S. (2013). Production of bioethanol from biomass: An overview. In *Biofuel technologies* (pp. 397–441). Springer Berlin Heidelberg. https://doi.org/10.1007/978-3-642-34519-7_16

Sarwono, R., Pusfitasari, E. D., & Ghozali, M. (2016). *Hydrothermal liquefaction of palm oil empty fruit bunch (EFB) into bio-oil in different organic solvents* (pp. 060015). https://doi.org/10.1063/1.4949322

SAS. (2014). SAS today operates a biofuel (cooking oil) flight from Stockholm. *World Airline News.*

Schaberg, P. W., Myburgh, I. S., Botha, J. J., & Khalek, I. A. (2000). *Comparative emissions performance of sasol Fischer-Tropsch diesel fuel in current and older technology heavy-duty engines.* https://doi.org/10.4271/2000-01-1912

Shell. (2021). *Military jet fuel specification. Aviation fuel—Military jet fuel.* https://www.shell.com/business-customers/aviation/aviation-fuel/military-jet-fuel-grades.html. Published 2021.

Silvester, L., Lamonier, J.-F., Faye, J., et al. (2015). Reactivity of ethanol over hydroxyapatite-based Ca-enriched catalysts with various carbonate contents. *Catalysis Science & Technology, 5*(5), 2994–3006. https://doi.org/10.1039/C5CY00327J

Singh, T., & Naveen, S. M. (2017). A review paper on production of linear alpha-olefins by undergoing oligomerization of ethylen. *International Journal of Engineering Science and Applied Science Technology, 2*(4), 83–86.

Singh, J., & Trivedi, J. (2017). Raw materials for biofuels production. In M. R. Riazi & D. Chiaramonti (Eds.), *Biofuels production and processing technology* (1st ed., pp. 127–152). CRC Press.

Sinha, A. K., Anand, M., & Farooqui, S. A. (2016). Aviation biofuels through lipid hydroprocessing. In *Biofuels for aviation* (pp. 85–108). Elsevier. https://doi.org/10.1016/B978-0-12-804568-8.00005-6

Siramon, P., Punsuvon, V., & Vaithanomsat, P. (2018). Production of bioethanol from oil palm empty fruit bunch via acid impregnation-steam explosion pretreatment. *Waste and Biomass Valorization., 9*(8), 1407–1414. https://doi.org/10.1007/s12649-017-9924-y

SkyNRG. (2021). *Sustainable aviation fuel.* https://skynrg.com/sustainable-aviation-fuel/saf/. Published 2021.

Snåre, M., Kubičková, I., Mäki-Arvela, P., Eränen, K., & Murzin, D. Y. (2006). Heterogeneous catalytic deoxygenation of stearic acid for production of biodiesel. *Industrial and Engineering Chemistry Research, 45*(16), 5708–5715. https://doi.org/10.1021/ie060334i

Solikhah, M. D., Pratiwi, F. T., Heryana, Y., et al. (2018). Characterization of bio-oil from fast pyrolysis of palm frond and empty fruit bunch. *IOP Conference Series: Materials Science and Engineering, 349*, 012035. https://doi.org/10.1088/1757-899X/349/1/012035

Sotelo-Boyas, R., Trejo-Zarraga, F., & de Jesus Hernandez-Loyo, F. (2012). Hydroconversion of triglycerides into green liquid fuels. In: *Hydrogenation*. InTech. https://doi.org/10.5772/48710

Stepacheva, A. A., Sapunov, V. N., Sulman, E. M. S. M., et al. (2016). Catalytic hydrodeoxygenation of fatty acids for biodiesel production. *Bulletin of Chemical Reaction Engineering & Catalysis, 11*(2), 125. https://doi.org/10.9767/bcrec.11.2.538.125-132

Sun, P., Heng, M., Sun, S., & Chen, J. (2010). Direct liquefaction of paulownia in hot compressed water: Influence of catalysts. *Energy, 35*(12), 5421–5429. https://doi.org/10.1016/j.energy.2010.07.005

Sutrisno, B., & Hidayat, A. (2018). Pyrolysis of palm empty fruit bunch: Yields and analysis of bio-oil. In: S. Ma'mun, H. Tamura, & M. R. A. Purnomo (Eds.), *MATEC Web Conference*, (Vol. 154, pp. 01036). https://doi.org/10.1051/matecconf/201815401036

Tabuchi, H. (2019). "Worse than anyone expected": Air travel emissions vastly outpace predictions. *The New York Times*.

Talebian-Kiakalaieh, A., Amin, N. A. S., & Mazaheri, H. (2013). A review on novel processes of biodiesel production from waste cooking oil. *Applied Energy, 104*, 683–710. https://doi.org/10.1016/j.apenergy.2012.11.061

Thangaraj, B., Solomon, P. R., Muniyandi, B., Ranganathan, S., & Lin, L. (2019). Catalysis in biodiesel production—a review. *Clean Energy, 3*(1), 2–23. https://doi.org/10.1093/ce/zky020

Thomas, G. K., Eric, D. L., Liu, G., & Robert, H. W. (2008). *Fischer-Tropsch fuels from coal and biomass*.

Timko, M. T., Herndon, S. C., de la Rosa, B. E., et al. (2011). Combustion products of petroleum jet fuel, a Fischer-Tropsch synthetic fuel, and a biomass fatty acid methyl ester fuel for a gas turbine engine. *Combustion Science and Technology, 183*(10), 1039–1068. https://doi.org/10.1080/00102202.2011.581717

Tiwari, R., Rana, B. S., Kumar, R., et al. (2011). Hydrotreating and hydrocracking catalysts for processing of waste soya-oil and refinery-oil mixtures. *Catalysis Communications, 12*(6), 559–562. https://doi.org/10.1016/j.catcom.2010.12.008

Toor, S. S., Rosendahl, L., & Rudolf, A. (2011). Hydrothermal liquefaction of biomass: A review of subcritical water technologies. *Energy, 36*(5), 2328–2342. https://doi.org/10.1016/j.energy.2011.03.013

Transesterification of waste cooking oil in biodiesel production utilizing CaO/Al_2O_3 heterogeneous catalyst. *Malaysian Jounral of Analytical Science, 22*(1). https://doi.org/10.17576/mjas-2018-2201-20

UAE. (2019). *Etihad Airways flies the world's first flight using fuel made in the UAE from plants grown in saltwater by Khalifa University*. Etihad Airways. https://www.etihad.com/en-my/news/etihad-airways-flies-the-worlds-first-flight--using-fuel-made-in-the-uae-from-plants-grown-in-saltwater-by-khalifa-university. Published 2019.

UK Defence Standardization. (2019). *Turbine fuel, kerosene type, Jet A-1*.

UNFCCC. *Adoption of Paris agreement*.

Varma, A. K., Thakur, L. S., Shankar, R., & Mondal, P. (2019). Pyrolysis of wood sawdust: Effects of process parameters on products yield and characterization of products. *Waste Management, 89*, 224–235. https://doi.org/10.1016/j.wasman.2019.04.016

Venderbosch, R. H., Ardiyanti, A. R., Wildschut, J., Oasmaa, A., & Heeres, H. J. (2010). Stabilization of biomass-derived pyrolysis oils. *Journal of Chemical Technology and Biotechnology, 85*(5), 674–686. https://doi.org/10.1002/jctb.2354

Verma, D., Kumar, R., Rana, B. S., & Sinha, A. K. (2011). Aviation fuel production from lipids by a single-step route using hierarchical mesoporous zeolites. *Energy & Environmental Science, 4*(5), 1667. https://doi.org/10.1039/c0ee00744g

Wang, W.-C. (2016). Techno-economic analysis of a bio-refinery process for producing hydro-processed renewable Jet fuel from Jatropha. *Renewable Energy, 95*, 63–73. https://doi.org/10.1016/j.renene.2016.03.107

Wang, W.-C., & Hsieh, C.-H. (2020). Hydro-processing of biomass-derived oil into straight-chain alkanes. *Chemical Engineering Research and Design, 153*, 63–74. https://doi.org/10.1016/j.cherd.2019.10.030

Wang, W.-C., & Tao, L. (2016). Bio-jet fuel conversion technologies. *Renewable and Sustainable Energy Reviews, 53*, 801–822. https://doi.org/10.1016/j.rser.2015.09.016

Wang, W.-C., Turner, T. L., Stikeleather, L. F., & Roberts, W. L. (2012a). Exploration of process parameters for continuous hydrolysis of canola oil, camelina oil and algal oil. *Chemical Engineering and Processing: Process Intensification, 57–58*, 51–58. https://doi.org/10.1016/j.cep.2012.04.001

Wang, W.-C., Turner, T. L., Roberts, W. L., & Stikeleather, L. F. (2012b). Direct injection of superheated steam for continuous hydrolysis reaction. *Chemical Engineering and Processing: Process Intensification, 59*, 52–59. https://doi.org/10.1016/j.cep.2012.04.003

Wang, W.-C., Thapaliya, N., Campos, A., Stikeleather, L. F., & Roberts, W. L. (2012c). Hydrocarbon fuels from vegetable oils via hydrolysis and thermo-catalytic decarboxylation. *Fuel, 95*, 622–629. https://doi.org/10.1016/j.fuel.2011.12.041

Wang, W.-C., Tao, L., & Markham, J. J. et al. (2016). *Review of biojet fuel conversion technologies.*

Weerachanchai, P., Tangsathitkulchai, C., & Tangsathitkulchai, M. (2011). Characterization of products from slow pyrolysis of palm kernel cake and cassava pulp residue. *Korean Journal of Chemical Engineering, 28*(12), 2262–2274. https://doi.org/10.1007/s11814-011-0116-3

Wei, H., Liu, W., Chen, X., Yang, Q., Li, J., & Chen, H. (2019). Renewable bio-jet fuel production for aviation: A review. *Fuel, 254*, 115599. https://doi.org/10.1016/j.fuel.2019.06.007

Winchester, N., McConnachie, D., Wollersheim, C., & Waitz, I. A. (2013). Economic and emissions impacts of renewable fuel goals for aviation in the US. *Transportation Research Part a: Policy and Practice, 58*, 116–128. https://doi.org/10.1016/j.tra.2013.10.001

Wu, C.-Y., & Wu, H.-S. (2017). Ethylene formation from ethanol dehydration using ZSM-5 catalyst. *ACS Omega, 2*(8), 4287–4296. https://doi.org/10.1021/acsomega.7b00680

Yaakob, Z., Mohammad, M., Alherbawi, M., Alam, Z., & Sopian, K. (2013). Overview of the production of biodiesel from Waste cooking oil. *Renewable and Sustainable Energy Reviews, 18*, 184–193. https://doi.org/10.1016/j.rser.2012.10.016

Yatish, T. S. (2015). *Energy and fuel systems integration* (1st ed.). CRC.

Yim, S. C., Quitain, A. T., Yusup, S., Sasaki, M., Uemura, Y., & Kida, T. (2017). Metal oxide-catalyzed hydrothermal liquefaction of Malaysian oil palm biomass to bio-oil under supercritical condition. *Journal of Supercritical Fluids, 120*, 384–394. https://doi.org/10.1016/j.supflu.2016.05.044

Zein, M., & Winter, R. (2000). Effect of temperature, pressure and lipid acyl chain length on the structure and phase behaviour of phospholipid–gramicidin bilayers. *Physical Chemistry Chemical Physics, 2*(20), 4545–4551. https://doi.org/10.1039/b003565n

Zhan, N., Hu, Y., Li, H., Yu, D., Han, Y., & Huang, H. (2010). Lanthanum–phosphorous modified HZSM-5 catalysts in dehydration of ethanol to ethylene: A comparative analysis. *Catalysis Communications, 11*(7), 633–637. https://doi.org/10.1016/j.catcom.2010.01.011

Zhang, X., Wilson, K., & Lee, A. F. (2016). Heterogeneously catalyzed hydrothermal processing of C5–C6 Sugars. *Chemical Reviews, 116*(19), 12328–12368. https://doi.org/10.1021/acs.chemrev.6b00311

Zhang, S., Yang, X., Zhang, H., et al. (2019). Liquefaction of biomass and upgrading of bio-oil: A review. *Molecules, 24*(12), 2250. https://doi.org/10.3390/molecules24122250

Zhang, L., Butler, T. L., & Yang, B. (2020). Recent trends, opportunities and challenges of sustainable aviation fuel. In *Green energy to sustainability* (pp. 85–110). Wiley. https://doi.org/10.1002/9781119152057.ch5

Oil Palm-Based Nanocellulose: From Extraction to Applications

Hong Jun Lim, Wai Kit Cheng, Khang Wei Tan, and Lih Jiun Yu

Abstract The palm oil industry is the largest contributor of biomass in Malaysia where vast amount of oil palm biomass waste is generated annually while only a small fraction is being converted into value-added products. The remaining is either burnt or left to decompose at the plantations, which will emit hazardous gases and often resulting in acute air pollution. Much effort has been invested for a more sustainable palm oil circular economy, in which the exploitation of lignocellulosic residues for nanomaterials production can be the golden answer. Nanocellulose may prove to be one of the most auspicious green materials for nanocomposite processing owing to its superb mechanical properties, abundance, renewability and biodegradability. Nanocellulose has garnered increasing attention over the last few decades due to its great potential in diverse applications including food industry, biomedical field, environmental remediation, construction composite materials, corrosion protection and catalysis. This chapter provided a comprehensive review of the recent advancements of oil palm-based nanocellulose from its initial isolation, characterizations to the final applications. Ultimately, the challenges and opportunities for future development associated with the commercialization of nanocellulose-based materials were also thoroughly canvassed.

Keywords Biomass · Nanocellulose · Nanocomposite · Food packaging · Biomedical · Effluent treatment

---

H. J. Lim · W. K. Cheng · K. W. Tan (✉)
School of Energy and Chemical Engineering, Xiamen University Malaysia, Selangor Darul Ehsan, 43900 Sepang, Malaysia
e-mail: khangwei.tan@xmu.edu.my

L. J. Yu
Faculty of Engineering, Technology and Built Environment, UCSI University, Kuala Lumpur Campus, No. 1, Jalan Menara Gading, UCSI Heights (Taman Connaught), Cheras, 56000 Kuala Lumpur, Malaysia

D. C. Y. Foo et al. (eds.), *Sustainable Technologies for the Oil Palm Industry*,
https://doi.org/10.1007/978-981-19-4847-3_3

1 Introduction

Biomass can be defined as organic matters consisting of carbon, oxygen, nitrogen and hydrogen originated from agricultural wastes and residues, forestry, as well as the biodegradable fractions of industrial and municipal wastes. Biomass is widely known for its potentials to supply myriads of different bioenergy such as biogas, briquettes and biofuels that contribute to the reduction of global greenhouse gas emissions. Biomass also plays an indispensable role in the production of renewable chemicals and biomaterial products including bio-fertilizers and bio-composites. Oil palm biomass refers to agricultural by-products originated from the palm oil industry during milling, replanting and pruning activities (Onoja et al., 2019). It can generally be classified into oil palm trunks (OPT), oil palm fronds (OPF), empty fruit bunches (EFB), mesocarp fiber (MF), palm kernel shell (PKS) and palm oil mill effluent (POME). OPT and OPF are typical examples of oil palm biomass produced at the plantation ; while EFB, MF, PKS and POME are generated at the oil palm processing mills (Onoja et al., 2019; Padzil et al., 2020).

Malaysia is well-endowed with oil palm and is the second largest producer after Indonesia. The palm oil industry is the fourth largest contributor to the Malaysian's Gross National Income that help generates billions of Ringgits (Onoja et al., 2019). Oil palm is the most important and cultivated agricultural crop in Malaysia, with a total of 5.87 million hectares of plantation area in the year 2020 (Parveez et al., 2021). From the year 2017 to 2019, the average amount of oil palm biomass generated in Malaysia was approximately 22.42 million tonnes of EFB, 7.13 million tonnes of PKS and 71.34 million tonnes of POME. As of February 2020, the amount of EFB, PKS and POME generated were around 2.80 million tonnes, 0.89 million tonnes and 8.92 million tonnes, respectively (Rubinsin et al., 2021). This indicated that oil palm biomass is readily available in Malaysia. Howbeit, only a small fraction of the produced biomass is being converted into value-added products, whereas the remaining large portion are left underutilized. Conventionally, oil palm biomass is either burnt or left at the plantations to decompose naturally, therefore contributing to environmental hazards due to the emission of harmful gases. In order to circumvent the issues raised by the improper disposal of oil palm biomass, future researches are salient to convert this abundant and renewable biomass into different value-added products with various prospective applications (Onoja et al., 2019) (Table 1).

1.1 Lignocellulosic Components of Oil Palm Biomass

Oil palm biomass is rich in lignocellulosic, except POME. As elucidated in Table 2, the solid oil palm biomass is mainly comprised of cellulose, hemicellulose and lignin in different proportion, depending on the biomass types. These lignocellulosic biomass materials can be regarded as renewable, sustainable and non-toxic (Onoja et al., 2019; Padzil et al., 2020). Cellulose represents one of the most abundant

Table 1 Different types of oil palm biomass (Onoja et al., 2019; Padzil et al., 2020)

Biomass type	Description	Site of production
 Oil palm trunk	Tree trunks of the oil palmtree	Plantation
 Oil palm frond	Leaves from the oil palmtree	Plantation
 Empty fruit bunch	Remains of fruit bunch after the removal of palm fruits	Mill
 Mesocarp fibre	Remains after crude palm oilextraction from mesocarp	Mill
 Palm kernel shell	Remains after palm kernel oil extraction	Mill
 Palm oil mill effluent	Liquid by-product produced from sterilization and other milling process of fresh fruit bunch	Mill

natural polymers on earth with the annual production of 1.5×10^{12} tons and consists of unique hierarchical structures made up from a millimeter-sized transverse section to angstrom-sized anhydroglucose units (Guzman-Puyol et al., 2019; Thomas et al., 2018). Generally, cellulose is made up of linear β-D-glucopyranose units linked together by β-1,4-glycosidic bonds (Mishra et al., 2018; Sinaga et al., 2018). One of the most prominent and functional characteristics of cellulose is its high degree of hydroxylation along the polymer chains in which each of the non-terminal monomer consists of 3 hydroxyl groups at C_2, C_3 and C_6 atoms (Fig. 1a) (Thomas et al.,

Table 2 Lignocellulosic content of different oil palm biomass

Biomasstype	Cellulose (%)	Hemicellulose (%)	Lignin (%)	References
Oil palm empty fruit bunch	34.10	31.90	25.90	Gan et al. (2020)
Oil palm frond	40.03	27.08	20.00	Azani et al. (2020)
Oil palm trunk	34.44	23.94	35.89	Onoja et al. (2019)
Oil palm mesocarp fiber	17.30	17.90	43.50	Campos et al. (2017)
Palm kernel shell	27.70	21.60	44.00	Onoja et al. (2019)
Oil palm leaf	32.49	22.97	26.00	Onoja et al. (2019)

2018; Vilarinho et al., 2018). Consequently, the hydrogen bonding arises between hydroxyl groups and oxygen atoms that are positioned within the same cellulose molecule (intramolecular) and between the adjacent cellulose chains (intermolecular). These intermolecular hydrogen bonding give rise to the fibrillar structures and semi-crystalline packing of cellulose which endow it with attractive physical properties including high strength and flexibility (Thomas et al., 2018). As elucidated in Fig. 1b, hemicellulose is made up of various monosaccharide units (mainly pentose sugars) and it exists as either homopolymer or heteropolymer. Hemicellulose can be hydrolyzed easily since it is highly soluble in alkalis. Conversely, lignin is an aromatic polymer that contains randomly substituted phenylpropane monomeric units such as syringyl, guaiacyl and p-hydroxyphenyl units (Fig. 1c) (Onoja et al., 2019). It plays a pivotal role in providing structural support in plants (Malucelli et al., 2017). Cellulose became a cynosure of all eyes within the research community due to its environmental friendliness and attractive features such as non-toxicity, low density, high biodegradability, biocompatibility, excellent thermal and mechanical properties, renewability and easy modification (Dai et al., 2018; Mohamed et al., 2015). The relatively high amount of cellulose in the oil palm biomass rendered it a very promising feedstock for the extraction of nanocellulose to produce a plethora of technologically advanced materials for various industrial applications.

1.2 Nanocellulose

Nanocellulose is a novel class of biopolymer in nanoscale dimensions that is creating a revolution in bio-based materials in the twenty-first century for myriads of interdisciplinary applications such as packaging, biomedical, pharmaceuticals, membrane, 3D printing, energy devices and flexible electronics (Fang et al., 2019; Yahya et al., 2015). This is because nanocellulose possesses distinctive properties including high reinforcing strength and stiffness which is often times comparable to Kevlar and steel, large surface area, remarkable optical properties, tailored crystallinity and easy surface functionalization. Nanocellulose is also biodegradable and renewable which

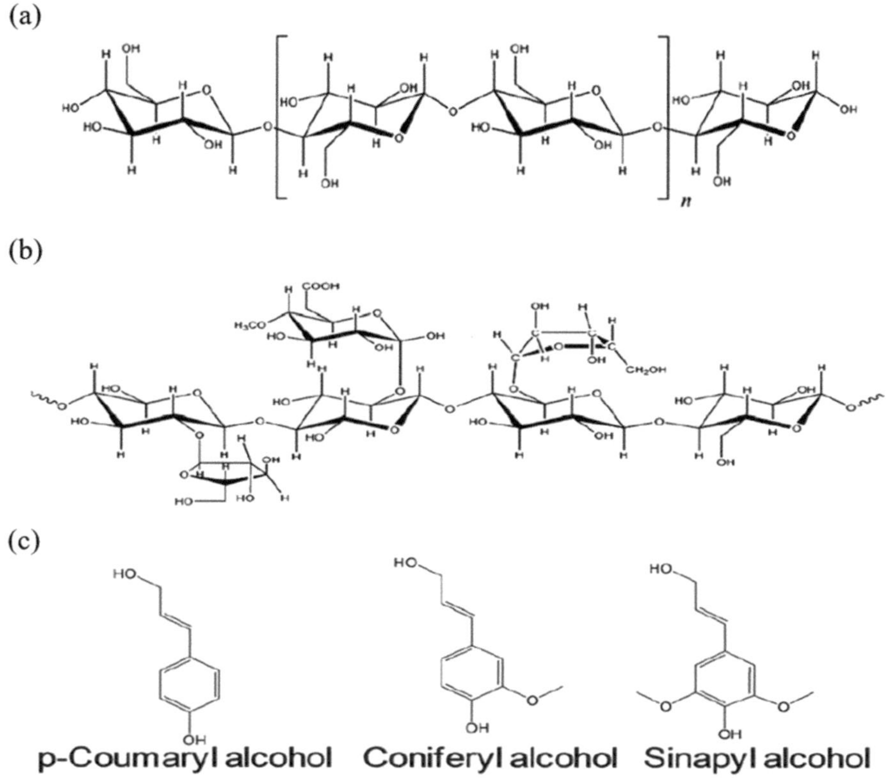

Fig. 1 Chemical structure of **a** cellulose, **b** hemicellulose and **c** lignin (Alonso et al., 2012)

renders it a propitious green alternative for further scrutiny by both the industry and academia (Yahya et al., 2015). Concerning the superior characteristics of nanocellulose, its global market demand stood at USD 87.5 million back in year 2016 and is expected to reach approximately USD 699.6 million in year 2023 with a staggering compound annual growth rate of 33.8%. Some of the key market players of nanocellulose are CelluForce Inc., Fpinnovations, CelluComp, Nippon Paper Group Inc. and Daicel with the current industrial production at tons per day scale. Withal, companies such as Borregaard, Stora Enso and Innventia also manufacture nanocellulose but in a much smaller laboratory scale (Tan et al., 2019).

Conventionally, nanocellulose can be classified into three different categories namely cellulose nanocrystals (CNCs), cellulose nanofibers (CNFs) and bacterial nanocellulose (BNC) according to their morphology and origins. Both CNCs and CNFs are obtained through a top-down approach by disintegrating plant matter through chemical or mechanical treatment. For instance, acid hydrolysis is one of the most common chemical synthesis approaches to produce the highly crystalline or needle-like fragments of CNCs. Contrarily, the mechanical shearing techniques (e.g., high pressure homogenization, grinding and cryocrushing) will disintegrate

Table 3 Classification of nanocellulose

Type of nanocellulose	Synonyms	Typical sources	Average size	Images
Cellulose nanocrystal	Nanocrystalline cellulose, cellulose nanowhiskers and cellulose nanorods	Wood, cotton, hemp, flex, wheat straw, rice straw, ramie, Avicel, tunicin, algae, bacteria etc.	Diameter: 5–70 nm Length: 100–250 nm (from plant); 100 nm-several micrometers (from cellulose of tunicates, algae, bacteria)	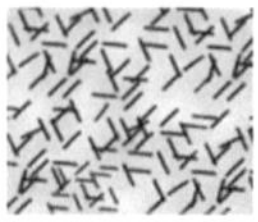
Cellulose nanofiber	Nanofibrils, nanofibrillated cellulose and microfibrillated cellulose	Wood, hemp, flax, sugar beet, potato tuber etc.	Diameter: 5–60 nm Length: several micrometers	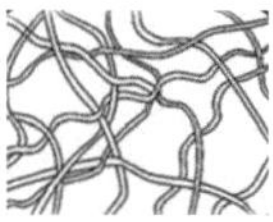
Bacterial nanocellulose	Microbial cellulose, biocellulose	Low molecular weight sugars and alcohols	Diameter: 20–100 nm	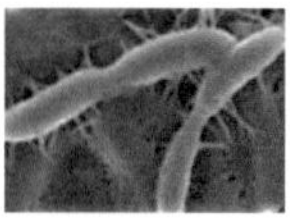

cellulose fibers into their sub-structural nanoscale units, producing longer CNFs that are micrometric in length. Furthermore, BNC is synthesized via the bottom-up approach by utilizing various bacteria cultures to produce the nanomaterial with desired properties (Bharimalla et al., 2017; Thomas et al., 2018). Table 3 shows the summary of the nomenclature used for the three main types of nanocellulose, their origins and average sizes. The next section further elucidates the different nanocellulose extraction methods and their distinct properties.

2 Extraction of Nanocellulose from Oil Palm Biomass

The techniques in which nanocellulose is extracted from oil palm biomass play a pivotal role on the morphology and properties of the resultant materials. This section will examine the main isolation methods used for the extraction of nanocellulose including mechanical treatment, chemical treatment, enzymatic treatment or any combination of these three strategies.

2.1 Pre-treatment

Conventionally, the oil palm biomass obtained from various sources should be milled or triturated to increase the surface area to enhance the subsequent treatments, followed by washing with deionized water to remove any soluble impurities that may be present on the raw materials before further processing. Milling to powder form is necessary to produce uniform particle size and to improve the swelling capacity in water. In addition, raw fibers containing excess impurities might reduce the efficiencies in the following extraction stages (Malucelli et al., 2017). Mazlita's group have performed dewaxing for the OPT fiber via Soxhlet extraction with a mixture of toluene and ethanol since normal water washing may not be efficient in removing the wax and pectin present on the raw fiber's surface. The treated samples are rinsed with deionized water for several times to obtain the wax-free fibers. This step is crucial to increase the accessibility of cellulose toward the subsequent chemical treatments (Mazlita et al., 2016).

2.2 Mechanical Treatment

In general, mechanical treatment refers to a vigorous fibrillation process that isolate CNFs from biomass sources by disintegrating the cellulose along their longitudinal axis. The cellulosic fibers will be broken down into their substructural fibrils with diameter ranging from 10 nm to a few hundred nanometers and length on micrometer scale (Thomas et al., 2018). The common mechanical techniques include high pressure homogenization, micro fluidization and grinding. In high pressure homogenization, the cellulose suspension is pumped into a homogenization machine under high pressure where it will be subjected to shearing and impact forces. These forces will lead to high degree of micro fibrillation of the cellulose slurry. Contrarily, unlike the homogenizer that operates at a constant pressure, the microfluidizer works at a constant shear rate and the cellulose suspension is pumped through a thin chamber with a specific geometry (either Z- or Y-shape). Furthermore, grinding can destroy the structure of plant cell wall by high shear force to produce individual nanofibrils. The cellulose fiber is forced through an ultrafine grinder with two specially grooved disks, one of which is constantly rotating while the other is static. The disk's rotation speed should be maintained at 1500 rpm to avoid clogging in the grinder. The samples must pass through the grinder until a gel is formed. It is noteworthy that the aforementioned mechanical processes should be repeated several times in order to enhance the degree of fibrillation (Ma et al., 2020). Electrospinning, high speed blending, cryocrushing, hammer milling and high intensity ultrasonication are the alternative mechanical treatments reported in the literature (Thomas et al., 2018).

Howbeit, the high energy consumption and production cost are some of the major cumbers to extract nanocellulose from lignocellulosic biomass via mechanical treatment alone. This is because a tremendous amount of energy is required to liberate the nanocellulose from the natural plant fibers due to the highly ordered hydrogen bond network of cellulose (Thomas et al., 2018). Ergo, numerous pre-treatment techniques have been introduced to reduce the energy consumption up to 98%. One of the archetypal pre-treatment methods is to introduce an electrostatic repulsion between the cellulosic fibers via 2,2,6,6-tetramethylpiperidine-1-oxyl (TEMPO)-mediated oxidation where anionic surface charge (i.e., carboxylate groups) can be incorporated onto the cellulose fibers (Mishra et al., 2018). Unlike any other approaches, TEMPO-mediated oxidation is highly selective and it will only react with the hydroxymethyl group located at C_6 on the glucosyl ring (Tan et al., 2019). Mishra and co-workers epitomized the pros and cons of the different mechanical isolation techniques including micro fluidization, high intensity ultrasonication, electrospinning and steam explosion to obtain the cellulosic nanoparticles (Mishra et al., 2018).

Solikhin and co-workers recently introduced the novel multi-mechanical stages (i.e., dry disk milling, vibrational milling and ultrasonication) for the isolation of lignocellulosic nanofibers (LCNFs) from EFB (Solikhin et al., 2017). The dry disk mill was used to pulverize EFB vascular bundles to produce a finer fraction of approximately 10–20% microsized particles (under 75 μm). The microfibers were then ground using a vibrational pressurized milling to reduce the particle size, destroy silica bodies imparted on the surface of EFB and increase the fibrillation degree of lignocellulosic fibrils. The fiber samples were ultrasonicated to further degrade the polysaccharide linkage via the microjets and shock waves generated during cavitation. This multi-mechanical approach is eco-friendly, inexpensive and easy to implement for large-scale production, indicating its potential to replace the traditional chemical and enzymatic processes.

2.3 Chemo-Mechanical Treatment

Chemical pre-treatments such as alkaline treatment (mercerization) and bleaching (delignification) are conducted to remove non-cellulosic materials such as hemicellulose, lignin and pectin present on the raw fibers (Gan et al., 2020; Malucelli et al., 2017). Alkaline treatment is necessary to partially solubilize the hemicellulose fraction from the oil palm biomass to expose the cellulose structure for further processing. Residual waxes, silica ash and natural fats will also be removed at this stage, while lignin and cellulose are barely affected (Malucelli et al., 2017). Alkaline treatment is usually carried out with 5 wt% NaOH at 80 °C for 1 h followed by washing with deionized water until neutrality (Supian et al., 2020; Thomas et al., 2018). The bleaching process usually takes place after alkaline treatment to remove lignin and residual hemicellulose from the fibers. Two orthodox treatments are well described in the literature archives, namely treatment with 5% sodium chlorite under acidic condition (more effective but hazardous) and 10% hydrogen peroxide under alkaline conditions (Gea et al., 2020; Supian et al., 2020). The bleaching process is often repeated several times to ensure higher level of lignin removal for subsequent hydrolysis treatment. Once cellulose is recovered, acid hydrolysis is employed to obtain nanocellulose. It is cheaper and more efficient compared to other methods such as enzymatic hydrolysis. During strong acid hydrolysis, the hydronium ions will penetrate the amorphous regions of cellulose chains to cleave the glycosidic bonds and produce highly crystalline cellulose nanoparticles (CNCs). The most commonly used hydrolysing agent is H_2SO_4 where it can react with surface hydroxyl groups of cellulose via esterification process, thus allowing the grafting of anionic sulfate ester groups (Dai et al., 2018; Thomas et al., 2018). The introduction of these negatively charged sulfate groups can promote better dispersion of CNCs in water by electrostatic repulsion (Dai et al., 2018). Other mineral acids including hydrochloric acid, phosphoric acid and nitric acid are also capable of yielding crystalline CNCs but with lower dispersibility in solutions since there is fewer or no charge incorporated on the nanoparticle's surface (Thomas et al., 2018). It is imperative to control the acid hydrolysis process to avoid excessive cleavage of the crystalline regions, which may lower the crystallinity of resultant CNCs and affect their mechanical properties (Malucelli et al., 2017).

After the hydrolysis process, the CNCs suspension is diluted with excess deionized water to quench the reaction. The excess acid is normally removed through a series of centrifugation and washing stages. The precipitate obtained is then subjected to dialysis for removal of non-reactive sulfate groups and soluble sugars. The surplus of sulfate groups may deteriorate the thermal stability of as-synthesized CNCs, although they may also help to prevent the aggregation of nanoparticles. Finally, the CNC suspension is ultrasonicated to ensure the uniform dispersion of nanocellulose followed by freeze drying to produce CNCs in powder form. Drying can reduce the transportation costs and also allows for easier processing in industrial applications. Oven drying and spray drying can also be used as cheaper alternatives besides lyophilization, but these methods may promote the aggregation of CNCs, therefore affecting its final properties (Malucelli et al., 2017).

The neoteric research conducted by Gan's group demonstrated the feasibility of utilizing the novel deep eutectic solvent (DES) as an environmentally friendly alternative for the pre-treatment of EFB to isolate CNCs (Gan et al., 2020). The DES can form hydrogen bonds with lignin and hemicellulose by accepting and donating the protons, thereby improving the solvation ability of EFB fibers. DES can be easily prepared from different mixtures of components at moderate temperature and atmospheric pressure. The conventional alkaline pre-treatment usually involves higher processing costs owing to the significant amount of water required to remove the salts from the treated biomass material. In the study, Gan et al. (2020) first pre-treated the raw EFB with alkaline DES mixture prepared by mixing potassium carbonate and glycerol at a mole ratio of 1:7, followed by bleaching with sodium chlorite solution. The microcellulose obtained was subsequently added into H_2SO_4 solution and centrifuged to obtain a CNC suspension. The suspension was then ultrasonicated to prevent nanocellulose aggregation. The optimum conditions determined for acid hydrolysis were 60 wt% H_2SO_4, hydrolytic temperature of 46.1 °C and reaction time of 58.5 min. The maximum yield of CNCs was 37.1% under the optimum conditions as predicted by the response surface model (Gan et al., 2020). A comprehensive review concerning the recent development of DES for nanocellulose extraction can be found in the work of Jiang's group (Jiang et al., 2021).

2.4 *Enzymatic-Mechanical Treatment*

As mentioned briefly in previous sections, the mechanical treatment for nanocellulose extraction requires high energy and high cost, thus it is normally employed as an additional treatment for both chemical and biological methods. Meanwhile, the conventional chemical treatment is highly corrosive and will generate toxic waste. Ergo, biological method such as enzymatic treatment is preferred due to its environmentally friendliness, low energy requirement and production of non-hazardous waste as it allows milder hydrolysis conditions compared to acid hydrolysis (Aditiawati et al., 2018; Thomas et al., 2018). For instance, xylanases are hydrolytic enzymes that can initiate random hydrolysis of β-1,4 non-reducing terminal regions located between the glycosidic linkages of glucose units. Enzymes can also modify or degrade specific components in the cellulosic fibers such as lignin and hemicellulose to yield the desired nanocellulose. Predominantly, the yields of nanocellulose obtained from enzymatic hydrolysis are much lower than those achieved via acid hydrolysis. However, the enzymatic synthesis route can be tuned to fulfil the societal demands on clean chemical processes for the production of state-of-the-art nanomaterials (Thomas et al., 2018).

An elegant example of this approach is the production of CNFs from EFB using the cellulase enzyme from *Trichoderma sp.* (Aditiawati et al., 2018). The non-cellulosic components were first removed from EFB by delignification process via inoculum of *Marasmius* sp. due to its high ligninolytic enzyme activity. Some of the common ligninolytic enzyme produced by *Marasmius* sp. were laccase, mangan peroxidase and lignin peroxidase. The delignified EFB was crushed with cryocrushing method followed by subsequent addition of sodium citrate buffer and crude cellulase ranging from 50 to 200% (v/w). The mixture was incubated for several days before terminating the enzymatic reaction to obtain the desired nanocellulose. Specifically, the addition of 50% (v/w) cellulase enzyme with 2 days incubation time produced almost 100% CNFs with maximum size distribution of 30 nm, indicating the high potential of enzymatic approach for nanocellulose isolation (Aditiawati et al., 2018).

3 Characterization of Nanocellulose from Oil Palm Biomass

Nanocellulose characterization is a crucial subject that had garnered more interest and discussion from experts in tandem with the advancement of nanotechnology and analytical techniques. The primary analytical methods for the characterization of nanocellulose involve microscopy, spectroscopy as well as thermal and rheological techniques. These types of characterization techniques are employed to determine the morphology of nanocellulose and other features such as crystallinity or mechanical performance (Mishra et al., 2018). It is generally agreed that the morphology and properties of nanocellulose depend on its raw material source and extraction methods. Therefore, it is indispensable to characterize the as-produced nanocellulose to understand its properties before utilizing it for different applications. The following sessions examine the typical characterization techniques employed for nanocellulose extracted from different oil palm biomass.

3.1 Fourier Transform Infrared (FTIR) Analysis

The changes in functional groups of nanocellulose at different treatment stages were analyzed using Fourier transform infrared spectroscopy (FTIR). The spectra were usually recorded in the wavenumber range of 4000–400 cm^{-1} with a resolution of 4 cm^{-1} (Foo et al., 2019; Szlapak Franco et al., 2020). As shown in Table 2, the oil palm biomass is mainly comprised of lignin, hemicellulose and cellulose, or more fundamentally by alkanes, aromatics, esters, ketones and alcohols with different oxygen containing functional groups. In short, the absorption peaks from 3300 to 2900 cm^{-1} were attributed to O–H stretching of hydroxyl groups, while peaks from 2900 to 2800 cm^{-1} can be associated with the aliphatic saturated C–H stretching vibration of cellulose derivatives (Shanmugarajah et al., 2019). Additionally, the peak observed at 1646–1630 cm^{-1} was corresponded to the adsorption of water on cellulose molecules (Gan et al., 2020). The adsorbed water was unable to be removed completely due to the formation of strong hydrogen bonding despite been subjected to the drying process. The peak at 1729 cm^{-1} was accredited to C–O stretching vibration of acetyl and uronic ester groups presented in the ester linkage of carboxylic group of lignin and/or hemicellulose. Furthermore, the peaks located at 1505 and 1592 cm^{-1} were attributed to the aromatic skeletal vibration in lignin (Foo et al., 2019). The typical small peak observed at 1235 cm^{-1} was associated with C–O–C stretching vibration of the aryl group in lignin (Gan et al., 2020). All these characteristic peaks usually disappeared after alkaline treatment and bleaching, indicating the successful removal of hemicellulose and lignin from the oil palm biomass. The peaks in the region of 1420–1430 cm^{-1} were ascribed to the symmetric bending of CH_2, whereas the peaks at 1058–1060 cm^{-1} were due to C–O stretching (Foo et al., 2019; Mazlita et al., 2016). Moreover, the presence of β-glycosidic linkages between glucose units of cellulose was illustrated by the peak at 896 cm^{-1} (Szlapak Franco et al., 2020). For the isolation of nanocellulose via sulfuric acid hydrolysis, the presence of negatively charged sulfate ester groups on the surface of CNCs can be identified by the distinct peak located at 1200 cm^{-1} due to S=O vibrations (Foo et al., 2019; Shanmugarajah et al., 2019). The FTIR spectra of cellulose and CNCs displayed similar patterns, suggesting that acid hydrolysis treatment do not significantly alter the chemical

Table 4 FTIR peaks analysis for nanocellulose

Wavenumber (cm^{-1})	Peak assignment	References
3300–2900	O–H stretching of hydroxyl groups	Shanmugarajah et al. (2019)
2900–2800	C–H stretching vibration	Shanmugarajah et al. (2019)
1729	C–O stretching vibration of acetyl and uronic ester groups	Foo et al. (2019)
1646–1630	adsorption of water on cellulose molecules	Gan et al. (2020)
1505	aromatic skeletal vibration in lignin	Foo et al. (2019)
1430–1420	symmetric bending of CH_2	Foo et al. (2019)
1235	C–O–C stretching vibration of the aryl group	Gan et al. (2020)
1200	S=O vibrations	Shanmugarajah et al. (2019)
1060–1058	C–O stretching	Mazlita et al. (2016)
896	β-glycosidic linkages between glucose units	Szlapak Franco et al. (2020)

structure of nanocellulose (Mazlita et al., 2016). Table 4 recapitulates the FTIR absorption peaks for various functional groups in the structure of nanocellulose.

3.2 X-ray Diffraction (XRD) Analysis

The crystallinity of extracted nanocellulose was evaluated by X-ray diffraction (XRD) analysis using the CuKα radiation ($\lambda = 0.154$ nm). The XRD pattern of the samples was typically recorded within the 2θ range from 10° to 40° (Elias et al., 2017; Gan et al., 2020). The crystallinity index (CrI) of samples was calculated based on the intensity between (002) and (101) lattice diffraction peaks using Segal's method (Eq. 1), where I_{002} represents both the crystalline and amorphous region of cellulose ($2\theta = 22°$) while I_{am} represents only the amorphous phase ($2\theta = 18°$) (Campos et al., 2017; Shanmugarajah et al., 2019).

$$\text{Cr}I(\%) = \frac{I_{002} - I_{am}}{I_{002}} \times 100\% \tag{1}$$

In the study conducted by Mazlita's group, the XRD patterns of the raw OPT biomass, extracted cellulose and CNCs showed major peaks at $2\theta = 15.6°$, 22.2° and 44.4°, which indicated the presence of cellulose I_β structure (Mazlita et al., 2016). The CrI increased in the order of OPT < cellulose < CNC with the values of 37.37%, 68.35% and 73.17%, respectively. The extracted cellulose displayed higher crystallinity compared to OPT due to the removal of non-cellulosic components (hemicellulose and lignin) during alkaline treatment and bleaching, thus exposing the cellulose phase for subsequent acid hydrolysis. The highest crystallinity recorded for CNCs can be attributed to the penetration of sulfuric acid molecules into the amorphous structure of cellulose fibers, causing hydrolytic cleavage of glycosidic bonds to release the individual crystallites (Campos et al., 2017; Elias et al., 2017). This lead to the rearrangement of the remaining crystalline regions into a more ordered structure, thus leading to enhance crystallinity of the as-synthesized CNCs (Mazlita et al., 2016). Foo and colleagues also recorded the increased crystallinity of CNCs after acid hydrolysis with CrI of 77.6%, which further substantiates the findings of Mazlita's group (Foo et al., 2019; Mazlita et al., 2016).

Solikhin and co-workers reported pronounced decreased of crystallinity for lignocellulose nanofibers (LCNFs) isolated from EFB via multi-mechanical stages (Solikhin et al., 2017). As compared to the CrI of raw EFB (41.4%), the CrI of LCNFs obtained with 7, 12 and 17 milling times were 24.02, 27.12 and 29.25%, respectively. This phenomenon can be attributed to the high forces involved during vibrational milling process, which damaged the crystalline structure of cellulose. The authors also observed the transformation of cellulose structure from cellulose I to cellulose II after vibrational milling. This might be due to the recrystalline nature of amorphous cellulose in the presence of water from EFB fibers and the heat caused by mechanical friction (Solikhin et al., 2017). Similarly, Franco and associates also observed the decline in CrI (31.5%) for the CNFs extracted from peach palm heart residues via mechanical defibrillation in a colloidal mill (Szlapak Franco et al., 2020). This is because mechanical fibrillation indiscriminately breaks apart both crystalline and amorphous region of cellulose, resulting in lower crystallinity (Campos et al., 2017; Szlapak Franco et al., 2020). Furthermore, the wide XRD peaks observed between 30° and 34° were corresponded to (040) crystallographic plane typical for type I and II cellulose, implying that the combination of chemical pulping and bleaching coupled with mechanical defibrillation transformed the crystalline structure of CNFs from cellulose I to cellulose II. The low crystallinity of cellulose II will contribute to its enhanced dispersibility in water as well as high compatibility with proteins and lipids, thus improving the functionality of the as-fabricated CNF as emulsifying or stabilizing agent in oil/water emulsions (Szlapak Franco et al., 2020).

3.3 Morphological Analysis

The surface morphology of nanocellulose was usually examined using scanning electron microscopy (SEM), field-emission scanning electron microscopy (FESEM) or atomic force microscopy (AFM). Conversely, transmission electron microscopy (TEM) was used to determine the size or dimensions of the extracted nanocellulose. Prior to the characterization, the samples were coated with a thin layer of platinum or gold to reduce the charging effect during analysis (Gea et al., 2020; Supian et al., 2020). The raw oil palm biomass materials typically displayed bundles of individual fibers containing encrusting substances such as lignin, hemicellulose, pectin and waxes. After the pre-treatment process (alkaline treatment and bleaching), the treated cellulose fibers will exhibit a smoother surface with reduced diameter since most non-cellulosic contents and impurities were removed (Campos et al., 2017). As illustrated in Fig. 2a, CNFs obtained via mechanical fibrillation consist of significant amounts of amorphous regions with long smooth fibrillar structure. This is because the high shear force and intensity produced during nano-grinding causes the cellulose to be broken into smaller sizes, thus yielding nano-dimensioned CNFs (Supian et al., 2020). Solikhin and colleagues observed the agglomerated, irregular and uneven external surface of LCNFs isolated from EFB via multi-mechanical stages with diameters below 100 nm (Solikhin et al., 2017). The aggregation of LCNFs depicted in Fig. 2b can be attributed to the hornification of fibers during oven-heated suspension, as well as strong hydrogen bonding and Van der Waals forces between the nanofibers. The amorphous cementing agents were still deposited on the nanoparticles since no chemical treatment was performed to purify the EFB fibrils. In addition, the high forces of vibrational milling contributed to the destruction of crystalline structure of cellulose, which resulted in highly amorphous and black-colored nanofibers (Solikhin et al., 2017).

On the contrary, CNCs exhibit crystalline rod-like or needle-like shapes with at least one of its dimensions equal to or less than 100 nm (Dai et al., 2018; Thomas et al., 2018). For instance, Gan's group obtained rod-shaped CNCs with

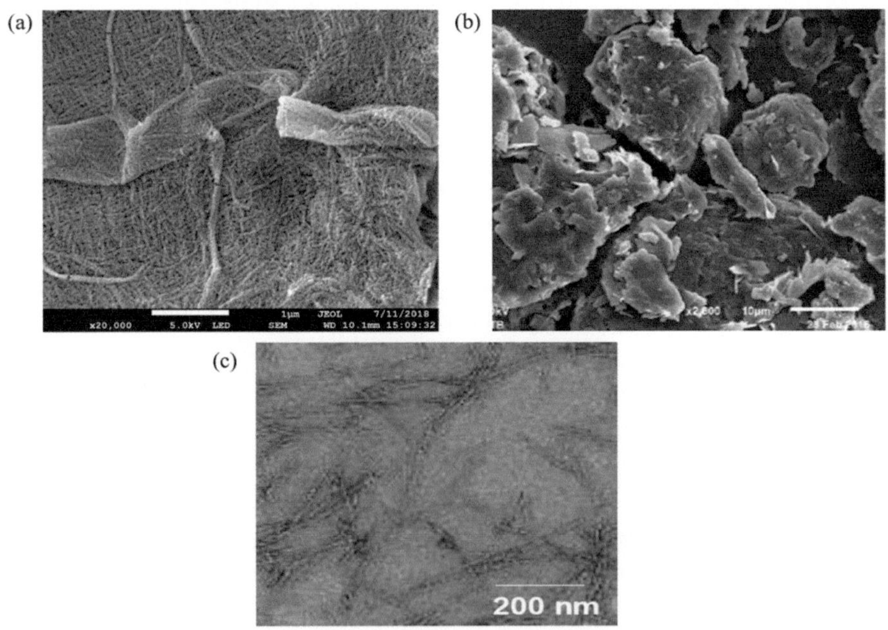

Fig. 2 **a** FESEM micrographs of CNFs isolated from EFB via nano-grinding with magnification × 20,000, **b** SEM micrographs of LCNFs extracted from EFB by multi-mechanical stages with magnification × 2800 and **c** TEM image of CNCs obtained from EFB (Gan et al., 2020; Solikhin et al., 2017; Supian et al., 2020)

6.78 ± 2.12 nm diameter and 160.06 ± 32.58 nm length from EFB hydrolyzed with 60 wt% H_2SO_4 after DES pre-treatment (Gan et al., 2020). Howbeit, aggregation of CNC was observed in the TEM micrograph of Fig. 2c. This can be accredited to the water evaporation step during sample preparation and the strong hydrogen bonding between CNCs. Interestingly, Campos et. al. (2017) discovered that the mechanical shearing (microfluidizer) would improve the homogenization of nanocellulose morphology in suspension, thereby minimizing the agglomeration of cellulose nanowhiskers without changing the dimensions of the nanoparticles. This novel finding provided additional insights for future researches where the implementation of mechanical shearing after acid hydrolysis could help to reduce the aggregation of nanocellulose, thus maintaining its good dispersion and mechanical performance.

4 Application of Nanocellulose from Oil Palm Biomass

Although the palm oil industry in Malaysia helps to boost the country's economy, it also generates a vast supply of biomass waste. Therefore, managing and processing this biomass waste in a greener approach remains one of the major predicaments by the research communities. One plausible solution is by converting the abundant oil palm waste into new value-added products such as nanocellulose-based composite materials to ensure the sustainable development of the country. Bio-nanocomposite had garnered increased research interest in the past decades due to its biocompatibility, biodegradability and unique functional properties. Bio-nanocomposite can be defined as a combination of a biopolymer and an inorganic moiety with one dimension in nanometer scale (Lamaming et al., 2020). For instance, bio-nanocomposite reinforced with nanocellulose could be used to produce biodegradable food packaging films to reduce the usage of synthetic plastic, whereas its biocompatibility property could be used in tissue engineering or drug carrier. The following sections abridged the current state-of-the-art applications of nanocellulose derived from various oil palm biomass in diverse fields or industries.

4.1 Food Industry

4.1.1 Food Packaging Film

Food insecurity is aggravated as a result of rapid population growth, decreased crop yields due to climate change and higher microbial growth rate at elevated temperature during transportation or storage. Therefore, the food price, trade flow and food access would likely jeopardize the development of a sustainable society in the next few decades (Tan et al., 2019). The ultimate function of food packaging is to ensure the food stability and quality as well as to extend the shelf life of food so that it is safe for human consumption (Azeredo et al., 2017). Nowadays, the usage of plastic packaging has dominated the food industry by replacing glass and metals (e.g., aluminium or tin) due to its superior flexibility, durability, low cost, ease of processing and broad applicability. In particular, the annual plastics production increased from 1.5 million tons in the 1950s to 359 million tons in 2018. The enormous amounts of plastics produced and its inability to degrade have caused "white pollution" that endangers the Earth's ecological environment. With the temperature rise at both north and south poles, in conjunction with the raging pandemic, protecting the environment and ecosystems has become more acute (Fang et al., 2019; Huang et al., 2020). Withal, plastics are usually derived from existing petroleum reserves, where the scarce resources are dwindling in recent years and cannot be renewed (Sinaga et al., 2018). Ergo, the growth of environmental concerns and public awareness toward the usage of petrochemical-based packaging materials has stimulated interest in biodegradable alternatives originating from replenishable agricultural feedstock (e.g., cellulose, guar gum and starch) or marine food processing industry waste such as chitosan. These types of materials demonstrate huge potential as the environment-friendly substitute for the synthetic plastics (Bhardwaj et al., 2020; Tang et al., 2018). Howbeit, the real-life utilizations of these biopolymers are still a mirage due to their poor mechanical properties, mediocre thermal stability and undesired barrier properties compared to their synthetic counterparts (i.e., plastics). In order to overcome these limitations, the incorporation of nanocellulose represents an efficacious approach to ameliorate the properties of biopolymers, which will expand their application in food packaging sector (El Miri et al., 2015; Tang et al., 2018).

Recently, Lamaming and co-workers prepared a PVA bio-nanocomposite film reinforced with CNCs isolated from OPT using various treatments as an auspicious material for food packaging applications (Lamaming et al., 2020). They reported that all the bio-nanocomposite films displayed higher Young's modulus and tensile strength compared to neat PVA films. The amelioration of mechanical performance can be attributed to the strong hydrogen bond interactions between CNCs and PVA, good compatibility between the fillers and polymer matrix, inherent stiffness of the nanocrystals and homogenous distribution of CNCs within the PVA polymer. The highest tensile strength (88.52 MPa) and Young's modulus (4.86 GPa) were observed for PVA bio-nanocomposite films blended with 5% CNCs with pre-hydrolysis treatment. Interestingly, the CNCs obtained with water pre-hydrolysis treatment had the lowest aspect ratio (L/D) but exhibited the highest tensile strength and Young's modulus (Lamaming et al., 2020). This is counterintuitive because higher aspect ratio of CNC forms more rigid filler network and has better capability to withstand mechanical stress uniformly over the matrix compared to CNC with lower aspect ratio. A plausible explanation might be due to the high zeta potential value of CNCs derived from pre-hydrolysis treatment (− 32.4 mV), thus endowing the good dispersion of nanoparticles in the PVA matrix. Conversely, the elongation at break was found to decrease with higher CNCs loading. This is because the formation of strong hydrogen bonds between the hydroxyl groups on CNC surface and PVA matrix will restrict the mobility of polymer chains, resulting in the decline of overall flexibility of bio-nanocomposite films. Furthermore, the onset degradation temperature and maximum decomposition temperature of all PVA/CNC bio-nanocomposite films increased by 1–7 °C compared to neat PVA film, suggesting the improvement of thermal stability upon addition of CNCs into PVA matrix (Lamaming et al., 2020).

Solikhin and colleagues also successfully developed PVA/chitosan nanocomposite films incorporated with different concentrations of LCNFs extracted from EFB (Solikhin et al., 2018). The tensile strength of neat PVA/chitosan film increased significantly from 32.19 to 65.65 MPa with only 0.5% LCNFs loadings. The enhancement of mechanical strength at low concentration of nanofillers can be ascribed to the homogeneous and good dispersion of LCNFs in PVA/chitosan matrix. Withal, the high level of compatibility and strong hydrogen bond interactions between LCNFs and PVA/chitosan might also contribute to the improved tensile strength of nanocomposite films, which is consistent with the findings of Lamaming's group (Lamaming et al., 2020; Solikhin et al., 2018). Nevertheless, the tensile strength decreased at higher LCNFs concentration (1, 2.5, 5, 7.5 and 10%) due to the formation of aggregated nanoparticles (Solikhin et al., 2018). A similar observation was also reported by Salehudin and associates where the starch-based biofilm reinforced with 10% CNFs derived from EFB displayed the tensile strength value of 2.90 MPa, which was lower compared to neat starch film (3.66 MPa) (Salehudin et al., 2014). This might be attributed to the agglomeration of nanoparticles at high concentrations that can act as stress concentration points, thereby deteriorating the mechanical performance of the nanocomposite films (Fig. 3a). Moreover, Solikhin and co-workers discovered that the addition of LCNFs induced lower transmittance at 300 nm compared to neat PVA/chitosan film, which is favorable for food packaging materials since it prevents photodegradation reaction between the packaged food and ultraviolet light (Solikhin et al., 2018). As depicted in Fig. 3b, the transparency of PVA/chitosan film reduced with the incorporation of nanocellulose and the effect was more conspicuous at higher concentration of LCNFs. This can be due to the aggregation and low dispersibility of LCNFs in the film matrix. Surprisingly, the as-synthesized PVA/chitosan nanocomposite films did not exhibit anti-bacteria and anti-fungal properties against gram-negative *Escherichia coli*, gram-positive *Staphylococcus aureus*, *Candida albicans* yeast and *Ganoderma* sp. fungi (Solikhin et al., 2018). This is counterintuitive because chitosan is known for its antimicrobial and antioxidant activities (Adel et al., 2019). The reasons postulated by the authors were: (1) water is needed to activate the chitosan as antimicrobial agent since the dried samples used could not release the energy stored inside the chemical bonds to initiate microbicidal reactions; (2) there was no interaction occurred between the cationic chitosan structure with the negatively charged cell membrane of microbes owing to the formation of intermolecular hydrogen bonds between amine groups of chitosan and hydroxyl groups of PVA and LCNFs (Solikhin et al., 2018).

4.1.2 Food Emulsion

Emulsions represent a system of two or more immiscible phases where one is dispersed as droplets in the other(s). The stability of emulsions depends heavily

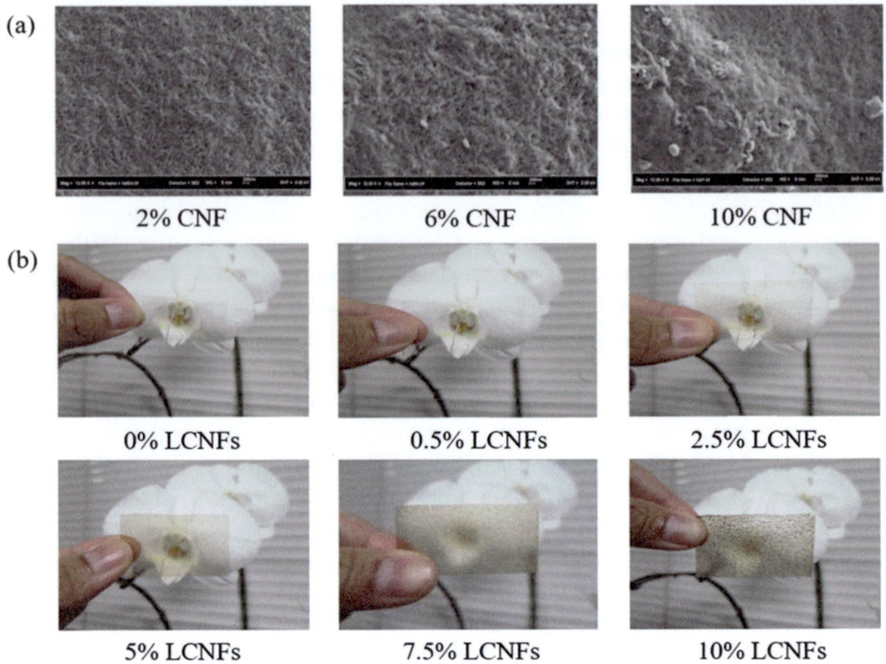

Fig. 3 **a** FESEM images of starch biofilm with different CNF concentration (×12,000 magnification and 200 nm scale) and **b** visual image of PVA/chitosan composite films at different LCNFs concentration (Salehudin et al., 2014; Solikhin et al., 2018)

on their formulation, emulsification techniques and emulsifier agent used that facilitate the interaction between different phases and deliver specific functional properties (Szlapak Franco et al., 2020). Emulsions can be used in plethora of industrial systems, especially in food, cosmetic and pharmaceutical products. Nonetheless, such systems are not thermodynamically stable and will often lead to phase separation to reduce the interfacial tension between the two phases. Ergo, amphiphilic molecules known as surfactants are added to stabilize these systems by reducing the interfacial energy to prevent the phase separation (Perrin et al., 2020). A wide variety of synthetic and natural additives that proportionate stability and emulsification can be legally used in food, including polysaccharides, proteins, phospholipids and small-molecule surfactants. The current consumer conscience about the relationship between personal health and food had stimulated the increased demand for functional and organic foods that are free of chemicals or harmful additives, indicating that the application of new natural emulsifiers is essential to fulfil the social requirements without compromising the sensorial attributes. These natural emulsifiers must display good emulsifying and stabilizing properties to maintain the emulsion characteristics and stability even under the extreme conditions applied during the production, storage and transporting stages (Szlapak Franco et al., 2020). Nanocellulose can be utilized as stabilizer for oil in water emulsions due to its amphiphilic nature and high aspect ratio. This type of particle stabilized emulsions shielded by a layer of solid particles are known as Pickering emulsions, where the formation of mechanical barrier stabilizes the emulsions against changes of temperature, pH, salt concentration and ionic strength (Perrin et al., 2020; Szlapak Franco et al., 2020).

In the neoteric research conducted by Franco's group, the authors proposed combining the functionality of CNF and health benefits of avocado oil to deliver a food emulsion that could be used without compromising the consumer blood fat levels and also to aid weight loss (Szlapak Franco et al., 2020). This is due to the potential of CNF as food additive or supplement that can reduce the adsorption of fat and limit the diffusion of glucose. They observed that the avocado oil/water emulsions prepared with CNF exhibited a negligent creaming index, indicating the capability of nanoparticles to maintain the emulsion format and their aggregation between the oil droplets even at a wide pH range. This phenomenon can be elucidated by the Pickering mechanism where the formation of CNF fibrillar network dispersed on the water and oil interface can prevent the coalescence of the emulsion droplets and mechanically stabilizing the system. Additionally, all the emulsions containing CNF also manifested resistance against interfacial disruption caused by gravity, collision and other effects during long-term storage. The hydrophilic CNFs with abundant hydroxyl groups have the tendency to overlap and join together into entangled networks and form macroscopic gels, acting as viscosity modifiers and thickeners in aqueous media at relatively low concentrations. The formation of this very high viscosity and gel-like behavior medium may trap the continuous phase between the droplets inside the network, thereby inhibiting gravity-induced creaming and serum separation during extended emulsion storage. Furthermore, the avocado oil/water emulsions formulated with 1% CNFs were able to remain its integrity after storage for 30 days, in tandem with heat treatment up to 80 °C and at extreme acidic (pH 2) or alkaline (pH 11) conditions (Szlapak Franco et al., 2020). The findings suggested that CNFs isolated from peach palm residues might be a propitious replacement for conventional surfactant (e.g., sorbitan monostearate) to deliver an edible emulsion with superior stability.

4.2 Biomedical Field

Nanocellulose is also commonly used in biomedical applications due to its superb mechanical properties, large specific surface area, availability for versatile surface functionalization, non-toxicity and biocompatibility (Mishra et al., 2018; Shazali et al., 2019). Nanocellulose membrane had captivated increased research interest for cell attachment and proliferation due to its unique 3D network, excessive porosity and low cytotoxicity. In this regard, the cell attachment can be enhanced via surface functionalization of nanocellulose such as the introduction of surface charges, plasma treatment method or protein coatings. Withal, the substantial water-holding capability, excellent flexibility and durability of nanocellulose membrane also rendered it a propitious material for wound dressing. Over the past decade, nanocellulose structured drug delivery systems have been adapted to regulate the drug release rates and subsequent drug levels in the bloodstream. One prominent example is the usage of nanocellulose-based hydrogel as a carrier for various kinds of drugs delivery (Mishra et al., 2018). The detailed summary of the most well-known nanocellulose hydrogel processing methods and their latest advanced applications are available in the work of Padzil's group (Padzil et al., 2020).

The short circulatory half-life and poor bioavailability of water-insoluble drug (e.g., anticancer drugs) have always been the major hurdles for efficient drug delivery inside the human body. Generally, rod-shaped particles can be regarded as a good drug carrier owing to its high cellular uptake in the body, long circulatory time and tumour accumulation. For instance, CNCs with rod-liked structure demonstrate great potential for drug delivery due to its high surface area and abundance of surface hydroxyl groups that allow its surface functionalization, thus facilitating a high level of drug loading. Recently, Foo and co-workers successfully attached curcumin onto CNC derived from EFB modified with tannic acid (TA) and decylamine (DA) as a superior and sustainable drug carrier (Foo et al., 2019). The modified CNC could replace the conventional surfactant, cetyl trimethylammonium bromide which might interact with the phospholipid bilayers of cells and lead to cell death. Curcumin is used because it possesses outstanding anti-inflammatory, antioxidant, antimicrobial, anticancer and antimutagenic properties. The authors reported that the curcumin binding capacity of surface-modified CNC was two-fold higher than pristine CNC, regardless of the concentration of curcumin used. This can be attributed to the increased hydrophobicity level of modified CNC that favored the hydrophobic interaction between the phenolic moieties of curcumin and long alkyl chain of DA, therefore yielding the remarkable curcumin binding efficiency ranging from 95 to 99%. Moreover, the formation of entangled and crosslinked network might also facilitate the binding of curcumin particles onto the surface of modified CNC. The schematics illustrating possible mechanisms for binding of curcumin onto surface of CNC modified with TA and DA are given in Fig. 4a (Foo et al., 2019). This work has proven that the surface modification using both TA and DA is a feasible approach to tailor the properties of CNC as an effective drug carrier.

In another similar study, Shazali and associates synthesized spherical CNC from EFB conjugated with fluorescein isothiocyanate (FITC) and investigated its cellular internalization into C6 (rat glioma) and NIH3T3 (normal murine fibroblast) cells for potential anticancer drug nanocarrier application (Shazali et al., 2019). Interestingly, the FITC-CNC showed poor cellular uptake into both normal and cancerous cell lines. Theoretically, the C6 and NIH3T3 take up nanoparticles via non-specific adsorptive endocytosis, which was governed by the surface charge, shape and hydrophobicity of the particles. Howbeit, the cellular accumulation in this study was inhibited due to the electrostatic repulsion between both negatively charged CNC surface and the cell membrane of fibroblasts (Fig. 4b). Consequently, the CNC failed to adhere to the cell surface in order to initiate membrane wrapping (Shazali et al., 2019). Therefore, FITC-CNC can be developed into targeted nanocarrier for the delivery of anticancer drugs by tuning its surface charge properties to enhance its cellular uptake

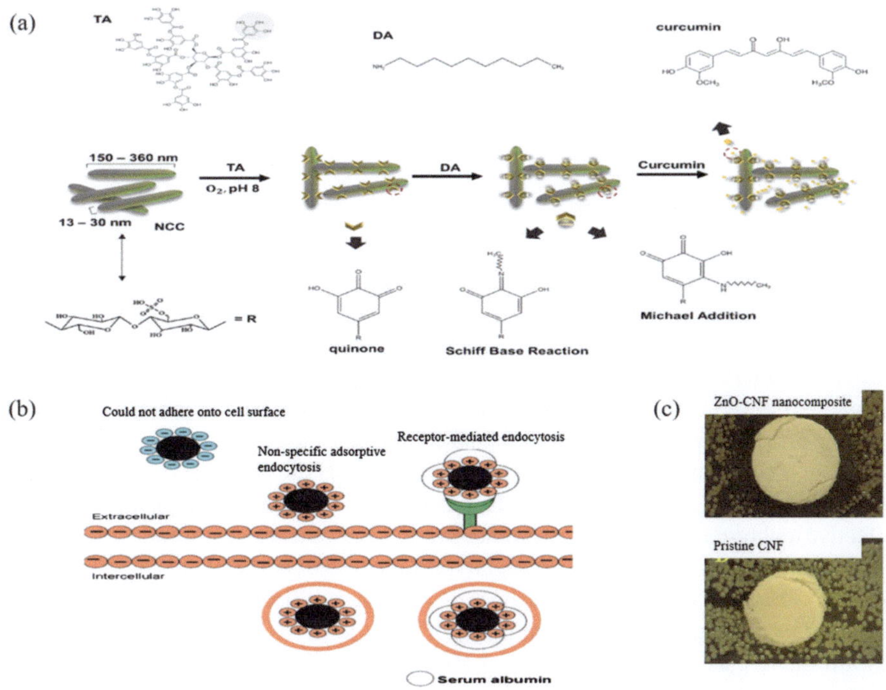

Fig. 4 **a** Schematic illustration of possible mechanism for binding of curcumin onto CNC surface modified with tannic acid (TA) and decylamine (DA), **b** illustration of possible uptake mechanism of FITC-CNC nanospheres with different surface charges and **c** enlarged image of ZnO-CNF and pristine CNF nanocomposite film in antibacterial test (Foo et al., 2019; Shazali et al., 2019; Supramaniam et al., 2020)

into cancerous cells. Furthermore, Supramaniam and colleagues fabricated oil palm biomass-derived CNFs loaded with zinc oxide (ZnO) nanocomposites for wound care application (Supramaniam et al., 2020). The wound disease caused by bacteria such as Methicillin-resistant *Staphylococcus aureus* (MRSA) can delay wound healing and leads to exudate development on the injured surface. ZnO has received considerable attention in nanofiller-related research due to its antibacterial activity, good photocatalytic performance, non-toxicity and high stability. They discovered that the ZnO-CNFs possessed higher swelling capacity (10%) compared to pristine CNF. This can be ascribed to the presence of interstitial pores in cellulosic polymeric network and charged ZnO nanoparticles that result in the seepage of water molecules to balance the increase of ion osmotic pressure. Based on Fig. 4c, the ZnO-CNF sample also exhibited 2 mm inhibition zone for MRSA antibacterial test, while no inhibition zone was observed for pristine CNF. The antibacterial property of ZnO might be attributed to the generation of reactive oxygen species (ROS) such as superoxide radicals, hydroxyl radicals and singlet oxygen. These ROS can then induce oxidative stress and interrupt the transmembrane electron transport, which will damage the DNA structure and ultimately lead to cell death (Supramaniam et al., 2020). The results from this study suggested that the as-prepared ZnO-CNF nanocomposites can be utilized as potential nanofillers for wound dressing application.

4.3 *Effluent Treatment*

The rapid development of petrochemical industries and urbanization is responsible for the massive deterioration in water quality due to the contamination from heavy metals and hazardous or toxic chemicals. These contaminants pose a severe threat to human health and ecological well-being (Septevani et al., 2020; Shanmugarajah et al., 2019). Hitherto, many techniques have been developed for both organic and heavy metal water remediation including electrochemical, physicochemical treatment, membrane filtration, photocatalysis, biological treatment and adsorption. Among all these methods, adsorption is the most promising and cost-effective technique due to its high efficiency, scalable processability and ease of operation (Septevani et al., 2020). Specifically, cellulose-based adsorbents have garnered interest for water remediation because of its hydrophilicity, high specific surface area, ease of chemical surface modification, non-toxicity, high adsorption affinity toward many classes of pollutants and 100% biodegradable with no adverse effect on the environment and human beings (Septevani et al., 2020; Thomas et al., 2018).

Some of the common organic contaminants in drinking water are natural oils, dyes, chemical pesticides and synthetic medicines materials (Mishra et al., 2018). For instance, methylene blue (MB) is a cationic dye that is commonly employed in the textile dyeing industry. It is poisonous since acute exposure could cause nausea, increased heartbeat and breathing difficulties. Recently, Shanmugarajah and co-workers managed to convert EFB into CNCs without further surface modification as a promising natural bio-sorbent for the removal of MB (Shanmugarajah et al., 2019). They found that the percentage of MB removal increased rapidly from 60 to 87% with increase of CNC dosage from 0.005 to 0.01 g, but the removal rate remained constant when the CNC dosage increased further from 0.01 to 0.05 g. Higher dose of CNC will increase the active sites available on the adsorbents to bind with more dye molecules, implying a quick initial adsorption with better removal efficiency. However, further increase in adsorbent dosage will cause particle aggregation and subsequently reduce the efficiency for dye removal. Withal, the optimal adsorption occurred in alkaline medium (pH 8) with removal rate of 88%. This is because the presence of OH^- ions in the dye solution would create better electrostatic attractions for more dye molecular binding, which is particularly important for cationic dyes such as MB. On the other hand, the adsorption rate at strong acidic condition (pH 2) was only 29% due to the excess H^+ ions competing with the cationic dye for adsorption sites on CNC surface (Fig. 5a). The negatively charged sulfate ions on the surface of CNCs were also protonated under acidic condition, hence reducing the binding efficiency with MB molecules. The authors also reported good adsorption capacity (50.91 mg/g) at very low pristine CNC dosage of 0.066 mg/ml attributed to the

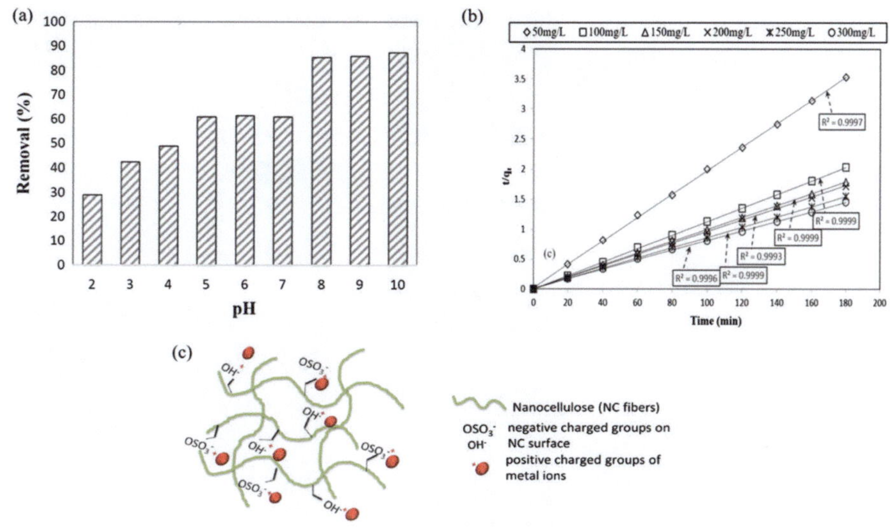

Fig. 5 **a** Removal of methylene blue by CNCs at different pH values, **b** pseudo-second order kinetic model for adsorption of methylene blue at various initial dye concentrations and **c** schematic of electrostatic interaction on the surface of nanocellulose bio-adsorbents (Septevani et al., 2020; Shanmugarajah et al., 2019)

abundant and highly functional methylol and hydroxyl groups on the nanocellulose. Figure 5b illustrates that the adsorption obeyed Langmuir isotherm and pseudo-second order kinetic model with the estimated maximum adsorption capacity of 144.93 mg/g, where chemisorption was the rate-controlling factor (Shanmugarajah et al., 2019).

In another similar study conducted by Septevani's group, EFB-based nanocellulose functionalized with activated carbon (AC) was successfully fabricated as a super-adsorbent for water remediation (Septevani et al., 2020). Surprisingly, the functionalization of AC on nanocellulose did not improve the metal adsorption, therefore confirming that the micron size AC will disrupt the network structure of nanocellulose and reduce the active adsorption sites, which yielded a lower heavy metal performance. The optimum dose for heavy metals removal is 2% of nanocellulose extracted via sulfuric acid hydrolysis (NCS) with removal efficiencies of 80%, 9% and 8% for Pb^{2+}, Cd^{2+} and Cr^{3+}, respectively (Septevani et al., 2020). The adsorption capacity toward Pb^{2+} is substantially higher than other metal ions due to the lower hydration energy of Pb^{2+}, which increased the ion interaction toward the activated sites of super-adsorbent. Additionally, the formation of sulfate groups on the surface of NCS during acid hydrolysis may enhance the heavy metal adsorption capacity since the sulfonated nanocellulose possess a large number of binding sites for the electrostatic interaction between the negatively charged sulfate groups and the positively charged metal ions (Fig. 5c). The authors also reported that the optimum adsorption efficiency for 2% NCS was 80% at initial metal ion concentration of

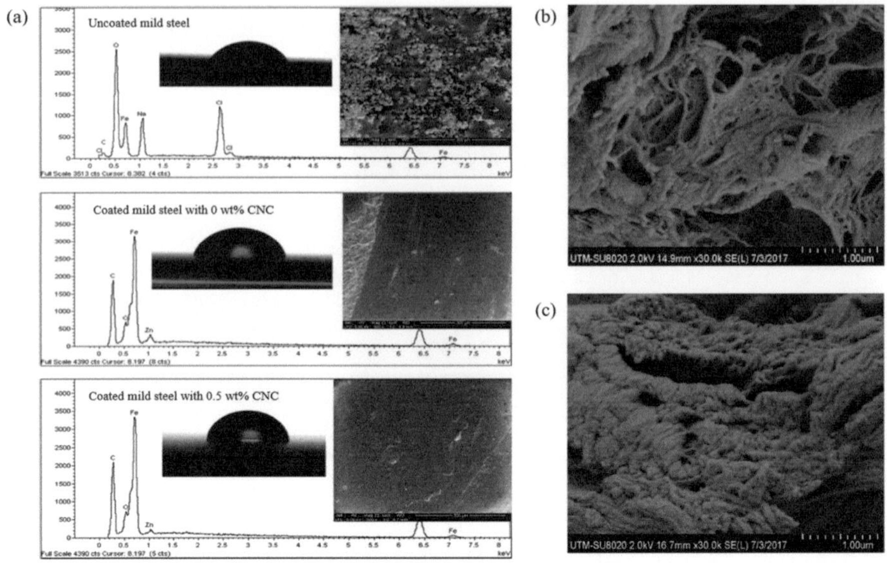

Fig. 6 **a** EDX spectra (×500 magnification), SEM image (×500 magnification and 300 μm scale) and water contact angle for mild steel reinforced epoxy-Zn rich coating with CNC, FESEM image of **b** lyophilized chitosan/nanocellulose and **c** lyophilized CRL/chitosan-nanocellulose biocatalysts (Azani et al., 2020; Elias et al., 2017)

30 ppm (Septevani et al., 2020). Increasing the initial metal ion concentration will accelerate the diffusion of metal ion onto the adsorbent until all the active sites on the nanocellulose adsorbent attain the saturation point. Once the saturation point was reached, the adsorption capacity remained constant while the adsorption efficiency gradually reduced. Furthermore, the optimum adsorption time was reported at 150 s, where the maximum adsorption efficiency and adsorption capacity for Pb^{2+} were 86% and 30.45 mg/g, respectively. The adsorption process obeyed the Langmuir isotherm model, where the adsorption occurred on the surface of solid following the monolayer sorption mechanism, which is consistent with the findings of Shanmugarajah and associates (Septevani et al., 2020; Shanmugarajah et al., 2019). The lead adsorption capacity reduced to 10.34 mg/g and 7.80 mg/g after the second and third cycles of nanocellulose adsorbent regeneration, respectively (Septevani et al., 2020). The significant decline of adsorption capacity might be due to the fact that the super-adsorbent was derived from easily degradable biomass sources such as EFB. The authors also examined the effect of nanocellulose super-adsorbent on the organic pollutant remediation and recorded the highest chemical oxygen demand removal of 93% after treatment with 0.5 wt% sulfonated nanocellulose functionalized with AC. Forbye, the treated water quality remained within the government specifications with pH and TDS level of 7.5 and 151 mg/L, respectively (Septevani et al., 2020). This indicated that nanocellulose derived from oil palm biomass can be utilized as robust and effective adsorbents for effluent treatment.

4.4 Miscellaneous

In addition to the food industry, biomedical field and effluent treatment, a recent shift toward the utilization of nanocellulose in building materials, corrosion protection and biocatalyst was observed in the research community. In particular, Mocktar and colleagues incorporated oil palm nanocellulose that acts as nanofillers into composite bricks to reduce the radon gas emanations in the indoor environment (Mocktar et al., 2020). Radon gas is notorious for causing lung cancer besides smoking. It is noteworthy that brick material is one of the significant sources of indoor radon emanations, in which the aggregates contain Ra-226 that will decay to Rn-222 and alpha particles. These alpha particles with high linear energy transfer property can eventually damage the DNA double bonding in humans, where it will be lodged in the lining of the lungs and release mass radiation energies. Nanocellulose can be incorporated into building materials to minimize the indoor radon concentration since the cellulosic materials will act as liquid fillers to substitute the radon sources such as gravel, cement and sand. The authors discovered that the radon concentration for composite bricks with 40 mL and 80 mL nanocellulose was 1.4 and 2.17 pCi per 1, respectively. Howbeit, further increase of nanocellulose contents (120 mL, 160 mL, 200 mL) led to higher radon concentration because more water was used to fabricate the bricks, thus resulting in the swelling of cellulose chain and increased moisture uptake. The high humidity environment will in turn increase the radon concentration due to the interactions of alpha particles with water molecules. The fibrillar networks of nanocellulose will also miniaturize the porosity of the composite bricks and reduce the formation of cracks, which indirectly help to mitigate the radon gas emission. The difference in radon concentrations between the control and composite bricks with 40 mL nanocellulose were 2.84 pCi per 1 and this could increase the probability of lung cancer. This is because for every 2.7 pCi per 1 radon exposure, the probability of lung cancer will be raised for approximately 16% in an indoor environment. In addition, the composite brick reinforced with 40 mL nanocellulose exhibited higher Young's modulus (27.8 MPa) compared to the control brick (27 MPa), indicating the high stiffness of composite brick materials (Mocktar et al., 2020). The results proved that this type of novel building materials can promote a healthy environment and contribute toward a better life.

Corrosion has become one of the biggest concerns in the marine, oil and gas, automotive and construction industries. Hence, corrosion protection is essential to ensure the production of high-quality products with relatively lower maintenance cost. The high toxicity of conventional chromate-based coating to the environment have urged researchers to develop state-of-the-art anticorrosive organic coating with high performance. In general, the epoxy-Zn rich coating has been widely used to protect the steel substrate in various industries. This is because the electrical contact of zinc particle in epoxy coating system offers excellent protection to the metal substrate against corrosive agents and ions such as H_2O_2, O_2 and Cl_2. The high crosslinking density of epoxy coating also provides good adhesion to the metal surface. Nevertheless, the epoxy coating might experience considerable mechanical damage when exposed to aggressive environment for an extended period of time. One possible solution is to introduce nanocellulose-based composites to enhance its adhesion to metal surface and durability in order to minimize the metal corrosions. For instance, Azani and associates successfully reinforced epoxy-Zn rich coating with CNCs derived from OPF to improve the corrosion protection of mild steel (Azani et al., 2020). Based on the results obtained from corrosion studies, the mild steel coated with epoxy-Zn rich at 0.5 wt% CNC was the optimum loading since it lessened the corrosion by 99%, exhibited highest hydrophobicity with water contact angle of $100.5 \pm 0.7°$ and a meagre amount of O element on the coating surface, indicating a low corrosion rate (Fig. 6a). The high surface area of CNCs (26.10 m^2/g) can simulate the growth of crosslinking within the epoxy coating network and the compact size of CNCs with diameter less than 100 nm can easily penetrate the coating surface, thus eliminating any damage or crack to prevent the penetration of any corrosive species into the metal surface. Nonetheless, when large amount of nanocellulose was added, the formation of strong hydrogen bonds between the nanocrystals might cause aggregation of epoxy coating. This might deteriorate the anticorrosion performance of coating due to the formation of small pores within the epoxy network, which will allow the contact of corrosive species with the metal surface (Azani et al., 2020). Therefore, small amount of CNCs (<1 wt%) was sufficient to ameliorate the mechanical strength and performance of epoxy coating in minimizing corrosion.

Furthermore, Elias's group utilized nanocellulose isolated from OPF leaves/chitosan nanocomposite as support material for immobilizing *Candida rugosa lipase* (CRL) to catalyze the production of butyl butyrate through esterification of butanol and butyric acid (Elias et al., 2017). Both nanocellulose and chitosan were chosen due to their biocompatibility, good mechanical strength, non-toxicity, amenability for chemical modification and environmentally friendliness. CRL as free lipases are known to be easily deactivated under harsh industrial processing conditions, thereby limiting their use in catalyzing chemical processes such as esterification, transesterification and interesterification. Hence, immobilizing CRL onto nanocellulose/chitosan hybrid composites might enhance the operational stability of the enzyme and improve its catalytic performance in aqueous and non-aqueous conditions, as well as enabling catalyst recovery and reuse for large-scale applications. The authors reported that the air-dried chitosan-nanocellulose support showed higher immobilized protein of 5.20 mg/g compared to 4.80 mg/g in the lyophilized chitosan-nanocellulose. This is because higher numbers of internally located pores were formed during lyophilization compared to the pores that are exposed to the outer environment (Fig. 6b). In other words, there were fewer surface cavities available on chitosan-nanocellulose support for the attachment of CRL and also lesser CRL with its active sites exposed to the surface for catalysis reaction (Fig. 6c). Therefore, 73.8% conversion of butyl butyrate was reported for CRL immobilized on air-dried chitosan-nanocellulose in 3 h, while only a mere 25% conversion of the ester was observed for CRL immobilized on lyophilized chitosan-nanocellulose. Moreover, the as-synthesized CRL/chitosan-nanocellulose yielded 76.3% of butyl butyrate within only 4 h. Longer immobilization durations beyond 4 h were futile and produced significantly fewer active batches of CRL/chitosan-nanocellulose biocatalysts.

This is because optimal immobilization duration is indispensable to allow formation of covalent bonds between the CRL and chitosan-nanocellulose support (Elias et al., 2017). This work had demonstrated the feasibility of using the highly functional CRL/chitosan-nanocellulose biocatalysts prepared from OPF leaves biomass to produce high yields of butyl butyrate.

5 Conclusion and Future Recommendations

The generation of enormous amounts of oil palm biomass in Malaysia caused a plethora of environmental issues despite the contribution of palm oil industry to the country's growth and development. The concept of waste valorization of oil palm biomass is an innovative idea that can bring imponderable economic and ecological benefits. The utilization of nanotechnology to produce nanocellulose from oil palm biomass has broad application prospects compared with the conventional disposal methods such as incineration and composting. Nanocellulose is abundant, non-toxic, renewable, biodegradable and has excellent mechanical properties, rendering it a highly propitious candidate to play a key role in the synthesis of many state-of-the-art nanocomposite materials. Nanocellulose has been scrutinized not only for its applications in food packaging films and emulsions, but also in much more surprising domains such as drug carriers, wound dressing, bio-sorbents for water remediation, building materials, corrosion protection and biocatalysts. However, there are several challenges that need to be addressed by the academia and industrial experts before the official roll-out of nanocellulose-based materials into the society. First, life cycle assessment must be conducted to identify and analyze the potential risk circumstances of nanocellulose-based composites. This is to ensure that nanocellulose does not pose any harm toward human health and the environment. Specifically, life cycle assessment should include detailed analysis during the extraction of raw materials, production, transportation, consumption and final disposal. In addition, the advanced 3D simulation software endorsed by machine learning and Industrial Revolution 4.0, coupled with atomistic and analytical modelling shall be utilized to further understand the structure-property relationship of the novel nanocellulose-based materials (e.g., packaging films, membranes or catalysts). This will provide an adequate guideline for the synthesis and design of nanocellulose composite products with desired properties or characteristics. The incorporation of artificial intelligence, Internet of Things and big data analytics presents an auspicious approach to achieve the fully automated manufacturing process of advanced nanocellulose-containing materials without human intervention in the coming years (Lim et al., 2022). Consequently, it represents a cornerstone in realizing the sustainable consumption and production patterns that aligned with the United Nations' Sustainable Development Goals.

Acknowledgements The authors are thankful to the Xiamen University Malaysia for the financial support through the Xiamen University Malaysia Research Fund (XMUMRF/2019-C3/IENG/0014).

References

Adel, A. M., Ibrahim, A. A., El-Shafei, A. M., & Al-Shemy, M. T. (2019). Inclusion complex of clove oil with chitosan/β-cyclodextrin citrate/oxidized nanocellulose biocomposite for active food packaging. *Food Packaging and Shelf Life, 20*, 100307.

Aditiawati, P., Dungani, R., & Amelia, C. (2018). Enzymatic production of cellulose nanofibers from oil palm empty fruit bunch (EFB) with crude cellulase of *Trichoderma* sp. *Materials Research Express, 5*(3), 034005. https://doi.org/10.1088/2053-1591/aab449

Alonso, D. M., Wettstein, S. G., & Dumesic, J. A. (2012). Bimetallic catalysts for upgrading of biomass to fuels and chemicals. *Chemical Society Reviews, 41*(24), 8075–8098.

Azani, N. F. S. M., Haafiz, M. K. M., Zahari, A., Poinsignon, S., Brosse, N., & Hussin, M. H. (2020). Preparation and characterizations of oil palm fronds cellulose nanocrystal (OPF-CNC) as reinforcing filler in epoxy-Zn rich coating for mild steel corrosion protection. *International Journal of Biological Macromolecules, 153*, 385–398. https://doi.org/10.1016/j.ijbiomac.2020.03.020

Azeredo, H. M., Rosa, M. F., & Mattoso, L. H. C. (2017). *Nanocellulose in Bio-Based Food Packaging Applications., 97*, 664–671.

Bhardwaj, A., Alam, T., Sharma, V., Alam, M. S., Hamid, H., & Deshwal, G. K. (2020). Lignocellulosic agricultural biomass as a biodegradable and eco-friendly alternative for polymer-based food packaging. *Journal of Packaging Technology and Research*, 1–12.

Bharimalla, A., Deshmukh, S., Vigneshwaran, N., Patil, P., & Prasad, V. (2017). Nanocellulose-polymer composites for applications in food packaging: Current status, future prospects and challenges. *Polymer-Plastics Technology and Engineering, 56*(8), 805–823.

Campos et al., 2017 Campos, A. D., Neto, A. R. D. S., Rodrigues, V. B., Kuana, V. A., Correa, A. C., Takahashi, M. C., Mattoso, L. H., & Marconcini, J. M. (2017). Production of Cellulose nanowhiskers from oil palm mesocarp fibers by acid hydrolysis and microfluidization. *Journal of Nanoscience and Nanotechnology, 17*(7), 4970–4976. https://doi.org/10.1166/jnn.2017.13451

Dai, H., Ou, S., Huang, Y., & Huang, H. (2018). Utilization of pineapple peel for production of nanocellulose and film application. *Cellulose, 25*(3), 1743–1756.

El Miri, N., Abdelouahdi, K., Barakat, A., Zahouily, M., Fihri, A., Solhy, A., & El Achaby, M. (2015). Bio-nanocomposite films reinforced with cellulose nanocrystals: Rheology of film-forming solutions, transparency, water vapor barrier and tensile properties of films. *Carbohydrate Polymers, 129*, 156–167. https://doi.org/10.1016/j.carbpol.2015.04.051

Elias, N., Chandren, S., Attan, N., Mahat, N. A., Razak, F. I. A., Jamalis, J., & Wahab, R. A. (2017). Structure and properties of oil palm-based nanocellulose reinforced chitosan nanocomposite for efficient synthesis of butyl butyrate. *Carbohydrate Polymers, 176*, 281–292. https://doi.org/10.1016/j.carbpol.2017.08.097

Fang, Z., Hou, G., Chen, C., & Hu, L. (2019). Nanocellulose-based films and their emerging applications. *Current Opinion in Solid State and Materials Science, 23*(4), 100764. https://doi.org/10.1016/j.cossms.2019.07.003

Foo, M. L., Tan, C. R., Lim, P. D., Ooi, C. W., Tan, K. W., & Chew, I. M. L. (2019). Surface-modified nanocrystalline cellulose from oil palm empty fruit bunch for effective binding of curcumin. *International Journal of Biological Macromolecules, 138*, 1064–1071. https://doi.org/10.1016/j.ijbiomac.2019.07.035

Gan, P. G., Sam, S. T., Abdullah, M. F., Omar, M. F., & Tan, L. S. (2020). An alkaline deep eutectic solvent based on potassium carbonate and glycerol as pretreatment for the isolation of cellulose nanocrystals from empty fruit bunch. *BioResources, 15*(1), 1154–1170.

Gea, S., Siregar, A. H., Zaidar, E., Harahap, M., Indrawan, D. P., & Perangin-Angin, Y. A. (2020). Isolation and Characterisation of Cellulose Nanofibre and Lignin from Oil Palm Empty Fruit Bunches. *Materials, 13*(10). https://doi.org/10.3390/ma13102290

Guzman-Puyol, S., Ceseracciu, L., Tedeschi, G., Marras, S., Scarpellini, A., Benítez, J. J., Athanassiou, A., & Heredia-Guerrero, J. A. (2019). Transparent and robust all-cellulose nanocomposite packaging materials prepared in a mixture of trifluoroacetic acid and trifluoroacetic anhydride. *Nanomaterials, 9*(3), 368. https://doi.org/10.3390/nano9030368

Huang, L., Zhao, H., Yi, T., Qi, M., Xu, H., Mo, Q., Huang, C., Wang, S., & Liu, Y. (2020). Preparation and properties of cassava residue cellulose nanofibril/cassava starch composite films. *Nanomaterials, 10*(4), 755.

Jiang, J., Zhu, Y., & Jiang, F. (2021). Sustainable isolation of nanocellulose from cellulose and lignocellulosic feedstocks: Recent progress and perspectives. *Carbohydrate Polymers*, 118188.

Lamaming, J., Hashim, R., Leh, C. P., Sulaiman, O., & Lamaming, S. Z. (2020). Bio-nanocomposite films reinforced with various types of cellulose nanocrystals isolated from oil palm biomass waste. *Waste and Biomass Valorization, 11*(12), 7017–7027. https://doi.org/10.1007/s12649-019-00892-7

Lim, H. J., Cheng, W. K., Tan, K. W., & Yu, L. J. (2022). Oil palm-based nanocellulose for a sustainable future: Where are we now? *Journal of Environmental Chemical Engineering, 10*(2), 107271. https://doi.org/10.1016/j.jece.2022.107271

Ma, T., Hu, X., Lu, S., Liao, X., Song, Y., & Hu, X. (2020). Nanocellulose: a promising green treasure from food wastes to available food materials. *Critical Reviews in Food Science and Nutrition*, 1–14. https://doi.org/10.1080/10408398.2020.1832440

Malucelli, L. C., Lacerda, L. G., Dziedzic, M., & da Silva Carvalho Filho, M. A. (2017). Preparation, properties and future perspectives of nanocrystals from agro-industrial residues: A review of recent research. *Reviews in Environmental Science and Bio/technology, 16*(1), 131–145. https://doi.org/10.1007/s11157-017-9423-4

Mazlita, Y., Lee, H. V., & Hamid, S. B. A. (2016). preparation of cellulose nanocrystals bio-polymer from agro-industrial wastes: Separation and characterization. *Polymers and Polymer Composites, 24*(9), 719–728. https://doi.org/10.1177/096739111602400907

Mishra, R. K., Sabu, A., & Tiwari, S. K. (2018). Materials chemistry and the futurist eco-friendly applications of nanocellulose: Status and prospect. *Journal of Saudi Chemical Society, 22*(8), 949–978. https://doi.org/10.1016/j.jscs.2018.02.005

Mocktar, F. A., Abdul razab, M. K. A., & Mohamed noor, A. A. (2020). Incorporating kenaf and oil palm nanocellulose in building materials for indoor radon gas emanation reduction. *Radiation Protection Dosimetry, 189*(1), 69–75. https://doi.org/10.1093/rpd/ncaa014 %J

Mohamed, M. A., Salleh, W. N. W., Jaafar, J., Asri, S. E. A. M., & Ismail, A. F. (2015). Physicochemical properties of "green" nanocrystalline cellulose isolated from recycled newspaper. *Rsc Advances, 5*(38), 29842–29849. https://doi.org/10.1039/C4RA17020B

Onoja, E., Chandren, S., Abdul Razak, F. I., Mahat, N. A., & Wahab, R. A. (2019). Oil palm (*Elaeis guineensis*) biomass in Malaysia: The present and future prospects. *Waste and Biomass Valorization, 10*(8), 2099–2117. https://doi.org/10.1007/s12649-018-0258-1

Padzil, F. N., Lee, S. H., Ainun, Z. M., Lee, C. H., & Abdullah, L. C. (2020). Potential of oil palm empty fruit bunch resources in nanocellulose hydrogel production for versatile applications: A review. *Materials, 13*(5). https://doi.org/10.3390/ma13051245

Parveez, G. K. A., Tarmizi, A. H. A., Sundram, S., Loh, S. K., Ong-Abdullah, M., Palam, K. D. P., Salleh, K. M., Idris, Z. (2021). Oil palm economic performance in Malaysia and R&D progress in 2020. *Journal of Oil Palm Research, 33*, 2.

Perrin, L., Gillet, G., Gressin, L., & Desobry, S. (2020). Interest of pickering emulsions for sustainable micro/nanocellulose in food and cosmetic applications. *Polymers, 12*(10). https://doi.org/10.3390/polym12102385

Rubinsin, N. J., Daud, W. R. W., Kamarudin, S. K., Masdar, M. S., Rosli, M. I., Samsatli, S., Tapia, J. F. D., Ghani, W. A. W. A. K., Hasan, A., & Lim, K. L. (2021). Modelling and optimisation of oil palm biomass value chains and the environment–food–energy–water nexus in peninsular Malaysia. *Biomass and Bioenergy, 144*, 105912.

Salehudin, M. H., Salleh, E., Muhamad, I. I., & Mamat, S. N. H. (2014). Starch-based biofilm reinforced with empty fruit bunch cellulose nanofibre. *Materials Research Innovations, 18*(sup6), S6-322–S326-325. https://doi.org/10.1179/1432891714Z.000000000977

Septevani, A. A., Rifathin, A., Sari, A. A., Sampora, Y., Ariani, G. N., Sudiyarmanto, & Sondari, D. (2020). Oil palm empty fruit bunch-based nanocellulose as a super-adsorbent for water remediation. *Carbohydrate Polymers, 229*, 115433. https://doi.org/10.1016/j.carbpol.2019.115433

Shanmugarajah, B., Chew, I. M., Mubarak, N. M., Choong, T. S., Yoo, C., & Tan, K. (2019). Valorization of palm oil agro-waste into cellulose biosorbents for highly effective textile effluent remediation. *Journal of Cleaner Production, 210*, 697–709. https://doi.org/10.1016/j.jclepro.2018.10.342

Shazali, N. A., Zaidi, N. E., Ariffin, H., Abdullah, L. C., Ghaemi, F., Abdullah, J. M., Takashima, I., & Nik Abd. Rahman, N. M. (2019). Characterization and cellular internalization of spherical cellulose nanocrystals (CNC) into normal and cancerous fibroblasts. *Materials, 12*(19). https://doi.org/10.3390/ma12193251

Sinaga, M. Z. E., Gea, S., Panindia, N., & Sihombing, Y. A. (2018). The preparation of all-cellulose nanocomposite film from isolated cellulose of corncobs as food packaging. *Oriental Journal of Chemistry, 34*(1), 562.

Solikhin, A., Hadi, Y. S., Massijaya, M. Y., & Nikmatin, S. (2017). Novel isolation of empty fruit bunch lignocellulose nanofibers using different vibration milling times-assisted multimechanical stages. *Waste and Biomass Valorization, 8*(7), 2451–2462. https://doi.org/10.1007/s12649-016-9765-0

Solikhin, A., Hadi, Y. S., Massijaya, M. Y., Nikmatin, S., Suzuki, S., Kojima, Y., & Kobori, H. (2018). Properties of poly(vinyl alcohol)/chitosan nanocomposite films reinforced with oil palm empty fruit bunch amorphous lignocellulose nanofibers. *Journal of Polymers and the Environment, 26*(8), 3316–3333. https://doi.org/10.1007/s10924-018-1215-6

Supian, M. A. F., Amin, K. N. M., Jamari, S. S., & Mohamad, S. (2020). Production of cellulose nanofiber (CNF) from empty fruit bunch (EFB) via mechanical method. *Journal of Environmental Chemical Engineering, 8*(1), 103024. https://doi.org/10.1016/j.jece.2019.103024

Supramaniam, J., Kiat Wong, S., Fen Leo, B., Teng Hern Tan, L., Hing Goh, B., & Ying Tang, S. (2020). Unravelling the swelling behaviour and antibacterial activity of palm cellulose nanofiber-based metallic nanocomposites. *IOP Conference Series: Materials Science and Engineering, 778*, 012027. https://doi.org/10.1088/1757-899x/778/1/012027

Szlapak Franco, T., Martínez Rodríguez, D. C., Jiménez Soto, M. F., Jiménez Amezcua, R. M., Urquíza, M. R., Mendizábal Mijares, E., & de Muniz, G. I. B. (2020). Production and technological characteristics of avocado oil emulsions stabilized with cellulose nanofibrils isolated from agroindustrial residues. *Colloids and Surfaces A: Physicochemical and Engineering Aspects, 586*, 124263. https://doi.org/10.1016/j.colsurfa.2019.124263

Tan, K., Heo, S., Foo, M., Chew, I. M., & Yoo, C. (2019). An insight into nanocellulose as soft condensed matter: Challenge and future prospective toward environmental sustainability. *Science of the Total Environment, 650*, 1309–1326. https://doi.org/10.1016/j.scitotenv.2018.08.402

Tang, Y., Zhang, X., Zhao, R., Guo, D., & Zhang, J. (2018). Preparation and properties of chitosan/guar gum/nanocrystalline cellulose nanocomposite films. *Carbohydrate Polymers, 197*, 128–136. https://doi.org/10.1016/j.carbpol.2018.05.073

Thomas, B., Raj, M. C., B, A. K., H, R. M., Joy, J., Moores, A., Drisko, G.L., & Sanchez, C. (2018). nanocellulose, a versatile green platform: From biosources to materials and their applications. *Chemical Reviews, 118*(24), 11575–11625. https://doi.org/10.1021/acs.chemrev.7b00627

Vilarinho, F., Sanches Silva, A., Vaz, M. F., & Farinha, J. P. (2018). Nanocellulose in green food packaging. *Critical Reviews in Food Science and Nutrition, 58*(9), 1526–1537.

Yahya, M. B., Lee, H. V., & Abd Hamid, S. B. (2015). Preparation of nanocellulose via transition metal salt-catalyzed hydrolysis pathway. *BioResources, 10*(4), 7627–7639.

Effect of Nanolignin and Nanocrystalline Cellulose on Thermal, Mechanical, and Water Barrier Properties of Starch Composites

Udari Prasadini Perera, Mei Ling Foo, and Irene Mei Leng Chew

Abstract In this work, a ternary biopolymer composite was obtained by reinforcing the thermoplastic starch (TPS) with nanocrystalline cellulose (NCC), and nanolignin (NL) originated from oil palm empty fruit bunch (EFB). The samples were characterized for optical, thermal, mechanical, and water barrier properties. UV-vis spectrometry and contact angle results revealed a synergistic effect between NL and NCC in improving transparency, UV light barrier and hydrophobicity of the composites. Furthermore, the ternary films showed better moisture absorbance and swelling performance than the neat and binary films. The glass transition temperature (T_g) of the film with 4 wt% of NL and 4 wt% of NCC increased from 40 to 61 °C compared to the neat film. In addition, the tensile strength was found to be approximately 150% higher than the neat film. This work suggested that incorporating NL and NCC into the starch composites could be a promising alternative to current packaging material with improved mechanical strength, thermal stability, water, and UV light barrier performance.

Keywords Nanocomposite film · Nanolignin · Packaging material · Empty fruit bunch · Biocomposites · Waste valorization

1 Introduction

Biopolymers are a sustainable alternative for synthetic polymers for many applications (e.g. biomaterial and packaging) in solving growing environmental problems (El Miri et al., 2015; Jumaidin et al., 2020). Biopolymers, such as starch, are abundant, non-toxic, biodegradable, and biocompatible compared to synthetic polymers. Starch occurs in nature in the form of semi-crystalline granules in plant seeds, fruits, roots, and tubers (Giroto et al., 2020; Pérez et al., 2009). Native starch can be processed to thermoplastic material, known as thermoplastic starch (TPS) with a plasticizer

U. P. Perera · M. L. Foo · I. M. L. Chew (✉)
Chemical Engineering Discipline, School of Engineering, Monash University Malaysia, 47500 Bandar Sunway, Selangor, Malaysia
e-mail: irene.chew@monash.edu

D. C. Y. Foo et al. (eds.), *Sustainable Technologies for the Oil Palm Industry*,
https://doi.org/10.1007/978-981-19-4847-3_4

such as polyols (i.e. glycerol and sorbitol) or amides, and water under high temperature and pressure (González et al., 2015). TPS is a renewable and flexible material that can be processed with existing plastic handling technologies for engineering and packaging applications. Nevertheless, the TPS is highly hydrophilic, and the mechanical performance of the material is low. Hence, many studies focused on the addition of bio fillers and biomaterials (e.g. nanocrystalline cellulose (NCC) and lignin) to starch composites to improve the mechanical, thermal, and water resistance properties (Area et al., 2019; Fazeli et al., 2018; Li et al., 2018).

NCC is a highly functional bio-nanomaterial, which can be prepared from a wide range of biomass sources, most commonly by acid hydrolysis (Wang et al., 2020a) NCC has been widely used in reinforcing natural (Huq et al., 2012; Lee et al., 2020) and synthetic (Hassanabadi et al., 2015; Li et al., 2014; Liu et al., 2020) polymer composites owing to their high crystallinity (54–88%), high tensile strength and Young's modulus (138–150 GPa), and biodegradability (Ng et al., 2015). Also, it was evident to improve the water-resistance of composite films (El Miri et al., 2015; Montero et al., 2017).

In nature, lignin presents in plant cell walls together with cellulose provides strength and structural rigidity. It is highly hydrophobic and possesses potent antioxidant activity (Kai et al., 2016; Liu et al., 2017). This makes lignin a suitable filler for biocomposites. However, its structural heterogeneity and processability of lignin limit the application as a filler (Garcia Gonzalez et al., 2017). Moreover, the high variability of the particle size distribution of lignin and uneven dispersion of the lignin particles in the matrix could negatively impact the mechanical properties of the composites (Yang et al., 2015). Hence, adding lignin nanoparticles (Nanolignin, NL) instead of unreduced lignin could be the potential solution. NL has better dispersity in polymer composites, increasing its interfacial area between NL and host polymer (Chrissafis & Bikiaris, 2011). This significantly improved the mechanical performance (Tian et al., 2017), thermal stability (Nair et al., 2014), and higher anti-oxidant activity (Yearla & Padmasree, 2016) upon adding NL to the polymer composites.

Reinforcing biopolymers with NL and NCC received limited attention in the literature (Yang et al., 2016a). Few studies have reported attempts on reinforcing Polyvinyl alcohol (PVA) and Polylactic acid (PLA) with NL and NCC (Yang et al., 2020) or with Cellulose Nanofibrils (CNF) (Bian et al., 2018), which are summarized in Table 1.

In this context, this study attempts to produce ternary biocomposites using NL, NCC and starch. In this study, oil palm empty fruit bunch (EFB), a common industrial waste from the oil palm industry was used to produce both NCC (Al-Dulaimi & Wanrosli, 2017; Foo et al., 2019) and lignin (Coral Medina et al., 2018; Tang et al., 2019). The synthesized lignin was converted to NL by ultrasonication (Ma et al., 2019). The properties of the composite films were analysed in terms of mechanical, thermal, UV light barrier, and water barrier performances. Also, the structural properties of the composite films were analysed using Fourier transform infrared (FTIR) spectroscopy. To our knowledge, this is the first work to study the performance of EFB-derived NCC and NL in starch composites (Fig. 1).

Table 1 Nanocomposites containing both NL and NCC/NFC

Polymer	Reinforcing material	Source biomass	Results	References
PLA	NL (1–3 wt%) NCC (1–3 wt%)	NL-*Arundo donax* L. NCC-industrial Microcrystalline cellulose (MCC)	Synergic effect of NL and NCC in composite transparency and UV blocking resulted in higher strength and modulus and better antibacterial activity	Yang et al. (2016a)
PLA	NL (0–4 wt%) CNF (1 wt%)	NL-birch wood chips CNF-bleached spruce kraft pulp	NL acted as nano-spaces which lead to enhanced viscoelasticity and thermal stability of hydrogel. When NL content increased, hydrogel showed 28- and 18-times higher storage modulus (8504 Pa) and loss modulus (3260 Pa), respectively	Bian et al. (2018)
PVA	NL (1–2 wt%) NCC (1–2 wt%)	NL-industrial lignin NCC-industrial MCC	Tensile strength increased from 26 (for neat) to 35.4 MPa of the composite containing 2 wt% and 1 wt%. PVA-CNC-LNP showed better water vapour barrier properties and UV irradiation shielding effect	Yang et al. (2020)
PVA	NL (1 wt%) CNF(2–10 wt%) boron nitride nanosheet (BNNS)	NL-Maple pulp CNF-bleached softwood pulp	CNF and LNP had synergistic actions on enhancing the film's tensile strength, thermal stability, and conductivity	Wang et al. (2020a, 2020b)

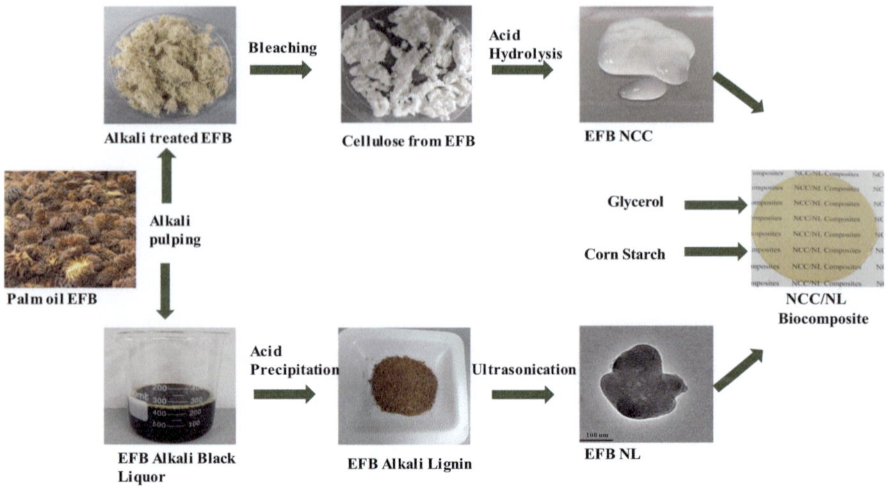

Fig. 1 The procedure to fabricate NL /NCC biocomposites

2 Materials and Methods

2.1 *Materials*

EFB fibres were complimentary given by Eureka Synergy Sdn. Bhd. Glycerol (99.5%) and glacial acetic acid (AR grade) were obtained from Fisher Scientific. Corn starch and sodium chlorite ($NaClO_2$, 80 w/w%) were supplied by Sigma Aldrich and Acros Organics, respectively. Sodium hydroxide (NaOH) and sulfuric acid (H_2SO_4, 95–97%) were purchased from Friendemann Schmidt Chemicals.

2.2 *Methods*

2.2.1 *NCC and NL Synthesis*

NCC preparation was adapted from our previous work Foo et al. (2019) with minor modifications. Firstly, EFB fibres were grounded using a rotor mill (Pulverisette 14, Fritsch) and sieved through a 250 μm mesh. Then, the fibres were washed three times with 50 °C water (fibre: liquid ratio of 1 g:30 ml) to remove the dirt and water-soluble substances and then oven-dried at 60 °C for 24 h. Subsequently, the fibres were treated with soda pulping using 4 w/v% NaOH solution at 80 °C for 6 h with continuous stirring. The fibres were then filtered and washed with deionized water until neutral pH is obtained, whereas the black liquor resulted from the soda pulping was collected for the isolation of lignin.

The fibres from soda pulping were treated with an acidic $NaClO_2$ solution (made with an equal volume of 2 w/v% of $NaClO_2$ and acetate buffer) at 80 °C for 6 h with continuous stirring. The fibres were water washed until neutral pH was obtained. The fibres were subsequently filtered with Whatman fibre paper (pore size = 11 μm) before freeze-drying for 24 h. Then the fibres were subjected to acid hydrolysis with 56 wt% H_2SO_4 for 30 min at 45 °C. After 30 min, cold water was added to the acid solution to stop the hydrolysis reaction. Then the mixture was centrifuged (10,000 rpm for 15 min) twice to collect the hydrolysed sample and remove the acid residue. The collected sample was dialysed (molecular weight cut off 12,000–14,000 Da) with deionized water until a fixed pH was reached. The solid suspension was sonicated at 60% amplitude for 10 min to obtain the dispersed NCC.

The black liquor was first filtered and then acidified to pH 2 under continuous stirring with 20 v/v% H_2SO_4 solutions to isolate lignin. The acidified black liquor was centrifuged at 11,000 rpm for 15 min to precipitate the lignin. The precipitated lignin was washed with H_2SO_4 solution (pH 2) and subsequently filtered using Whatman filter paper (pore size = 11 μm) before being oven-dried at 45 °C for 24 h. The isolated lignin was dispersed in deionized water at 0.5 wt% and adjusted to pH 7 using NaOH solution. The lignin suspension was sonicated for 30 min using the sonicator (QSONICA Q700, 20 kHz, 700 W) at 60% amplitude to obtain NL. The sample was kept inside an ice bath during the sonication to present excessive heating.

2.2.2 *NL and NCC Characterization*

The particle size distribution of NL and NCC was examined using the dynamic light scattering (DLS) method (Malvern Nano-ZS Zetasizer, Malvern Instruments, UK). The morphology of NCC and NL was characterized using field emission scanning electron microscope (FE-SEM; SU8010, Hitachi) with an acceleration voltage of 15 kV. The diluted sample (~0.001 w/v%) was dropped onto a carbon-coated copper grid and air-dried, followed by the staining with a drop of 2 w/v% uranyl acetate and air-dried before analysis.

2.2.3 Bionanocomposites Preparation

The solvent casting method was used to prepare three types of films, namely neat, binary, and ternary. Neat film (i.e. control sample) consists of starch, glycerol, and water. For the neat film, a 10 g of starch and a 5 g of glycerol were dispersed in 185 g of deionized water. The mixture was heated to 80 °C for 5 min with continuous stirring. The resulting gelatinized suspension was sonicated at 60% amplitude for 10 min. The suspension was then poured into petri dishes and oven dried at 45 °C for 24 h. Binary films (contain one nanoparticle, i.e. NCC, lignin or NL) and ternary films (contain two nanoparticles, NL and NCC) were prepared following the same procedure as neat film. The quantity of nanoparticles added to the composite films

Table 2 Composition of prepared films

Sample	Lignin (%)	NCC (%)	NL (%)
Neat	0	0	0
2NCC	0	2	0
2L	2	0	0
2NL	0	0	2
1NL1NCC	0	1	1
2NL2NCC	0	2	2
4NL4NCC	0	4	4

during the mixing stage are given in Table 2. The nanoparticles percentages were calculated with respect to the starch dry mass and glycerol weight.

2.2.4 Fourier Transform Infrared Spectroscopy (FTIR) Analysis

FTIR analysis was performed using Thermo Nicolet iS10 spectrometer (Thermo Fisher Scientific) equipped with an ATR diamond probe accessory. The spectra were recorded in the transmittance mode with an accumulation of 64 scans over the range of 500–4000 cm^{-1}.

2.2.5 UV Absorbance

UV absorbance of composite films were measured using a UV-vis spectrometer (Cary 100, Agilent) in the range of 200–800 nm with triplicate measurements for each sample.

2.2.6 Contact Angle Measurements

The hydrophobicity of the composite films was determined using a goniometer (Ramehert) equipped with a computer-automated program. The films were fixed on glass plates with double sided adhesive tape. The distilled water droplet of 4 μl was placed on the surface of the film with images captured immediately. The contact angle of water was determined with the aid of DROPimage software based on the sessile drop technique. First ten contact angles were taken for at least three different positions of the film and the mean value was reported.

2.2.7 Moisture Content

The moisture content of the nanocomposite films was determined through weight measurement. The samples of nanocomposite films were cut into rectangular shapes (1 mm × 2 mm) and oven dried at 50 °C for 48 h (W_1, g). Then, the samples were conditioned at 100% RH for 24 h (W_2, g) in a desiccator. The measurement of films was performed in triplicate. The moisture content was evaluated using Eq. (1) as follows:

$$\text{Moisture content}(\%) = \frac{W_2 - W_1}{W_1} \times 100\% \tag{1}$$

2.2.8 Swelling Ratio

The swelling ratio of samples was analysed as suggested in previous studies (López-Córdoba et al., 2019; Shankar et al., 2015). The samples (1 mm × 2 mm) were dried at 50 °C for 48 h to determine the initial dry weight (W_0, g). Then the films were submerged into 30 ml of distilled water at room temperature for 24 h. Subsequently, the water was removed r and weighted after eliminating surface water with blotting paper (W_t, g). The swelling ratio percentage was calculated as below (Eq. (2):

$$\text{Swelling Ratio}(\%) = \frac{W_t - W_0}{W_0} \times 100 \tag{2}$$

2.2.9 Thermogravimetric Analysis

TGA was performed using a thermogravimetric analyser (TGA 550, TA instruments). Film samples (approximately 10 mg) were heated under nitrogen environment from 25 to 800 °C at a ramping rate of 10 °C/min. The weight loss rate and maximum thermal degradation temperature (T_{max}) were determined from derivative thermogravimetric (DTG). Important thermal characteristics, namely peak temperatures and residue wt% were obtained from thermogravimetric analysis.

2.2.10 Differential Scanning Calorimetry (DSC)

DSC measurements were carried out with TA instruments DSC Q200 (TA instruments). Approximately 5 mg of each film sample was placed in a sealed sample pan and heated from −25 to 210 °C (1st scan) at a heating rate of 10 °C/min under the nitrogen environment. Then the sample was cooled with the same heating rate to −25 °C and reheated to 210 °C (2nd scan).

2.2.11 Mechanical Properties

The mechanical properties of the films were evaluated using a texture analyser TA-XT2 (Stable Micro Systems). The samples (40 mm × 12.5 mm) were tested with crosshead speed of 1.00 mm/s at a distance of 40 mm. A 5 kg load cell was used, and the initial gauge length of 25 mm was set. Three specimens from each sample were tested, and the results were reported as the average of these replications.

2.2.12 Statistical Analysis

Analysis of variance (ANOVA) was performed using GraphPad Prism 8 software at the significance level of $p < 0.05$.

3 Results and Discussion

3.1 NCC and NL Characterization

The STEM and DLS measurements showed NCC is 21–33 nm in width and 260–374 nm in length. The NL was observed to have spherical shapes from the SEM image in Fig. 2. The particle size and particle distribution of NL were obtained from the DLS measurements. NL was estimated to be 220 ± 1.6 nm with a polydispersity index of 0.2.

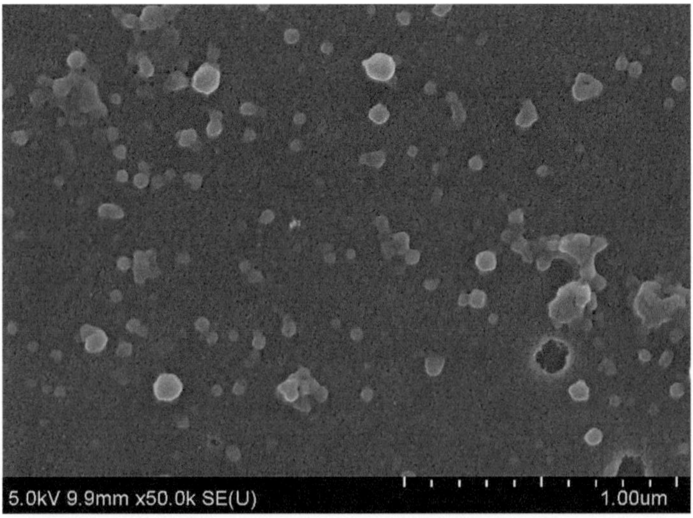

Fig. 2 SEM image of NL

3.2 Bionanocomposite Characterization

3.2.1 Fourier Transform Infrared (FTIR) Spectroscopy

FTIR examined the structural components and intermolecular interactions between starch, NCC, lignin, and NL. Figure 3a shows the FTIR spectrum of all composites, and Fig. 3b demonstrates the enlarged section of the range 600 to 1440 cm^{-1}. All the spectrums showed the following peaks: broad peak from 3000 to 3600 cm^{-1} (stretching vibrations of hydrogen-bonded OH groups of the free inter- and intramolecular bound OH groups in the composite), 2854 cm^{-1} (referred to the presence of C–H stretching of aldehydes), 2921 cm^{-1} (C–H stretching of alkyl chain), 1739 cm^{-1} (C=O carbonyl stretching bands), 1640 cm^{-1} (O–H bending of absorbed water of composites), and 1450 cm^{-1} (C–H bending and wagging of CH_2) (Gilfillan, 2015; Mano et al., 2003; Meneguin et al., 2017). Similarly, many peaks were located from 1200 to 1400 cm^{-1} in all spectrums, owing to the overlapping of C–H in the plane and O–H bonding. Plasticized starch has few characteristic peaks between 900 and 1200 cm^{-1} regions, corresponding to C–O stretching vibration of primary, secondary, and tertiary alcohols (Gilfillan, 2015). The peaks near 1104 and 1149 cm^{-1} are related to C–O bond stretching of the C–O–H group. The two peaks around 1020 cm^{-1} were attributed to C–O bond starching of C–O–C of the anhydroglucose ring of native starch (Fang et al., 2002).

3.2.2 UV Absorbance and Optical Properties

Figure 4a shows the UV spectra of biocomposites films. The 2NL film has excellent UV absorbance in the 250–350 nm area compared to the 2L film. The high absorbance could be ascribed to nanostructured lignin's enhanced UV barrier properties (Yearla & Padmasree, 2016; Zimniewska et al., 2008). Both 1NL1NCC and 2NL2NCC films showed similar level of absorbance values in the UV area. The highest absorbance of 4NL4NCC film could possibly be due to the high NL load, which suggested adding NL could improve the UV blocking of the ternary film. This indicated that the NL-NCC composites could be used in high-end packaging applications when UV protection is required.

These results are in good agreement with the digital images captured for neat and composite films in Fig. 4b. Both neat and 2NCC films exhibited the highest optical transparency. With the addition of lignin/ NL, colour of the films became darker. The 4NL4NCC film showed the brownest colour compared to other samples, which was attributed to the high NL load. The non-homogeneous colour on the 2L film indicated the uneven distribution of lignin particles in the starch matrix. However, 2NL, 1NL1NCC, 2NL2NCC, and 4NL4NCC films presented a uniform colour, owing to the homogeneous distribution of NL and NCC in the starch matrix (Peng et al., 2018).

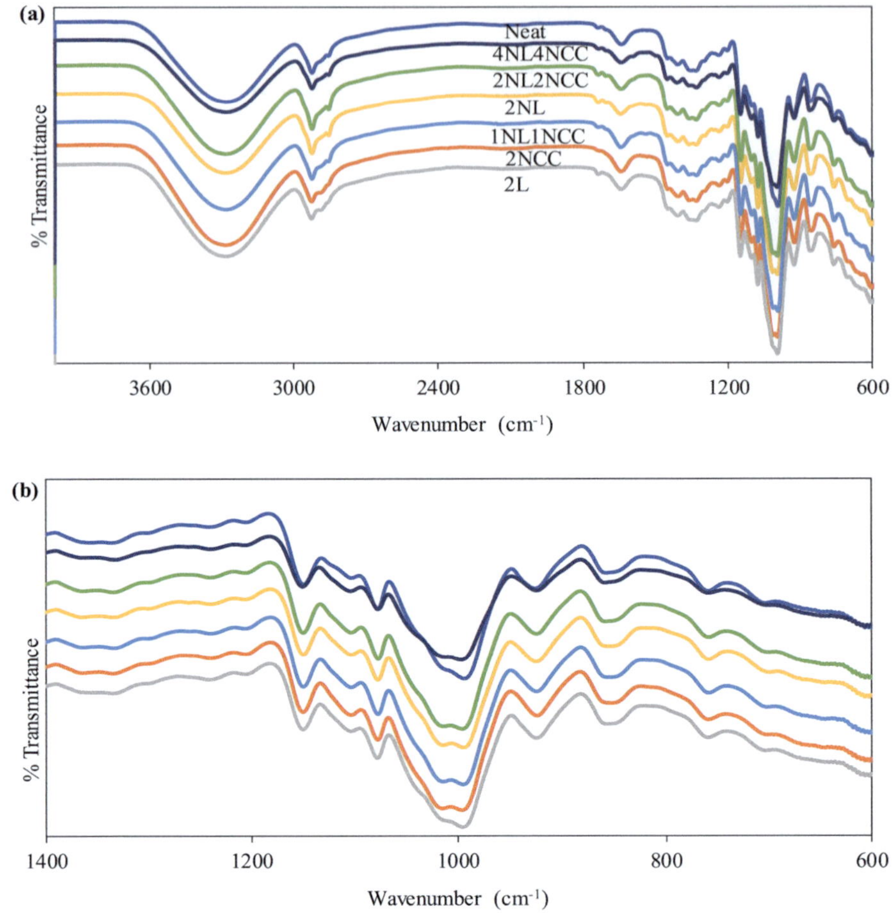

Fig. 3 **a** FTIR spectra on neat and composite films, and **b** enlarged FTIR spectra of neat and composite films for the 1400–600 cm^{-1} region

3.2.3 Water Contact Angle

The contact angle measurements are helpful to study the wetting behaviour and surface hydrophobicity of the polymer composites. Figure 5 presents the images of contact angle measurements of composite films, where their respective data are summarized in Table 3. The contact angle of the neat film was 42°, which indicated the hydrophilic nature of starch due to the abundant hydroxyl groups present in starch (Balakrishnan et al., 2019). The water contact angle of 2NCC was found similar to the neat film, owing to its hydrophilic dominant property. The hydrophobic nature of lignin has resulted in a water contact angle of 55° with the addition of lignin. Generally, surface textures with micro/nanoparticles are favourable to promoting surface hydrophobicity (Owais et al., 2016), which was confirmed by the relatively

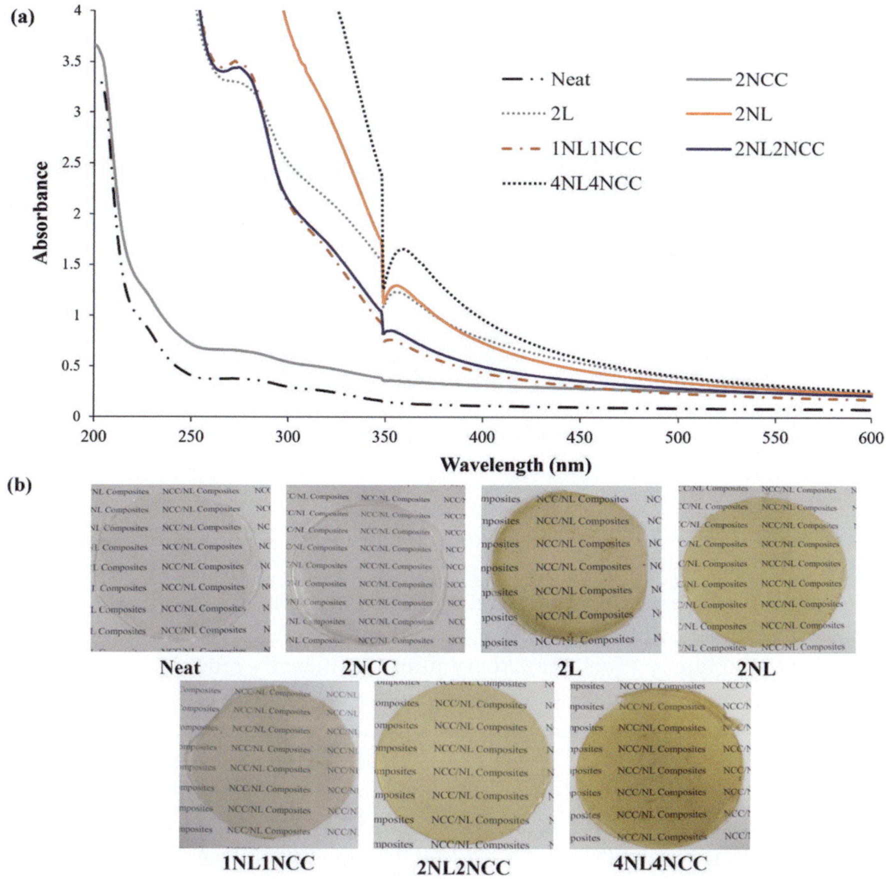

Fig. 4 **a** UV-vis spectra of neat and composite films, and **b** digital images of neat and composite films

higher water contact angle of 2NL film (77°) compared to 2L. On the other hand, the ternary composite films had led to a higher level of hydrophobicity. The water contact angle of 1NL1NCC, 2 NL2NCC, and 4NL4NCC were 79°, 83°, and 83°, respectively. It is noteworthy that the water contact angle obtained from the ternary films was higher than 2NL, suggesting that the blending of NL and NCC could also enhance the hydrophobicity of starch composite films.

3.2.4 Moisture Content and Swelling Ratio

One of the significant disadvantages of starch and nanocellulose-based biocomposites is the high moisture absorption due to its hydrophilic nature (Syafri et al., 2019). Table 3 presents the moisture content of composites, and the neat films showed the

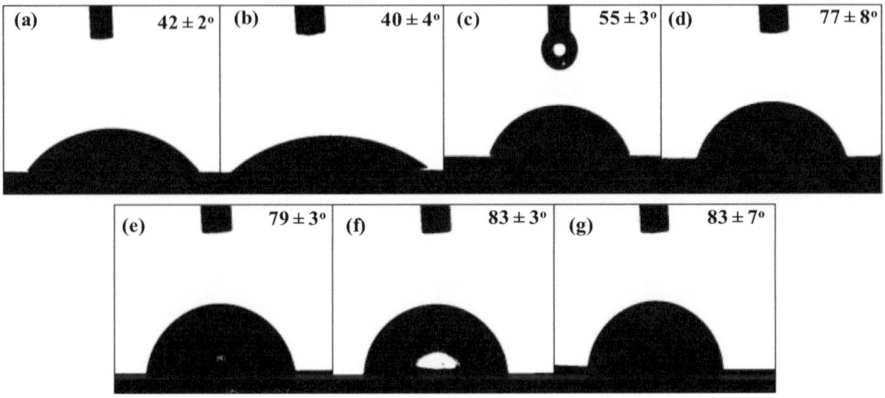

Fig. 5 Contact angle of **a** neat, **b** 2NCC, **c** 2L, **d** 2NL, **e** 1NL1NCC, **f** 2NL2NCC, and **g** 4NL4NCC films

highest moisture content compared to other composite films. The addition of particles caused a decrease in the moisture content in 2L and 2NCC films. However, those values were not statistically significant as shown in Table 3 where 2NCC and 2L samples are denoted with the same [a].

Compared to binary films, the ternary films significantly reduced ($p < 0.05$) the moisture content. Lignin and NL are hydrophobic polymers owing to their aromatic skeleton, and it could reduce the hydrophilicity of a composite once added. (Duval et al., 2013; Yang et al., 2015; Zhang et al., 2014;) As reported in earlier literature, NCC has better performance in moisture absorbance than starch due to its higher degree of molecular order (González et al., 2015). When mixed, NCC caused the starch networks to be more rigid and reduced the mobility of water molecules across the starch matrix (Montero et al., 2017). On the other hand, NL created a hydrophobic surface on the film, thereby limiting the formation of hydrogen bonding between

Table 3 Contact angle, moisture content, and swelling ratio of prepared films

Sample	Contact angle (°)	Moisture content (%)	Swelling ratio (%)
Neat	42 ± 2^{a}	39.6 ± 1.4^{a}	45.6 ± 2.1^{a}
2NCC	40 ± 4^{b}	37.9 ± 2.1^{a}	43.6 ± 1.3^{a}
2L	55 ± 3^{c}	38.4 ± 0.6^{a}	44.0 ± 0.9^{a}
2NL	77 ± 8^{d}	35.1 ± 0.8^{b}	27.0 ± 4.9^{b}
1NL1NCC	79 ± 3^{e}	34.7 ± 1.9^{b}	49.2 ± 3.2^{a}
2NL2NCC	83 ± 3^{f}	34.9 ± 0.4^{b}	$32.2 \pm 0.5^{b,c}$
4NL4NCC	83 ± 7^{f}	34.6 ± 1.0^{b}	$31.1 \pm 0.8^{b,c}$

Average value of triplicate measurements was presented as means±standard deviation. Different letters ([a] to [f]) indicate statistically significant differences ($p<0.05$) between means

starch and water molecules (Zhang et al., 2020). The synergic effect of NL and NCC in lowering the moisture absorbance would potentially be beneficial to the applications where moisture absorbance is critical for food packaging.

The swelling ratio of samples is given in Table 3. The swelling ratio of 2NCC and 2L did not significantly differ from the swelling ratio of the neat film ($p > 0.05$). This indicates that adding small amounts of NCC or lignin does not substantially influence the composite's water resistance properties. On the contrary, a remarkable reduction in swelling ratio (40.9% lower than the neat film) was observed for 2NL film, suggesting that the NL has a higher resistance to water penetration. NL may act as an impermeable medium for moisture and shield the spreading of water in the starch-based composite (Nair et al., 2017; Spiridon et al., 2011). On the other hand, blending of NCC and NL only showed a significant reduction in swelling ratio when NCC and NL in the ternary film is > 1%. This signified that the NL played a prominent role against swelling.

3.2.5 Thermal Properties

TG and DTG curves of the samples are presented in Fig. 6a, b, respectively, where the thermal parameters are reported in Table 4. Three main peaks were observed from the TG curve of samples, which were attributed to moisture evaporation (60–100 °C), decomposition of glycerol (160–210 °C), and decomposition of starch (240–280 °C) (Area et al., 2019; Yang et al., 2016a, 2016b). Generally, pure lignin/NL and NCC possess a T_{max} at 285 °C and 293 °C, respectively (Foo et al., 2019; Sen et al., 2015). In the composites, T_{max} values of lignin/NL and NCC intersect with the T_{max} region of starch. Hence, no discernible distinct degradation peaks were observed from the lignin, NL, and NCC TG curves. The temperature of water evaporation (T_{peak1}) of 1NL1NCC and 2NL2NCC films increased with NL-NCC loading as the NL and NCC occupied more in the starch matrix and thus, hindered the evaporation of water and glycerol (Yang et al., 2016a, 2016b). However, 4NL4NCC film has a lower T_{peak1} (78 °C) than 1NL1NCC and 2NL2NCC, which could be due to the agglomerations occurring from the high loading of NCC (Yang et al., 2016a, 2016b). No significant change on T_{max} with the addition of NCC, lignin or NL, whilst the T_{max} value of the ternary composite film was slightly lower as compared to the neat and binary films.

The residue amount of neat film obtained from the DTG curve at 800 °C was 7%. The resulting char could be ascribed to the partial carbonization of starch (Area et al., 2019). The addition of NCC caused the residual char amount to increase to 11% (2NCC film) (González et al., 2015; Montero et al., 2017). The comparative increase in char for 2NCC film was attributed to less volatile crystalline contents that resist complete degradation. The 2L and 2NL composites had lower final residue amounts (8% and 9%, respectively) compared to 2NCC film, owing to the less crystalline structure. The residue amount of the ternary films was found to be higher with the increased loading of nanoparticles. The highest residue was obtained from 4NL4NCC film due to the highest nanomaterial load.

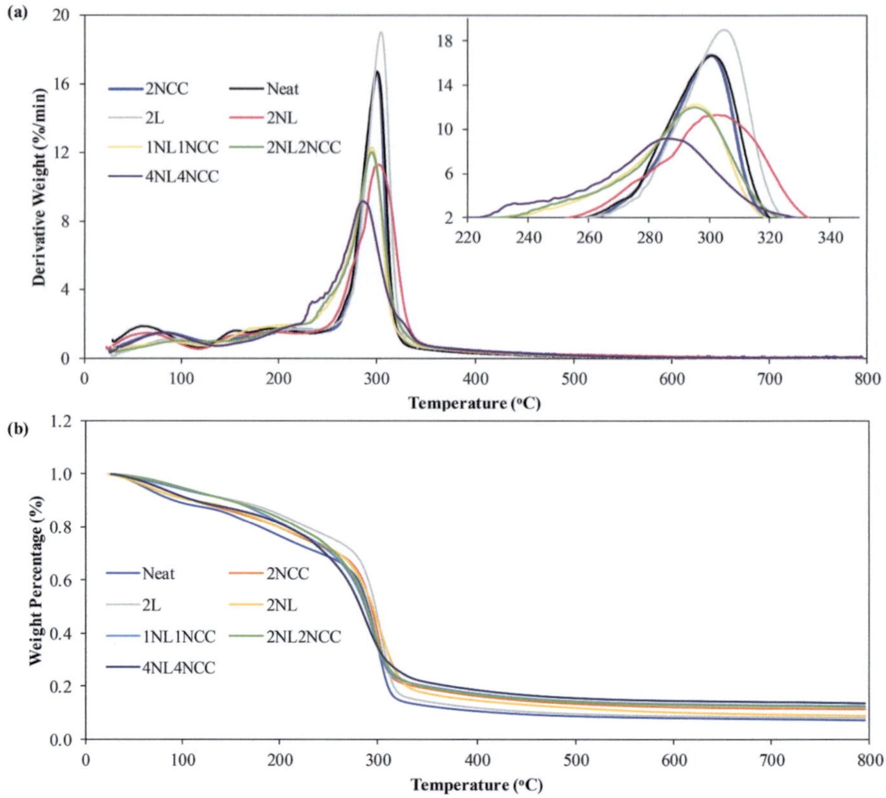

Fig. 6 **a** Derivative weight curves, and **b** residual mass curves from TGA

Table 4 Thermal properties of neat and composite films

Film	T_{peak1} (°C)	T_{max} (°C)	Residue (%)	T_g (°C)
Neat	60	302	7	40
2NCC	85	302	11	54
2L	86	305	8	42
2NL	67	304	9	52
1NL1NCC	95	297	12	55
2NL2NCC	99	297	12	60
4NL4NCC	78	286	14	61

T_g of the samples determined using DSC analysis are shown in Table 4. The T_g was found higher with the addition of nanoparticles. Adding NCC had promoted the T_g from 40 °C (neat film) to 54 °C. The 2NCC film has a higher T_g as compared to 2L and 2NL. The higher degree of crystallinity of NCC and the strong hydrogen bonding formed between starch and NCC could be the reason for improved thermal stability (Li et al., 2018). In addition, the T_g of 2NL film (52 °C) was higher than 2L film (42 °C), which could be ascribed to the strong interfacial bonding between NL and starch matrix. Furthermore, T_g of the composites increased with the addition of nanomaterial. The synergistic effect between NL and NCC has improved T_g compared to the neat and binary film.

3.2.6 Mechanical Properties

The effect of incorporating nanomaterials to the starch matrix on the mechanical properties (tensile strength and elongation at break) was shown in Fig. 7. All composites showed significantly improved tensile strength compared to the neat film except for 2L, where the highest value was obtained from the ternary composites (Fig. 7a). The 2L film showed similar tensile strength to the neat film due to the larger particle size and heterogeneous dispersion of lignin compared to other additives. The tensile strength of 2NL was higher compared to the 2L, due to the nano reinforcement and homogeneous dispersion of NL. On the other hand, the 2NCC showed better enhancement in tensile strength compared to 2NL owing to the crystalline structure and the good interfacial interaction between NCC and starch (Chang et al., 2010). For the ternary systems, the tensile strength was gradually improved with the increased loading of nanoparticles (both NL and NCC). The highest tensile strength was obtained from 4NL4NCC film, which was threefold increment compared to the neat film. The high loading of nanoparticles (8 wt% in total) and the compatibility between nanoparticles and starch had contributed to the high tensile strength of 4NL4NCC film (Shankar & Rhim, 2017). However, the elongation at break of the composite films was lower than the neat film as the addition of NL and NCC made the films more brittle. The lowest elongation at break was reported for 4NL4NCC.

Table 5 summarizes the performance of UV barrier, water barrier, surface hydrophobicity, thermal and mechanical properties of binary and ternary films compared to neat film. The 2L film in comparison to neat film showed a better UV barrier, surface hydrophobicity, and slightly improved T_g. However, the water barrier property and tensile strength of 2L films were still in the same range as the neat film. Nevertheless, 2NL showed better performance in all categories than both neat and 2L films, implying the effectiveness of NL counter to lignin in reinforcing biocomposites. 2NCC film also showed enhanced performance in water barrier, mechanical, and thermal properties relative to neat and 2L films. Also, the mechanical and thermal performance of 2NCC is better than 2NL. However, 2NCC was lacking in UV barrier and surface hydrophobicity in comparison to both 2NL and 2L. 1NL1NCC film showed intermediate properties between binary and ternary films, which could be due to the lower loading levels of NL and NCC. 2NL2NCC and 4NL4NCC films showed overall improved properties compared to binary films, and the most desirable performance was demonstrated 4NL4NCC film owing to combined properties of NL and NCC.

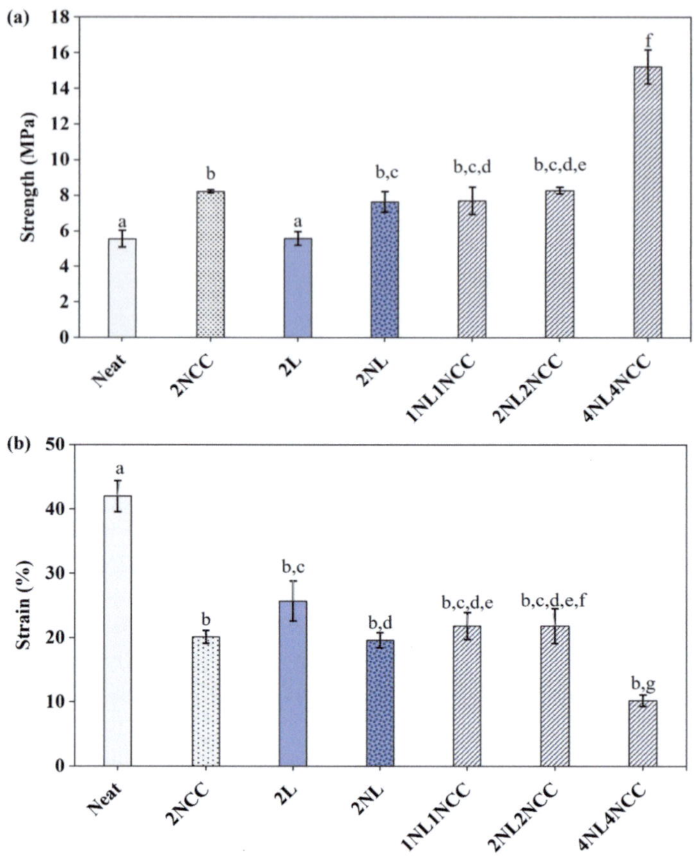

Fig. 7 **a** mechanical strength and **b** elongation at break of neat and composite films. Average values of measurements were presented as mean ± standard deviation. Different letters (a to g) denote statistically significance ($p < 0.05$) between means

4 Conclusion

NL and NCC obtained from oil palm empty fruit bunch were blended to starch composites to produce a series of biocomposites using the solvent casting method. Binary films (containing only lignin/NL or NCC in 2 wt%) were made to gauge the performance of ternary composites. The 2L (film containing 2 wt% lignin) was casted in comparison to 2NL film. The 2NL film showed improved performance in all aspects related to 2L film due to smaller particle size and uniform dispersion of lignin. The ternary composites demonstrated good UV blocking abilities and hydrophobicity. In addition, the 2NL2NCC and 4NL4NCC films showed a statistically significant reduction in moisture absorbance and swelling ratio, suggesting a positive effect of combing NL and NCC with starch. The 4NL4NCC film resulted in a 143% increment in tensile strength in comparison to neat film. Furthermore, the

Table 5 Summsary of the performance of different composites

Film	UV barrier	Water barrier	Contact angle (°)	T_g 40 (°C)	Tensile strength (MPa)
Neat	Weak	MC 40%, SR 46%	42	40	5.42
2L	Good	MC 38%, SR 44%	55	42	5.72
2NL	Excellent	MC 35%, SR 27%	78	52	6.88
2NCC	Weak	MC 38%, SR 44%	40	54	7.40
1NL1NCC	Good	MC 35%, SR 49%	79	55	7.71
2NL2NCC	Good	MC 35%, SR 33%	83	60	8.29
4NL4NCC	Excellent	MC 34%, SR 31%	83	61	13.15

Note MC represents moisture content and *SR* represents Swelling Ratio

thermal properties of ternary films were also improved, as evidenced by the increase of T_g. In summary, the crystalline structure of NCC could enhance the thermal and mechanical properties of ternary films, whilst NL improved the UV blocking ability and surface hydrophobicity of the films. Considering the performance of ternary films and their biodegradable nature, these films could be a sustainable alternative to petroleum-based packaging materials for the food industry. Further studies towards the antibacterial, antimicrobial properties of composites as well as the optimization of NL and NCC loads shall be conducted to explore the potential benefits of the NL-NCC biocomposites at a cost-effective usage.

References

Al-Dulaimi, A. A., & Wanrosli, W. D. (2017). Isolation and characterization of nanocrystalline cellulose from totally chlorine free oil palm empty fruit bunch pulp. *Journal of Polymers and the Environment, 25*(2), 192–202. https://doi.org/10.1007/s10924-016-0798-z

Area, M. R., Rico, M., Montero, B., Barral, L., Bouza, R., López, J., & Ramírez, C. (2019). Corn starch plasticized with isosorbide and filled with microcrystalline cellulose: Processing and characterization. *Carbohydrate Polymers, 206*, 726–733. https://doi.org/10.1016/j.carbpol.2018.11.055

Balakrishnan, P., Sreekala, M. S., Geethamma, V. G., Kalarikkal, N., Kokol, V., Volova, T., & Thomas, S. (2019). Physicochemical, mechanical, barrier and antibacterial properties of starch nanocomposites crosslinked with pre-oxidised sucrose. *Industrial Crops and Products, 130*, 398–408. https://doi.org/10.1016/j.indcrop.2019.01.007

Bian, H., Jiao, L., Wang, R., Wang, X., Zhu, W., & Dai, H. (2018). Lignin nanoparticles as nano-spacers for tuning the viscoelasticity of cellulose nanofibril reinforced polyvinyl alcohol-borax hydrogel. *European Polymer Journal, 107*. https://doi.org/10.1016/j.eurpolymj.2018.08.028

Chang, P. R., Jian, R., Zheng, P., Yu, J., & Ma, X. (2010). Preparation and properties of glycerol plasticized-starch (GPS)/cellulose nanoparticle (CN) composites. *Carbohydrate Polymers, 79*(2), 301–305. https://doi.org/10.1016/j.carbpol.2009.08.007

Chrissafis, K., & Bikiaris, D. (2011). Can nanoparticles really enhance thermal stability of polymers? Part I: An overview on thermal decomposition of addition polymers. *Thermochimica Acta*. https://doi.org/10.1016/j.tca.2011.06.010

Coral Medina, J. D., Woiciechowski, A. L., Filho, A. Z., Brar, S. K., Magalhães Júnior, A. I., & Soccol, C. R. (2018). Energetic and economic analysis of ethanol, xylitol and lignin production using oil palm empty fruit bunches from a Brazilian factory. *Journal of Cleaner Production, 195*, 44–55. https://doi.org/10.1016/j.jclepro.2018.05.189

Duval, A., Molina-Boisseau, S., & Chirat, C. (2013). Comparison of Kraft lignin and lignosulfonates addition to wheat gluten-based materials: Mechanical and thermal properties. *Industrial Crops and Products*. https://doi.org/10.1016/j.indcrop.2013.04.027

El Miri, N., Abdelouahdi, K., Barakat, A., Zahouily, M., Fihri, A., Solhy, A., & El Achaby, M. (2015). Bio-nanocomposite films reinforced with cellulose nanocrystals: Rheology of film-forming solutions, transparency, water vapor barrier and tensile properties of films. *Carbohydrate Polymers, 129*, 156–167. https://doi.org/10.1016/j.carbpol.2015.04.051

Fang, J. M., Fowler, P. A., Tomkinson, J., & Hill, C. A. S. (2002). The preparation and characterisation of a series of chemically modified potato starches. *Carbohydrate Polymers*. https://doi.org/10.1016/S0144-8617(01)00187-4

Fazeli, M., Keley, M., & Biazar, E. (2018). Preparation and characterization of starch-based composite films reinforced by cellulose nanofibers. *International Journal of Biological Macromolecules, 116*, 272–280. https://doi.org/10.1016/j.ijbiomac.2018.04.186

Foo, M., Tan, C., Lim, P., Ooi, E. C., Tan, K., & Chew, I. M. (2019). Surface-modified nanocrystalline cellulose from oil palm empty fruit bunch for effective binding of curcumin. *International Journal of Biological Macromolecules*. https://doi.org/10.1016/j.ijbiomac.2019.07.035

Garcia Gonzalez, M. N., Levi, M., Turri, S., & Griffini, G. (2017). Lignin nanoparticles by ultrasonication and their incorporation in waterborne polymer nanocomposites. *Journal of Applied Polymer Science, 134*(38). https://doi.org/10.1002/app.45318

Gilfillan, W. N. (2015). *Developing starch-based polymer composites*. Queensland University of Technology (QUT). Retrieved from https://eprints.qut.edu.au/86612/6/William_Gilfillan_Thesis.pdf

Giroto, A. S., Garcia, R. H. S., Colnago, L. A., Klamczynski, A., Glenn, G. M., & Ribeiro, C. (2020). Role of urea and melamine as synergic co-plasticizers for starch composites for fertilizer application. *International Journal of Biological Macromolecules*. https://doi.org/10.1016/j.ijbiomac.2019.12.094

González, K., Retegi, A., González, A., Eceiza, A., & Gabilondo, N. (2015). Starch and cellulose nanocrystals together into thermoplastic starch bionanocomposites. *Carbohydrate Polymers, 117*, 83–90. https://doi.org/10.1016/j.carbpol.2014.09.055

Hassanabadi, H. M., Alemdar, A., & Rodrigue, D. (2015). Polypropylene reinforced with nanocrystalline cellulose: Coupling agent optimization. *Journal of Applied Polymer Science, 132*(34). https://doi.org/10.1002/app.42438

Huq, T., Salmieri, S., Khan, A., Khan, R. A., Le Tien, C., Riedl, B., Lacroix, M., Fraschini, C., Bouchard, J., Uribe-Calderon, J., Kamal, M. R., Lacroix, M., & Lacroix, M. (2012). Nanocrystalline cellulose (NCC) reinforced alginate based biodegradable nanocomposite film. *Carbohydrate Polymers, 90*(4), 1757–1763. https://doi.org/10.1016/j.carbpol.2012.07.065

Jumaidin, R., Khiruddin, M. A. A., Asyul Sutan Saidi, Z., Salit, M. S., & Ilyas, R. A. (2020). Effect of cogon grass fibre on the thermal, mechanical and biodegradation properties of thermoplastic cassava starch biocomposite. *International Journal of Biological Macromolecules*. https://doi.org/10.1016/j.ijbiomac.2019.11.011

Kai, D., Tan, M. J., Chee, P. L., Chua, Y. K., Yap, Y. L., & Loh, X. J. (2016). Towards lignin-based functional materials in a sustainable world. *Green Chemistry*. https://doi.org/10.1039/c5gc02616d

Lee, H., You, J., Jin, H. J., & Kwak, H. W. (2020). Chemical and physical reinforcement behavior of dialdehyde nanocellulose in PVA composite film: A comparison of nanofiber and nanocrystal. *Carbohydrate Polymers*. https://doi.org/10.1016/j.carbpol.2019.115771

Li, J., Song, Z., Li, D., Shang, S., & Guo, Y. (2014). Cotton cellulose nanofiber-reinforced high density polyethylene composites prepared with two different pretreatment methods. *Industrial Crops and Products, 59*, 318–328. https://doi.org/10.1016/j.indcrop.2014.05.033

Li, M., Tian, X., Jin, R., & Li, D. (2018). Preparation and characterization of nanocomposite films containing starch and cellulose nanofibers. *Industrial Crops and Products, 123*, 654–660. https://doi.org/10.1016/j.indcrop.2018.07.043

Liu, D., Li, Y., Qian, Y., Xiao, Y., Du, S., & Qiu, X. (2017). Synergistic antioxidant performance of lignin and quercetin mixtures. *ACS Sustainable Chemistry and Engineering*. https://doi.org/10.1021/acssuschemeng.7b02282

Liu, Y., Zhang, S., Wang, X., Pan, Y., Zhang, F., & Huang, J. (2020). Mechanical and aging resistance properties of polypropylene (PP) reinforced with nanocellulose/attapulgite composites (NCC/AT). *Composite Interfaces*. https://doi.org/10.1080/09276440.2019.1600971

López-Córdoba, A., Estevez-Areco, S., & Goyanes, S. (2019). Potato starch-based biocomposites with enhanced thermal, mechanical and barrier properties comprising water-resistant electrospun poly (vinyl alcohol) fibers and yerba mate extract. *Carbohydrate Polymers*. https://doi.org/10.1016/j.carbpol.2019.03.105

Ma, M., Dai, L., Si, C., Hui, L., Liu, Z., & Ni, Y. (2019). A facile preparation of super long-term stable lignin nanoparticles from black liquor. *Chemsuschem*. https://doi.org/10.1002/cssc.201902287

Mano, J. F., Koniarova, D., & Reis, R. L. (2003). Thermal properties of thermoplastic starch/synthetic polymer blends with potential biomedical applicability. *Journal of Materials Science: Materials in Medicine*. https://doi.org/10.1023/A:1022015712170

Meneguin, A. B., Stringhetti, B., Cury, F., Aline, M., Faza, D., Barud, H. S., & Filho, S. (2017). Resistant starch/pectin free-standing films reinforced with nanocellulose intended for colonic methotrexate release. *Carbohydrate Polymers, 157*, 1013–1023. https://doi.org/10.1016/j.carbpol.2016.10.062

Montero, B., Rico, M., Rodríguez-llamazares, S., & Barral, L. (2017). Effect of nanocellulose as a filler on biodegradable thermoplastic starch films from tuber, cereal and legume. *Carbohydrate Polymers, 157*, 1094–1104. https://doi.org/10.1016/j.carbpol.2016.10.073

Nair, S. S., Kuo, P. Y., Chen, H., & Yan, N. (2017). Investigating the effect of lignin on the mechanical, thermal, and barrier properties of cellulose nanofibril reinforced epoxy composite. *Industrial Crops and Products*. https://doi.org/10.1016/j.indcrop.2017.02.032

Nair, S. S., Sharma, S., Pu, Y., Sun, Q., Pan, S., Zhu, J. Y., Ragauskas, A. J., & Deng, Y. (2014). High shear homogenization of lignin to nanolignin and thermal stability of nanolignin-polyvinyl alcohol blends. *Chemsuschem, 7*(12), 3513–3520. https://doi.org/10.1002/cssc.201402314

Ng, H. M., Sin, L. T., Tee, T. T., Bee, S. T., Hui, D., Low, C. Y., & Rahmat, A. R. (2015). Extraction of cellulose nanocrystals from plant sources for application as reinforcing agent in polymers. *Composites Part b: Engineering, 75*, 176–200. https://doi.org/10.1016/j.compositesb.2015.01.008

Owais, A., Khaled, M., & Yilbas, B. S. (2016). Hydrophobicity and surface finish. *Comprehensive Materials Finishing*. https://doi.org/10.1016/B978-0-12-803581-8.09172-4

Peng, Y., Nair, S. S., Chen, H., Yan, N., & Cao, J. (2018). Effects of lignin content on mechanical and thermal properties of polypropylene composites reinforced with micro particles of spray dried cellulose nanofibrils. *ACS Sustainable Chemistry and Engineering, 6*(8), 11078–11086. https://doi.org/10.1021/acssuschemeng.8b02544

Pérez, S., Baldwin, P. M., & Gallant, D. J. (2009). Structural features of starch granules I. *Starch*. https://doi.org/10.1016/B978-0-12-746275-2.00005-7

Sen, S., Patil, S., & Argyropoulos, D. S. (2015). Thermal properties of lignin in copolymers, blends, and composites: A review. *Green Chemistry*. https://doi.org/10.1039/c5gc01066g

Shankar, S., Reddy, J. P., & Rhim, J. W. (2015). Effect of lignin on water vapor barrier, mechanical, and structural properties of agar/lignin composite films. *International Journal of Biological Macromolecules, 81*, 267–273. https://doi.org/10.1016/j.ijbiomac.2015.08.015

Shankar, S., & Rhim, J. W. (2017). Preparation and characterization of agar/lignin/silver nanoparticles composite films with ultraviolet light barrier and antibacterial properties. *Food Hydrocolloids, 71*, 76–84. https://doi.org/10.1016/j.foodhyd.2017.05.002

Spiridon, I., Teaca, C. A., & Bodirlau, R. (2011). Preparation and characterization of adipic acid-modified starch microparticles/plasticized starch composite films reinforced by lignin. *Journal of Materials Science*. https://doi.org/10.1007/s10853-010-5210-0

Syafri, E., Jamaluddin, Wahono, S., Irwan, A., Asrofi, M., Sari, N. H., & Fudholi, A. (2019). Characterization and properties of cellulose microfibers from water hyacinth filled sago starch biocomposites. *International Journal of Biological Macromolecules*. https://doi.org/10.1016/j.ijbiomac.2019.06.174

Tang, P. L., Hassan, O., Yue, C. S., & Abdul, P. M. (2019). Lignin extraction from oil palm empty fruit bunch fiber (OPEFBF) via different alkaline treatments. *Biomass Conversion and Biorefinery*. https://doi.org/10.1007/s13399-019-00413-5

Tian, D., Hu, J., Bao, J., Chandra, R. P., Saddler, J. N., & Lu, C. (2017). Lignin valorization: Lignin nanoparticles as high-value bio-additive for multifunctional nanocomposites. *Biotechnology for Biofuels, 10*(1). https://doi.org/10.1186/s13068-017-0876-z

Wang, X., Tang, Y., Zhu, X., Zhou, Y., & Hong, X. (2020a). Preparation and characterization of polylactic acid/polyaniline/nanocrystalline cellulose nanocomposite films. *International Journal of Biological Macromolecules*. https://doi.org/10.1016/j.ijbiomac.2019.09.233

Wang, X., Bian, H., Ni, S., Sun, S., Jiao, L., & Dai, H. (2020b). BNNS/PVA bilayer composite film with multiple-improved properties by the synergistic actions of cellulose nanofibrils and lignin nanoparticles. *International Journal of Biological Macromolecules, 157*. https://doi.org/10.1016/j.ijbiomac.2020.04.178

Yang, W., Fortunati, E., Dominici, F., Giovanale, G., Mazzaglia, A., Balestra, G. M., Kenny, J. M., & Puglia, D. (2016a). Synergic effect of cellulose and lignin nanostructures in PLA based systems for food antibacterial packaging. *European Polymer Journal*. https://doi.org/10.1016/j.eurpolymj.2016.04.003

Yang, W., Owczarek, J. S., Fortunati, E., Kozanecki, M., Mazzaglia, A., Balestra, G. M., Kenny, J. M., Torre, L., & Puglia, D. (2016b). Antioxidant and antibacterial lignin nanoparticles in polyvinyl alcohol/chitosan films for active packaging. *Industrial Crops and Products, 94*, 800–811. https://doi.org/10.1016/j.indcrop.2016.09.061

Yang, W., Kenny, J. M., & Puglia, D. (2015). Structure and properties of biodegradable wheat gluten bionanocomposites containing lignin nanoparticles. *Industrial Crops and Products, 74*, 348–356. https://doi.org/10.1016/j.indcrop.2015.05.032

Yang, Weijun, Qi, G., Kenny, J. M., Puglia, D., & Ma, P. (2020). Effect of cellulose nanocrystals and lignin nanoparticles on mechanical, antioxidant and water vapour barrier properties of glutaraldehyde crosslinked PVA films. *Polymers, 12*(6). https://doi.org/10.3390/POLYM12061364

Yearla, S. R., & Padmasree, K. (2016). Preparation and characterisation of lignin nanoparticles: Evaluation of their potential as antioxidants and UV protectants. *Journal of Experimental Nanoscience, 11*(4), 289–302. https://doi.org/10.1080/17458080.2015.1055842

Zhang, C. W, Nair, S. S., Chen, H., Yan, N., Farnood, R., & Li, F. yi. (2020). Thermally stable, enhanced water barrier, high strength starch bio-composite reinforced with lignin containing cellulose nanofibrils. *Carbohydrate Polymers*. https://doi.org/10.1016/j.carbpol.2019.115626

Zhang, M., Ahmad, M., Lee, S. S., Xu, L. H., & Ok, Y. S. (2014). Sorption of polycyclic aromatic hydrocarbons (PAHs) to lignin: Effects of hydrophobicity and temperature. *Bulletin of Environmental Contamination and Toxicology*. https://doi.org/10.1007/s00128-014-1290-x

Zimniewska, M., Kozłowski, R., & Batog, J. (2008). Nanolignin modified linen fabric as a multifunctional product. *Molecular Crystals and Liquid Crystals*. https://doi.org/10.1080/15421400801903395

Process Simulation and Scheduling of Bio-succinic Acid Production from Palm Biomass

Dominic C. Y. Foo, Steve Z. Y. Foong, Denny K. S. Ng, and Jian Ping Tan

Abstract This chapter describes process modelling of bio-succinic acid production from palm biomass. Batch scheduling is carried out to determine the time bottleneck of the process, i.e. equipment that limits the annual production due to its long cycle time. In this case, the fermenter is identified as the time bottleneck, as it has the longest cycle time. Therefore, debottlenecking strategy is applied where additional fermenters are added to reduce the cycle time. Doing so leads to increased batches and annual production of the bio-succinic acid. With deployment of five fermenters, an increase of 340% of bio-succinic acid was resulted, as compared to that of the base case process.

Keywords Process simulation · Process scheduling · Palm biomass · Succinic acid · SuperPro designer · Debottlenecking

D. C. Y. Foo (✉)
Department of Chemical and Environmental Engineering/Centre for Green Technologies, University of Nottingham Malaysia, Broga Road, 43500 Semenyih, Selangor, Malaysia
e-mail: dominic.foo@nottingham.edu.my

S. Z. Y. Foong · D. K. S. Ng
School of Engineering and Physical Sciences, Heriot-Watt University Malaysia, No. 1, Jalan Venna P5/2, Precinct 5, Wilayah Persekutuan Putrajaya, 62200 Putrajaya, Malaysia
e-mail: stevefoong92@gmail.com

D. K. S. Ng
e-mail: denny.ng@hw.ac.uk

J. P. Tan
School of Energy and Chemical Engineering, Xiamen University Malaysia, 43900 Sepang, Selangor, Malaysia
e-mail: jianping.tan@xmu.edu.my

College of Chemistry and Chemical Engineering, Xiamen University, Xiamen 361005, China

D. C. Y. Foo et al. (eds.), *Sustainable Technologies for the Oil Palm Industry*,
https://doi.org/10.1007/978-981-19-4847-3_5

1 Introduction

With the increasing consciousness on environmental impact of the fossil-based petrochemical industries, research on biochemical production was emerging in recent years. The US Department of Energy (US DOE) listed 12 chemical building blocks which have the potential to be commercially produced through biological pathway (Werpy & Peterson, 2004). In this regard, succinic acid appeared to be among the top rank in the list. Succinic acid is also known as amber acid or butanedioic acid, which is a dicarboxylic acid with molecular formula of $C_4H_6O_4$. Such acid is an intermediate in the tricarboxylic acid cycle produced during aerobic metabolism, and an end product of anaerobic metabolism (Tan et al., 2014). This chemical can be produced via various strict and facultative anaerobic microorganisms and used as a precursor in many industries, such as pharmaceutical, food, agriculture, polymer, cosmetic and textile industries. In the pharmaceutical industries, for instance, succinic acid is used in the preparation of active calcium succinate, active pharmaceutical ingredients, vitamin A, anti-inflammatory-erthyrodiol derivates, sedative, antipasmer, etc. Succinic acid is utilized in a polyester synthesis and acts as an intermediate in the manufacturing of plasticizer, resins, biodegradable solvents, engineering plastics, surfactants, detergents, etc. The major market potential of succinic acid lies as a precursor for the production of other chemicals (Allied Market Research, 2020).

Compared to the fossil-based succinic acid production, bio-succinic acid production is a benign environmental route (Xu et al., 2021). The usability of renewable raw materials and its carbon dioxide (CO_2) fixation are the major advantages of bio-succinic acid production. Apart from being carbon neutral, it requires 30–40% lower energy consumption as compared to conventional fossil-based succinic acid. These factors have led to the commercialization of bio-based succinic acid in recent years (Kuenz et al., 2020). In fact, production of bio-succinic acid could serve as a downstream process for CO_2-producing biofuel industries such as bio-methane, bio-ethanol and bio-hydrogen plants as it could absorb CO_2 in the process, reducing CO_2 concentration while purifying the biogas in the process. (Tan et al., 2018).

In this chapter, a commercial process simulation tool, i.e. SuperPro Designer (SPD) v12 (www.intelligen.com) is used to model a newly developed bio-succinic acid production process from palm biomass, i.e. oil palm frond (OPF) that is collected from the oil palm plantation. The process was developed based on laboratory-scale experimental work (Luthfi et al., 2016).

2 Process Simulation

Figure 1 shows the simulation flowsheet developed in SPD for a bio-succinic acid production process. The process is operated in batch mode, with an annual operating time of 7920 h. The feed stream of the process consisted of 1000 kg of OPF. Note that OPF is a lignocellulose material, and is modelled as a stock mixture in SPD.

The properties of OPF are given in Table 1, obtained from OPF sample taken from the oil palm estate of Universiti Kebangsaan Malaysia, located at Bangi, Selangor, Malaysia (Tan et al., 2017).

As shown in Fig. 1, the OPF is sent to the ball mill where its juice is pressed from the fresh OPF, before being sent for centrifugation. Apart from juice removal, the ball mill also serves as physical treatment that enlarges the surface area of the bagasse, so to allow more effective pretreatment at later stage. The OPF juice from the centrifuge is transferred to the fermenter, while the separated bagasse from ball mill and centrifuge are sent to the Pretreatment Section. This follows the same process as reported by Luthfi et al. (2017).

In the Pretreatment Section, the OPF bagasse is first treated with sodium hydroxide (NaOH) of 4 wt%; the latter disrupts the crystalline structure of the lignocellulosic bagasse, so to recover its fermentable sugars. The liquid effluent from the pretreatment unit is then cooled to 40 °C before it is sent to a rotary siever. The latter removes the black liquor, which consists of hydrolyzed xylan and lignin. Effluent from the rotary siever is transferred into the enzymatic reactor. In the enzymatic reactor, hydrochloric acid (HCl) is first used to neutralize the NaOH content, while enzyme (consisting of cellulase and xylanase) is later used to break down the structural carbohydrate of pretreated bagasse into fermentable sugars (Luthfi et al., 2019). The hydrolysate from the enzymatic reactor is then sent for membrane filtration to remove all suspended solids. The filtrate (OPF sugar) is sent to the fermenter.

In the fermenter, OPF juice (extracted in ball mill) and OPF sugar (from pretreatment section) are first transferred in from their operations. Mineral mixture (consisting 50 wt% of potassium dihydrogen phosphate (KH_2PO_4) and sodium chloride) are next added into the fermenter (Tan et al., 2016), which is then followed by sterilization-in-place (SIP). The SIP ensures no other microbe presence in the fermenter prior to the charging of the necessary microbes, i.e. *Actinobacillus succinogenes*. Fermentation process is next initiated upon the charging of inoculum *Acinobacillus succinonogenes*. Throughout the fermentation process which takes a total duration of 24 h, CO_2 is charged to fermenter, while its temperature is maintained at 37 °C (Luthfi et al., 2018). Upon the completion of fermentation process, product from the fermenter is sent to the Downstream Processing Section.

In the Downstream Processing Section, several processing units are used for the purification of bio-succinic acid. Two units of membrane filtration are used to remove the suspended solids and liquid content from the fermentation broth. The suspended solid are mainly due to microbe and other traces of impurities resulted from enzymatic hydrolysis and OPF juice. Filtrate from the membrane unit is sent to a flash evaporator, where the product is further concentrated in the medium to allow a more efficient crystallization process (Luthfi et al., 2020). Effluent from the evaporator is then mixed with concentrated HCl in order to reduce its pH value, before being sent to the crystallization, where succinic acid solids are formed. Concentrated HCl is used to lower the pH for the optimum succinic acid crystallization. Products from the crystalizer are dried in a drum dryer. The latter removes remaining moisture of the crystals. Dried crystals from the drum dryer are ready for sale. Detailed setting for all units are shown in Table 2.

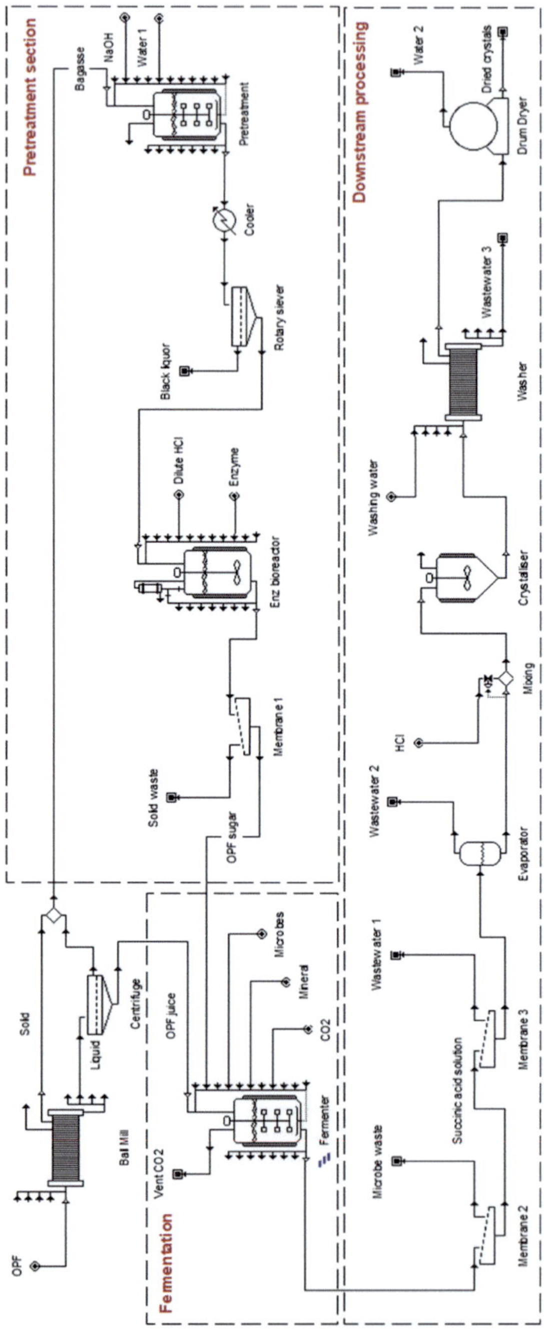

Fig. 1 Simulation flowsheet for bio-succinic acid production in SuperPro Designer (SPD) v12

Table 1 Properties for OPF feed stream

Properties	Value
Pressure	1 bar
Temperature	25 °C
Mass flow	1000 kg
Composition (wt%)	
Ash	2.21
Fructose	0.02
Glucan	29.31
Glucose	1.25
Lignin	15.43
Sucrose	0.09
Water	41.67
Xylan	10.02

3 Process Scheduling

Details of scheduling for each process units, such as duration and start time are given in the last two column of Table 2 (all units are assume to have zero set-up time). The operation Gantt chart of the process is shown in Fig. 2. SPD determines that the overall process has a batch time of 62 h, while its minimum cycle time is identified as 40 h. With the annual operation time of 7920 h, the minimum cycle time determines that the process has a yearly production of 197 batches, which is equivalent to an annual production of 6878 kg of bio-succinic acid products.

4 Options for Debottlenecking

SPD next identifies that the fermenter is the *time bottleneck* of this process (see Fig. 2 for its longest cycle time among all units, i.e. 40 h). To improve the overall annual productivity, the minimum cycle time is to be reduced. One potential solution is to make use of multiple units of fermenter. This may be done using the staggered mode function of SPD. Results of the latter are shown in Table 3. As shown, with five fermenters, the annual production is increased to 30,303 kg/y, i.e. 340% of the base case process. Note however that the use of multiple fermenters will increase the capital expenditure of the process. However, during the course of this process simulation work, economic feasibility study was not carried out due to the lack of process economic data.

Table 2 Important setting and scheduling data for all units (set-up time is assumed as zero)

Processes	Important setting	Duration	Start time
Ball mill	Removal of ash, glucan, lignin, water and xylan (99% removal for all)	4 h	Beginning of the batch
Centrifuge	Removal of ash, glucan, lignin and xylan (100% removal for all)	30 min	After completion of ball mill procedure
Pretreatment	Transfer in bagasse	15 min	After completion of ball mill procedure
	Charge-in NaOH (28.1 kg)	15 min	After completion of previous operation
	Charge-in water (607.7 kg)	1 h	After completion of previous operation
	Agitation	1 h	After completion of previous operation
	Heating (to 98 °C)	1 h	Start with agitation operation
	Transfer-out bagasse	30 min	After completion of previous operation
Cooler	Exit temperature: 40 °C	30 min	Start with transfer-out of Pretreatment procedure
Rotary siever	Split %: ash (30.18), glucan (4.27), NaOH (79.5), water (79.5), xylan (73.3)	1	After completion of cooler procedure
Enzymatic bioreactor	Transfer-in	30 min	After completion of rotary siever procedure
	Charge HCl (3539.3 kg, with 0.0364% HCl and 99.9636% water)	10 min	After completion of previous operation
	Reaction 1 (neutralization), with mass stoichiometry: 36.46 HCl + 40 NaOH $\rightarrow$ 58.44 NaCl + 18.02 H_20 (100% conversion) Final temperature: 40 °C	20 min	Start with charging of HCl
	Agitation	20 min	Start with Reaction 1
	Charge enzyme (0.18 kg)	5 min	After completion of previous operation
	Heating (final temperature: 40 °C)	6 h	Start with charging of enzyme

(continued)

Table 2 (continued)

Processes	Important setting	Duration	Start time
	Reaction 2 (glucan conversion), with mass stoichiometry: 180.16 glucan → 180.16 glucose (88.4% conversion) Final temperature: 40 °C	6 h	Start with Heating
	Reaction 3 (xylan conversion), with mass stoichiometry: 150.13 xylan → 150.13 xylose (89.93% conversion) Final temperature: 40 °C	6 h	Start with Reaction 2
	Transfer-out	20 min	After completion of previous operation
Membrane 1	Removal of enzyme (99%)	1 h	Start with transfer-out of Enzymatic bioreactor procedure
Fermenter	Transfer-in OPF juice	30 min	After completion of centrifuge procedure
	Transfer-in OPF sugar	1 h	Start with Membrane 1 operation
	Charge mineral (6.27 kg for both KH_2PO_4 and sodium chloride)	10 min	After completion of previous operation
	SIP (100 kg/h m^3)	30 min	After completion of previous operation
	Charge CO_2 (10.87 kg)	24 h	After completion of previous operation
	Charge microbe (0.519 kg)	15 min	Start with charging of CO_2
	Agitate	24 h	Start with charging of microbe
	Reaction (fermentation)	24 h	Start with agitation
	Transfer-out	30 min	After completion of previous operation
Membrane 2	Removal of microbe (99%)	1 h	After fermentation procedure completed
Membrane 3	Removal of NaCl (100%), NaOH (100%) and water (32.26%)	1 h	After Membrane 2 procedure completed
Evaporator	Removal of water (80%), and acetic acid (70%) Temperature: 70 °C	1 h	After Membrane 3 procedure completed
Mixer	Mixing of HCl (input mass ratio of 0.01)	10 min	After completion of Evaporation procedure

(continued)

Table 2 (continued)

Processes	Important setting	Duration	Start time
Crystallizer	Evap. Data: Water in vapor phase (82.18%) Crystal. Data: succinic acid	6 h	Start with Mixing procedure
Washer	Filtration: removal of succinic acid solid (100%) and glucose (0.1%) Loss on drying (LOD): 10%	2 h	After completion of Cyrtalization procedure
	Cake wash	30 min	After completion of previous operation
Drum dryer	Filtration: removal of water (calculated based on final LOD) Final LOD: 1%	6 h	After completion of Washer procedure

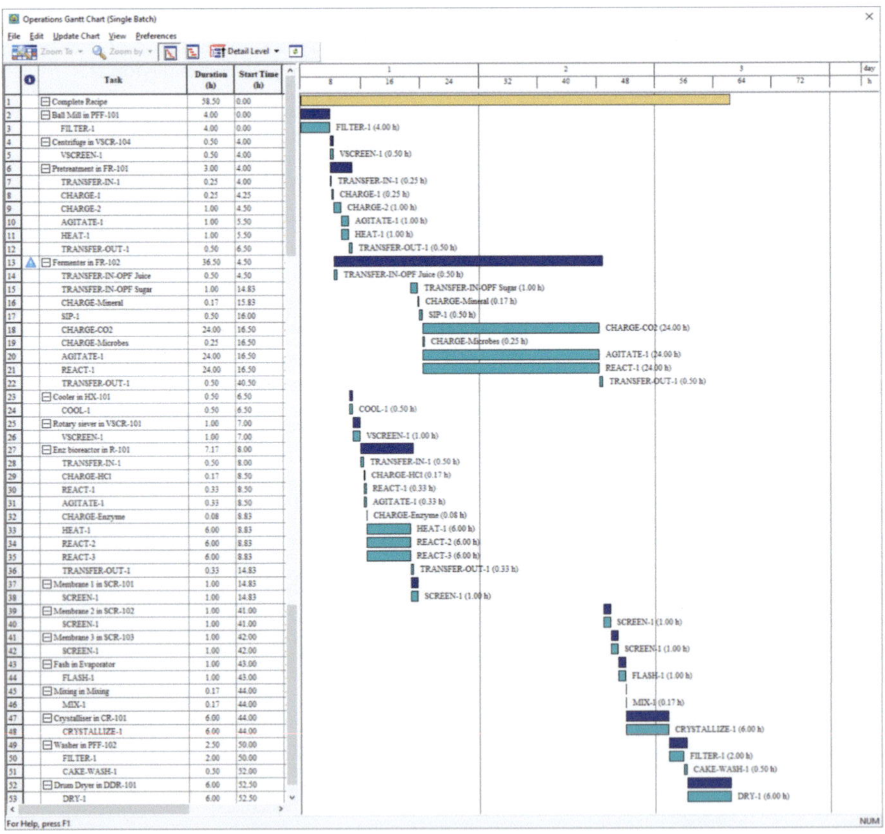

Fig. 2 Operational Gantt chart for bio-succinic acid production

Table 3 Summary of production with different units of fermenters

Numbers of fermenter	Number of annual batches	Annual production (kg/y)
1	97	6878
2	393	13,722
3	590	20,600
4	786	27,444
5	1097	30,303

5 Conclusion

In this chapter, a production process of bio-succinic acid crystal is modelled using SuperPro Designer v12. Scheduling is also performed to identify the time bottleneck of the process, which limits the annual production. The fermenter which has 40 h cycle time is determined as the time bottleneck. More fermenters are hence added to reduce the cycle time, and hence increase the number of batches and annual production of the bio-succinic acid product. Doing this leads to 340% increase of production, as compared to that of the base case process. Detailed economic evaluation should be carried out next in order to identify economic hotspots of the process.

Acknowledgements This research is supported by the Ministry of Higher Education Malaysia through the Fundamental Research Grant Scheme (project number "FRGS/1/2020/TK0/XMU/03/2"), and Long Term Research Grant Scheme (Project Code: LRGS/2013/UKM-UNMC/PT/05). The support of Xiamen University Malaysia Research Fund (project number "XMUMRF/2020-C5/IENG/0027") is also acknowledged.

References

Allied Market Research. (2020). *Bio succinic acid market by application (1-butanediol,4-butanediol (BDO), polyester polyols, PBS, plasticizers, solvents & lubricants, alkyd resins, resins, coatings, pigments, De-icer solutions)—Global opportunity analysis and industry forecast, 2013–2020*. https://www.alliedmarketresearch.com/bio-succinic-acid-market. Accessed December, 2021

Kuenz, A., Hoffmann, L., Goy, K., Bromann, S., & Prüße, U. (2020). High-level production of succinic acid from crude glycerol by a wild type organism. *Catalysts, 10*(5), 470.

Luthfi, A. A. I., Jahim, J. M., Harun, S., Tan, J. P., & Mohammad, A. W. (2016). Biorefinery approach towards greener succinic acid production from oil palm frond bagasse. *Process Biochemistry, 51*(10), 1527–1537.

Luthfi, A. A. I., Jahim, M. J., Harun, S., Tan, J. P., & Mohammad, A. W. (2017). Potential use of coconut shell activated carbon as an immobilisation carrier for high conversion of succinic acid from oil palm frond hydrolysate. *7*, 49480–49489.

Luthfi, A. A. I., Jahim, M. J., Harun, S., Tan, J. P., Manaf, S. F. A., & Shah, S. S. M. (2018). Kinetics of the bioproduction of succinic acid by actinobacillus succinogenes from oil palm lignocellulosic hydrolysate in a bioreactor. *BioResources, 13*(4).

Luthfi, A. A. I., Tan, J. P., Isa, Harun, S., Manaf, S. F. A., & Jahim, M. J. (2019). Homogeneous solid dispersion (HSD) system for rapid and stable production of succinic acid from lignocellulosic hydrolysate. *Bioprocess and Biosystems Engineering*, *42*, 117–130.

Luthfi, A. A. I., Tan, J. P., Isa, N. F. A. M., Bukhari, N. A., Shah, S. S. M., Mahmod, S. S., & Jahim, J. M. (2020). Multiple crystallization as a potential strategy for efficient recovery of succinic acid following fermentation with immobilized cells. *Bioprocess and Biosystems Engineering, 43*, 1153–1169.

Tan, J. P., Luthfi, A. A. I., Abdul Manaf, S. F., Wu, T. Y., & Jahim, J. M. (2018). Incorporation of CO_2 during the production of succinic acid from sustainable oil palm frond juice. *Journal of CO_2 Utilization, 26*, 595–601.

Tan, J. P., Jahim, M. J., Harun, S., Wu, T. W., Harun, S., Kim, B. H., & Mohammad, A. W. (2014). Insight into biomass as a renewable carbon source for the production of succinic acid and the factors affecting the metabolic flux toward higher succinate yield. *Industrial Engineering and Chemistry Research, 53*(42), 16123–16134.

Tan, J. P., Jahim, M.J., Wu, T. W., Harun, S., & Mumtaz, T. (2016). Use of corn steep liquor as an economical nitrogen source for biosuccinic acid production by Actinobacillus succinogenes. *IOP Conference Series: Earth and Environmental Science,* 36.

Tan, J. P., Jahim, M. J., Harun, S., & Wu, T. W. (2017). Overview of the potential of bio-succinic acid production from oil palm fronds. *Journal of Physical Science, 28*(1), 53–72.

Xu, C., Alam, M. A., Wang, Z., Peng, Y., Xie, C., Gong, W., Yang, Q., Huang, S., Zhuang, W., & Xu, J. (2021). Co-fermentation of succinic acid and ethanol from sugarcane bagasse based on full hexose and pentose utilization and carbon dioxide reduction. *Bioresource Technology, 339*, 125578.

Werpy, T., & Peterson, G. (2004). *Top value added chemical from biomass (Vol. 1)—Results of screening for potential candidates from sugars and synthesis gas*. US Department of Energy. www.nrel.gov. Accessed December, 2021.

Stochastic Modeling for Palm Biomass Supply Chain

Bing Shen How, Shirleen Lee Yuen Lo, Karen Gah Hie Kong, and Sin Yong Teng

Abstract Oil palm industry is one of the key contributors to the Gross Domestic Product (GDP) and Comprehensive National Strength (CNS) of Malaysia. According to the Department of Statistical Malaysia (2021), oil palm industries had contributed about 2.7% or RM 38.26 billion (equivalent to 9.45 billion USD) to Malaysia's GDP in 2019. An abundant amount of palm-based biomass has been generated through oil palm harvesting and palm oil production. Therefore, conceptual design and modeling of the *palm biomass supply chain* are deemed necessary to ensure the sustainability of the oil palm industry. Nevertheless, to enhance the model reliability and robustness, *stochastic modeling* should be opted to incorporate various uncertainties into the supply chain model. Keeping this in mind, this chapter presents an overview of the key supply chain uncertainties that should be incorporated into the supply chain model. It is followed by three illustrative examples which cover (i) biomass selection decision, (ii) facility location decision, and (iii) policy selection decision.

Keywords Biomass supply chain · Uncertainty · Stochastic modeling · Decision-making · Monte Carlo

---

B. S. How (✉) · S. L. Y. Lo · K. G. H. Kong
Biomass Waste-to-Wealth Special Interest Group, Research Centre for Sustainable Technologies, Faculty of Engineering, Computing and Science, Swinburne University of Technology, Jalan Simpang Tiga, 93350 Kuching, Sarawak, Malaysia
e-mail: bshow@swinburne.edu.my

S. Y. Teng
Institute for Molecules and Materials, Radboud University, P.O. Box 9010, 6500 GL Nijmegen, The Netherlands
e-mail: sinyong.teng@ru.nl

D. C. Y. Foo et al. (eds.), *Sustainable Technologies for the Oil Palm Industry*,
https://doi.org/10.1007/978-981-19-4847-3_6

Nomenclature

Abbreviations

AP	Action plan
AP-1	Engage in demand contract
AP2-HS	Engage in supply contract
AP3-FI	Introduce new financing incentive
AP4-SF	Substitute fossil fuel with biodiesel
AP5-TI	Introduce new tax incentive
AP6-RF	Revise Feed-in-Tariff (FiT) rate
AP7-CM	Introduce carbon management system
BOTA	Bottleneck Tree Analysis
CAPEX	Capital expenditure
CNS	Comprehensive National Strength
DoE	Design of Experiment
EFB	Empty fruit bunch
FiT	Feed-in-tariff
GDP	Gross domestic product
GTFS	Green Technology Financing Scheme
LNG	Liquefied natural gas
MF	Mesocarp fiber
MILP	Mixed Integer Linear Programming
NPV	Net present value
OPEX	Operating expenditure
PBP	Payback Period
PCA	Principal component analysis
PKS	Palm kernel shell

Indices

g	Index for synthetic gas products
m	Index for months
t	Index for years
u	Index of units

Parameters

$\text{Biomass}_{m,t}^{\text{AVAILABLE}}$	Biomass availability at month m in year t (t)

$\text{Biomass}_{\text{EFB}}^{\text{QUALITY}}$	Biomass quality for EFB
$\text{Biomass}_{\text{PKS}}^{\text{QUALITY}}$	Biomass quality for PKS
$\text{Biomass}_{\text{MF}}^{\text{QUALITY}}$	Biomass quality for MF
c	Specific heat capacity of water (J kg/K)
$\text{Cap}^{\text{TRUCK}}$	Vehicle load limit (t)
$C_{m,t}^{\text{BIOMASS}}$	Unit cost of biomass at month *m* in year *t* (USD/t)
C_{t}^{BIOMASS}	Unit cost of biomass in year *t* (USD/t)
C_{u}^{CAPEX}	Capital cost for unit *u* (USD)
C^{CAPEX}	Total capital cost for the conversion process (USD)
$C_{m,t}^{\text{COAL}}$	Unit cost of coal at month *m* in year *t* (USD/t)
C_{t}^{COAL}	Unit cost of coal in year *t* (USD/t)
C^{CO2}	Compensation cost required per unit of carbon emission (USD/kg)
C^{ELEC}	Unit cost of imported power (USD/kWh)
C^{FIT}	Feed-in-tariff (USD/kWh)
$C_{m,t}^{\text{FUEL}}$	Fuel price at month *m* in year *t* (USD/L)
C_{t}^{FUEL}	Fuel price in year *t* (USD/L)
$C_{u,m,t}^{\text{OPEX}}$	Operating cost for unit *u* at month *m* in year *t* (USD)
C_{t}^{OPEX}	Total operating cost for the conversion process in year *t* (USD)
$C_{m,t}^{\text{OIL}}$	Bio-oil price at month m in year t (USD/L)
$\text{Cap}^{\text{TRUCK}}$	Vehicle load limit (t)
d^{D}	Distance between polygeneration plant and the demand point (km)
d^{S}	Distance between biomass supply and the polygeneration plant (km)
$\text{Elec}_{m,t}^{\text{REQ}}$	Power demand at month *m* in year *t* (kWh)
$F_{m,t}^{\text{OIL\_DEMAND}}$	Bio-oil demand at month *m* in year *t* (L)
HV^{coal}	Heating value of coal (MJ/kg)
in	Discount rate (%)
ITA	Tax exemption indicator (USD)
LHV_{g}	Low heating value of gas *g* (kJ/mol)
LHV^{CHAR}	Low heating value of biochar (MJ/kg)
LHV^{COAL}	Low heating value of coal (MJ/kg)
$\text{MC}_{m,t}^{\text{IN}}$	Moisture content of biomass before drying (%)
$\text{MC}_{t}^{\text{IN}}$	Moisture content of biomass before drying (wt%)
MC^{OUT}	Moisture content of biomass after drying (%)
Q^{FC}	Weight composition of fixed carbon of biomass (wt%)
Q^{VM}	Weight composition of volatile matter of biomass (wt%)
Q^{A}	Weight composition of ash of biomass (wt%)
Q^{MC}	Weight composition of moisture content of biomass (wt%)
Q^{H}	Weight composition of hydrogen of biomass (wt%)
Q^{C}	Weight composition of carbon of biomass (wt%)
Q^{O}	Weight composition of oxygen of biomass (wt%)
Q^{S}	Weight composition of sulfur of biomass (wt%)

Q^{N}	Weight composition of nitrogen of biomass (wt%)
TAX	Corporate tax rate (%)
$\text{Thermal}_{m,t}^{\text{REQ}}$	Heat demand at month m in year t (kWh)
y^{CHAR}	Biochar yield (%)
$y^{\text{CO2\_COGEN}}$	CO_2 emitted during co-generation unit (kg CO_2/kWh)
$y^{\text{CO2\_PY}}$	CO_2 emitted during pyrolysis process (kg CO_2/kg biomass)
$y^{\text{CO2\_TR}}$	CO_2 emitted during transportation (kg CO_2/L fuel)
y^{GAS}	Syngas yield (%)
y_{g}^{PY}	Molecular composition of gas g (%)
y^{OIL}	Bio-oil yield (%)
ξ^{COGEN}	Conversion efficiency of the co-generation unit (%)
ξ^{DRY}	Drying efficiency (%)
ψ^{FUEL}	Fuel consumption rate (L/km)
ψ^{PY}	Thermal energy required to try per unit mass of biomass (kWh/t)
ψ^{THERMAL}	Thermal energy required to try per unit mass of moisture (kWh/t)

Variables

$\text{Biomass}_{m,t}^{\text{DRY}}$	Flowrate of dried biomass fed into the pyrolyser at month m in year t (t)
$\text{Biomass}_{m,t}^{IN}$	Flowrate of biomass fed into the plant at month m in year t (t)
$\text{Biomass}_{t}^{\text{IN}}$	Flowrate of biomass fed into the plant in year t (t)
$\text{Biomass}_{t}^{\text{SUPPLY}}$	Biomass supply in year t (t)
$C_{m,t}^{\text{PENALTY}}$	Carbon penalty at month m in year t (USD)
$C_{m,t}^{\text{PROCURE}}$	Procurement cost at month m in year t (USD)
C_{t}^{PROCURE}	Procurement cost in year t (USD)
C_{t}^{SYNGAS}	The selling price of syngas in year t (USD)
$C_{m,t}^{\text{TR}}$	Transportation cost at month m in year t (USD)
C_{t}^{TR}	Transportation cost in year t (USD)
$CF_{m,t}^{\text{IN}}$	Input cash flow (USD)
CF_{t}^{IN}	Input cash flow in year t (USD)
$CF_{m,t}^{\text{OUT}}$	Output cash flow (USD)
CF_{t}^{OUT}	Output cash flow in year t (USD)
$\text{Elec}_{m,t}^{\text{EXP}}$	Exported power at month m in year t (kWh)
$\text{Elec}_{m,t}^{\text{GEN}}$	Generated power at month m in year t (kWh)
$\text{Elec}_{m,t}^{\text{IMP}}$	Imported power at month m in year t (kWh)
$F_{m,t}^{\text{CHAR}}$	Biochar production at month m in year t (t)
$F_{m,t}^{\text{COAL}}$	Coal consumption at month m in year t (t)
F_{t}^{COAL}	Coal consumption in year t (t)
$F_{m,t}^{\text{GAS}}$	Syngas production at month m in year t (t)

$F_{m,t}^{\mathrm{OIL}}$	Bio-oil production at month m in year t (L)
F_{t}^{SYNGAS}	Syngas demand in year t (MWth)
Q^{coal}	Energy required to reduce the moisture content of biomass (MJ)
S^{BIOMASS}	Specific syngas yield (kg/kg)
T_{Final}	Temperature of biomass after drying (°C)
T_{Initial}	Temperature of biomass before drying (°C)
$\mathrm{NCF}_{m,t}$	Net cash flow at month m in year t (USD)
NCF_{t}	Net cash flow in year t (USD)
$\mathrm{Thermal}_{m,t}^{\mathrm{GEN}}$	Generated heat at month m in year t (kWh)

1 Introduction

Laying the agriculture foundation in Malaysia, the cultivation of the oil palm industry has been emerging ever since the first commercial planting was materialized in Tennamaran Estate, Selangor, in 1917 (The Oil Palm, 2021). Oil palm plantation was first introduced by the Malaysian government to diversify the country's agricultural products and at the same time diminish the nation's economic dependency on rubber and tins in the early 1960s (The Oil Palm, 2021).

Over the years, Malaysia has been showered and fueled with oil palm biomass resulted as the world's second-largest palm oil producer (around 19.14 million tonnes of crude palm oil produced in 2020) (Malaysia Palm Oil Board, 2020). Oil palm biomass exists as one of the most appealing substituents for energy generation feedstock whereas many nations are still exploring diversifying the country's energy profile. In order to appeal to more potential investors into biomass-based industries, a sustainable biomass supply chain (from harvesting to distribution) is required to be systemized and well-ordered. To this end, Hong et al. (2016) had proposed four critical elements in developing a sustainable biomass supply chain, which consists of (i) biomass harvesting and management, (ii) integrated biorefinery, (iii) product distribution, and (iv) logistics management. Nonetheless, Lo et al. (2021) stated that almost 95% of the researchers used deterministic models (Mixed Integer Linear Programming (MILP)) in evaluating biomass supply chain models. However, various uncertainties related to the biomass supply chain are yet to be considered in most of the supply chain optimization models. The relevant works mainly utilize deterministic optimization options in performing their studies and the outcomes may be too ideal which are not realistic in real life. For instance, Zakaria et al. (2020) did mention that deterministic optimization models would opt to generate ideal results which some of the systems are often uncertain in real life, causing the generated results to be impractical and imprecise. The addition of uncertainties and risk parameters into the supply chain model may shift it from deterministic to stochastic, where the model robustness can be enhanced and the outcomes can be more accurate and reliable. Stochastic optimization models with uncertainties integrated are able to develop random-probability-based distribution results (Kieffer et al., 2016), in which the generated outcomes would be more solid (Zakaria et al., 2020). One acknowledged example is the work conducted by Kristianto and Zhu (2017), where the stochastic optimization is implemented in the biomass-to-bioethanol supply chain model, had successfully generated promising results in minimizing emission, energy, and water utilization.

Hence, this chapter focus on four key components as the uncertainty variables to be considered in the stochastic biomass supply chain model:

(i) Biomass quality
(ii) Biomass supply uncertainties
(iii) Biomass demand variation
(iv) Pricing fluctuation.

Considering uncertainties in supply-chain, optimization models may be able to perform probability-based distribution outcomes in which the results might be closer to the real-world scenarios. Though, the details of the uncertainties need to be further discussed to prevent unexpected consequences that would lead to process failure or profit loss. A schematic figure of the basic idea for the uncertainties to be considered in stochastic biomass supply chain optimization models is shown in Fig. 1. The manipulating variables are then demonstrated with three different case scenarios, where the formulated models and their corresponding impacts are further discussed in the later sections.

One of the concerns in developing a sustainable supply chain would be the biomass source of supply (i.e., biomass availability). Various research works were reported to include biomass supply as one of the uncertainties in the biomass supply chain due to different harvesting seasons and logistics arrangements as described by Martinkus et al. (2018) and Lim et al. (2019). In the case of biomass shortage, imported biomass is mandatory, where additional operating costs (transportation cost and importing cost) may be required, which directly affects the profitability and efficiency of the biomass supply chain. Extended from that, various biomass qualities (or characteristics) should also be incorporated into the stochastic biomass supply chain model. It has been reported that the quality of the biomass is the most sensitive parameter

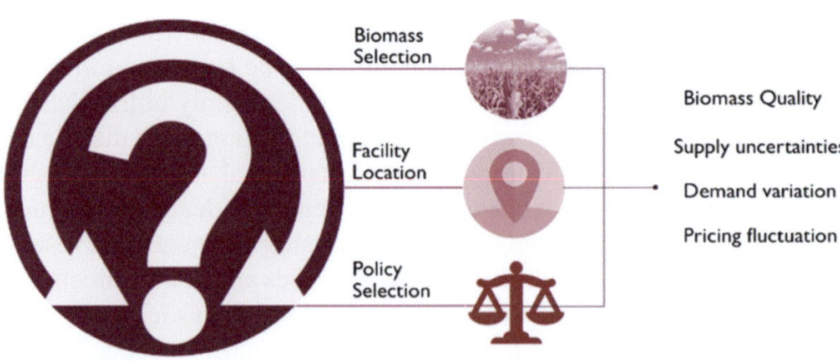

Fig. 1 Key uncertainties considered in stochastic biomass supply chain

in influencing the biorefinery economics, where it has been justified by Baral et al. (2019) that better biomass quality eventually contributes to higher profitability even though higher purchase cost for biomass is required. Another example from How et al. (2019), proved that higher moisture content and low density of biomass quality requires higher logistics cost which increases the prime cost of the biomass and the overall biomass supply chain.

Likewise, the distinctive quality of biomass has a significant impact on their respective conversion routes (e.g. gasification, pyrolysis, anaerobic digestion, etc.). Lim et al. (2019) found that a broader quality range of biomass will lead to higher fluctuation of conversion efficiency. Several research works have discussed the impact of biomass quality on the overall economic feasibility of the biomass supply chain (Bussemaker et al., 2017; Tanzer et al., 2019). For example, the lowering of moisture content of biomass (60% to 40%) would help in reducing the drying cost along with the logistics cost. Similarly, biomass with different compositions would favor the generation of different end products (biochar, bio-oil, etc.), given the various decomposition rate of the respective components (Bussemaker et al. (2017). Hence, the uncertainty of biomass quality plays a crucial role in developing a feasible biomass supply chain.

The quality of biomass not only influences the choice of conversion route and products yielded but also the demand variability and setting the market price which can further affect the profitability and feasibility of the biomass supply chain (Lin et al., 2013). The demand variability of the biomass-derived products is often impacted by several factors (e.g. quality of biomass, price of the products, availability of the raw material, competitive market, etc.). For example, investing in biogas production industry would be a very risky act in Malaysia, as the nation's liquefied natural gas (LNG) had dominated the gas market (Malaysia is the fifth largest exporter of LNG in the world in 2019) (Energy Information Administration (EIA), 2021). In fact, the low market demand in Malaysia results in the hindrance of the development of the biogas-related industry. Therefore, to meet the challenge of demand variability, stochastic biomass supply chain models will be needed to forecast and predict demand response.

The instability of biomass-derived product price is attributed to the demand and quality of the products. At a regional level, the fluctuating prices of products have been disproportionate with the product fluctuations, causing the pricing of the yielded products to become inconsistent. Khatiwada et al. (2016) stated that the fluctuation of biomass-derived electricity price would eventually impact the feasibility of the overall biomass supply chain substantially. Typically, the unit costs for raw material and the biomass-derived products are expressed as fixed variables in techno-economic analysis which, therefore cause the results to be unrealistic and less reliable. It has been indicated that the feasibility of the biomass supply chain strongly depends on the biomass market where its inconsistency will eventually lead to a high level of risk management. As such, the sensitive correlation between price and other manipulative variables needs to be paid attention as they pose a great impact on developing a feasible biomass supply chain.

Beyond technological challenges and their uncertainties in the supply chain, policy selection could also, be one of the decisive criteria to be included in the stochastic biomass supply chain. The induction of nation's relevant regulations and policies (i.e., feed-in tariff system (FIT) from Renewable Energy Act, Renewable Energy Policy and Action Plan, Green Technology Financing Scheme (GTFS), etc. (International Energy Agency (IEA), 2019)), would cause the biomass industries to face potential compliance issues and operating cost increment (How et al., 2019). The failure of finding common ground for cooperation on policy-related matters between government and industries may lead to serious consequences such as potential social scandal, higher costing, or project delay (Yatim et al., 2016). Given that the policy selection will continue to play important role in the biomass industries to drive the development of renewable energy in Malaysia, the strategic model formulation that considers policy selection decisions for the sustainable biomass supply chain is highlighted in this chapter later.

The aforementioned research works had shown how biomass availability, biomass quality, and fluctuating price can significantly affect the feasibility of the supply chain. This reveals the necessity of incorporating these uncertainties into the stochastic biomass supply chain model. This chapter aims to provide an overview of the key uncertainties to be considered in the stochastic biomass supply chain model, then followed by three case studies that deal with (i) biomass selection decision, (ii) facility location decision, and (iii) policy selection decision, to demonstrate the application of the developed stochastic model.

2 Biomass Selection Decision

The Monte Carlo model can be utilized to perform decision-making such as the selection of biomass for the conversion process. Apart from biomass or product pricing, biomass quality is also one of the criteria that will affect the feasibility of the conversion process. Few researchers have considered numerous biomass qualities (e.g., moisture content (Ngan et al., 2020); moisture and ash content (Aboytes-Ojeda et al., 2019)) in their developed model. This section demonstrates the usage of the Monte Carlo model to select the optimal type of palm-based biomass (i.e., palm kernel shell (PKS), empty fruit bunches (EFB), and mesocarp fiber (MF)) for biomass gasification process with the consideration of their respective biomass qualities. The methodology to perform decision-making for biomass selection involves six general steps as follows:

(i) Data collection and pre-processing
(ii) Process simulation and validation of biomass conversion process
(iii) Generation of design matrix from Design of Experiment (DoE) software
(iv) Perform simulation based on the design matrix extracted from DoE software
(v) Extraction of the generic correlation equation
(vi) Formulation of Monte Carlo model

Note that the explanations of each step is presented in Sect. 2.1 (step (ii) to step (v)) and Sect. 2.2 (step (vi)).

2.1 Formulation of Generic Correlation Equation

The integrated use of a *process simulation* software and *Design of Experiment (DoE)* software is to formulate the generic correlation equation to be integrated into the Monte Carlo evaluation model. Firstly, the biomass conversion process can be simulated or modified *via* process simulation software. In this work, Aspen Plus and Design Expert software are used. Before simulating the desired process in Aspen Plus, a few information is required:

(i) Process flowsheet diagram
(ii) Process stream information (i.e., temperature, pressure, and other relevant information)
(iii) Operating conditions of the equipment.

For this case, the process flowsheet diagram and other required information are adapted from Han et al. (2017) whereby the authors simulated a downdraft biomass gasification process. The simulated flowsheet is illustrated in Fig. 2, whereas the significant operating conditions used are tabulated in Table 1. Before proceeding into the simulation environment, setting up of the simulation environment is significant. Two conventional components were input into the components that are biomass and ash. The enthalpy and density model selected for the two aforementioned conventional components are HCOALGEN and DCOALIGT. On top of that, due to the combination of non-conventional and conventional components in the process stream, the stream class MCINCPSD were defined. After defining the required physical property method, the simulation model can be developed based on the process flowsheet. The flowsheet is distributed into three sections (i.e., pre-treatment, gasification, and syngas clean up). The pre-treatment of biomass begins with the mixing of 80 kg of ambient wet biomass with ambient air (comprise of 78% nitrogen and 21% oxygen). The air to fuel ratio is 1.38, the amount of ambient air to be mixed with the wet biomass is 110.4 kg. Subsequently, the mixture of ambient air and wet biomass is transferred into the drier to undergo drying process to decrease the biomass moisture content to 8.91 wt%. The drier selected in Aspen Plus is **RStoic** block. The reduction of biomass moisture content is performed using FORTRAN calculator to be able to reduce varying moisture content of biomass into the system to 8.91 wt%. Subsequently, the moist air will be separated from the biomass *via* liquid–gas separation process in the moist air separator (default ID as **Flash2** in Aspen Plus environment). The dry biomass will then be transferred to the decomposer that is the **RYield** reactor to be broken down into its conventional elemental state (i.e., carbon, hydrogen, nitrogen, sulfur, oxygen, ash, volatile matter, and fixed carbon) using another FORTRAN calculator. It is worthy to note that the energy consumed by the gasifier is provided by the heat released during decomposition process. This is achieved by using the energy stream, QDECOMP, to link the decomposer to the gasifier. After undergoing decomposition process, the dry biomass stream that has been broken down into their conventional elemental state will enter a separator (default ID **Sep2** in Aspen Plus environment). The separator is used to separate a portion of the carbon to be treated as the unreacted char, whereby the split fraction used is 0.1. The solid dry biomass stream exiting the separator will be mixed with heated air (comprised of 78% nitrogen and 21% oxygen) of temperature 150 °C using Mixer_2 before transferred to the gasifier to undergo gasification process. The gasifier reactor's default ID in Aspen Plus simulation is **RGibbs** reactor. The downdraft biomass gasification process can be split into three zones (i.e., pyrolysis, oxidation, and reduction) as shown in Table 2. Different reactions will occur at different zones. The first zone, pyrolysis zone, dried biomass is pyrolyze into volatiles and char. Several processes occur in the next zone, oxidation zone (i.e., hydrogen oxidation, carbon monoxide oxidation, light hydrocarbon oxidation, heavy hydrocarbon oxidation, and char partial oxidation). Last zone, reduction zone will have the water gas shift reaction, Boudouard reaction, water gas reaction, methanation, and steam methane reforming. The reactions occurring at the reduction zone were input as restricting chemical reactions. At the same time, the previously separated portion of unreacted char will be heated to the same temperature (856.17 °C) as GASOUT stream from the gasifier to mix the two streams using Mixer_3. The outlet gas stream from Mixer_3, MGAS, is the final gas product stream of the gasification process with unreacted char being a minor constituent in the stream. The last stage will be the syngas clean-up. The stream MGAS leaving Mixer_3 will move into the cyclone (default ID in Aspen Plus environment is **SSplit**) to remove the unreacted char. The main product stream leaving the cyclone is HSYNGAS (syngas containing water). The syngas stream will be cooled to ambient temperature using a cooler before heading into Separator_2 (default ID is **Sep2** in Aspen Plus). The function of Separator_2 is to remove the water content from the clean syngas stream.

Validation of the simulated process with literature sources is required to ensure the accuracy of the simulated process. Subsequently, the design matrix can be extracted from DoE software. The design matrix represents the required number of simulation runs to be fulfilled in order to extract the formulated generic correlation equation. The selection of design mode is the first step in the DoE software to extract the design matrix. There are several available design modes within the DoE software, i.e., (i) factorial design mode, (ii) response surface design mode, (iii) mixture design mode, and (iv) custom design mode.

In this case, a full factorial design mode has been adopted (Teng et al., 2019). Six factorial was selected based on the number of input biomass quality variables. The biomass quality variables selected for input in the design matrix are (i) moisture content, (ii) ash content, (iii) carbon content, (iv) hydrogen content, (v) oxygen content, and (vi) sulfur content.

The DoE software will then require an input of the maximum and minimum value for each biomass quality. Subsequently, the response or result intended will be required to be defined in the model. In this case, the specific syngas yield, $S^{BIOMASS}$ (kg syngas/kg biomass) was defined as the desired response. Once the required input

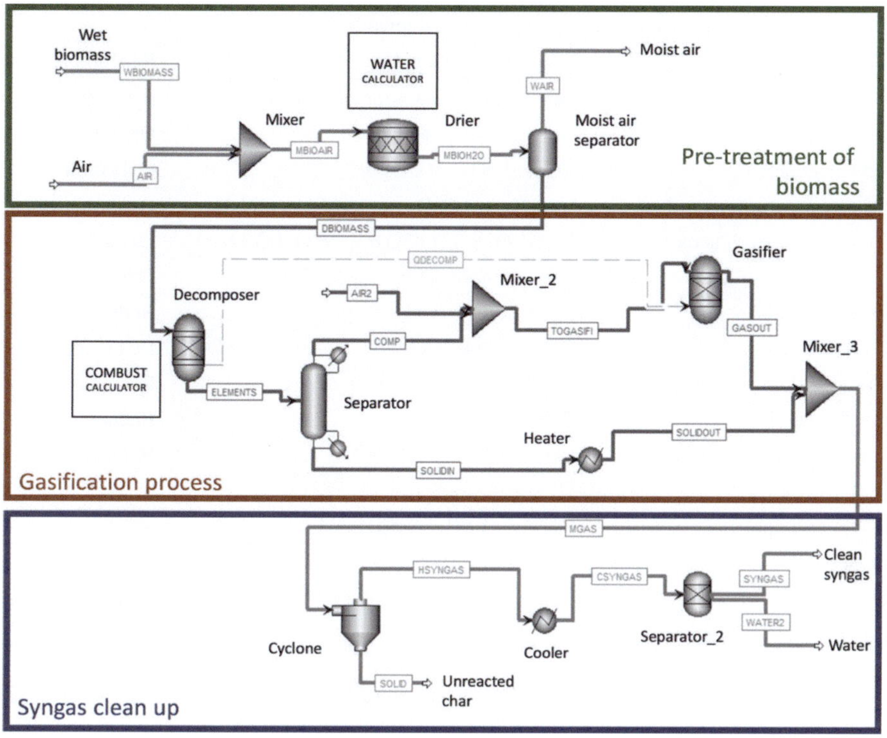

Fig. 2 Process simulation of biomass gasification process

Table 1 Fixed parameters used for this case study (Han et al., 2017)

Parameter	Unit	Value
Split fraction for unreacted carbon	–	0.1
Air inlet to gasifier	kg/hr	110
Air to fuel ratio	kg/kg	1.38
Temperature of inlet air	°C	150
Wet biomass into the system	kg/hr	80
Moisture content after drying	wt%	8.91
Outlet temperature of gasifier	°C	856.17

has been entered, the design matrix is then generated by the software. It is worthy to note that all the simulated runs of the design matrix require manual screening to ensure they meet the constraints shown in Eqs. (1) and (2). Both constraint equations require ensuring the summation of both proximate analysis and ultimate analysis to be at 100%, once the design matrix has been fulfilled, the generic correlation equation relating biomass quality to specific syngas yield can be extracted.

Table 2 Biomass gasification process reactions (adopted from Han et al., 2017)

Reaction name	Reaction equation	Reaction zone
Pyrolysis	Dried biomass → volatiles + char	Pyrolysis
H_2 oxidation	$H_2 + 0.5O_2 \rightarrow H_2O$	Oxidation
CO oxidation	$CO + 0.5O_2 \rightarrow CO_2$	Oxidation
Light hydrocarbon oxidation	$C_{1.16}H_4 + 1.58O_2 \rightarrow 1.16CO + 2H_2O$	Oxidation
Heavy hydrocarbon oxidation	$C_6H_{6.2}O_{1.2} + 4.45O_2 \rightarrow 6CO + 3.1H_2O$	Oxidation
Char partial oxidation	$C + 0.5O_2 \rightarrow CO$	Oxidation
Water gas shift	$CO + H_2O \leftrightarrow CO_2 + H_2$	Reduction
Boudouard	$C + CO_2 \leftrightarrow 2CO$	Reduction
Water gas	$C + H_2O \leftrightarrow CO + H_2$	Reduction
Methanation	$C + 2H_2 \leftrightarrow CH_4$	Reduction
Steam methane reforming	$CH_4 + H_2O \leftrightarrow CO + 3H_2$	Reduction
H_2S formation	$H_2 + S \rightarrow H_2S$	–
NH_3 formation	$0.5N_2 + 1.5H_2 \leftrightarrow NH_3$	–

$$Q^{\mathrm{FC}} + Q^{\mathrm{VM}} + Q^{A} + Q^{\mathrm{MC}}\ \mathrm{wt\%} \quad (1)$$

$$Q^{H} + Q^{C} + Q^{O} + Q^{S} + \mathrm{Q}^{\mathrm{N}} = 100\mathrm{wt\%} \quad (2)$$

Q^{FC}, Q^{VM}, Q^{A}, and Q^{MC} denote the weight composition of fixed carbon, volatile matter, ash, and moisture content of the biomass; whileQ^{H},Q^{C},Q^{O}, Q^{S} and Q^{N} denote the weight composition of hydrogen, carbon, oxygen, sulfur, and nitrogen of the biomass.

2.2 *Monte Carlo Model Formulation for Biomass Selection*

Net present value (NPV) (expressed in Eq. (3)) is used as an indicator to compare the feasibility of utilization of each type of biomass in consideration, where NCF_t refers to the net cash flow in year t, while in refers to the discount rate.

$$\mathrm{NPV} = \sum\nolimits_t \frac{\mathrm{NCF}_t}{(1 + in)^t} \quad (3)$$

Equation (4) is used to determine NCF_t, where the input and output cash flows are denoted as $\mathrm{CF}_t^{\mathrm{IN}}$ and $\mathrm{CF}_t^{\mathrm{OUT}}$ respectively.

$$\mathrm{NCF}_t = (\mathrm{CF}_t^{\mathrm{IN}} - \mathrm{CF}_t^{\mathrm{OUT}}) \quad (4)$$

The input cash flows, CF_t^{IN} depicts the revenue obtained from the sales of value-added products from the biomass conversion process. In this case, it would be the sales of syngas produced from the biomass gasification process. CF_t^{IN} can then be obtained from the multiplication of three variables, specific syngas yield, $S^{BIOMASS}$, the amount of biomass required, $Biomass_t^{IN}$ and the selling price of syngas, C_t^{SYNGAS}.

$$CF_t^{IN} = S^{BIOMASS} \times Biomass_t^{IN} \times C_t^{SYNGAS} \quad (5)$$

CF_t^{OUT}, on the other hand, is contributed by the capital expenditure (CAPEX) and operating expenditure (OPEX) (C^{CAPEX} and C_t^{OPEX}, respectively); and procurement cost for biomass ($C_t^{PROCURE}$) and coal ($C_t^{PROCURE}$) as shown in Eq. (6).

$$CF_t^{OUT} = C^{CAPEX} + C_t^{OPEX} + C_t^{PROCURE} \quad (6)$$

On the other hand, $C_t^{PROCURE}$ can be determined by multiplying the amount of the material to their respective unit cost, where $C_t^{BIOMASS}$ and C_t^{COAL} refers to the acquisition cost of biomass and coal, respectively:

$$C_t^{PROCURE} = (Biomass_t^{IN} \times C_t^{BIOMASS}) + (F_t^{COAL} \times C_t^{COAL}) \quad (7)$$

F_t^{COAL} can be calculated using Eqs. (8) and (9) whereby Q^{coal} denotes the energy required to reduce the moisture content, MC_t^{IN} and MC_t^{OUT} represents the initial and final moisture content of biomass, respectively. On the other hand, T_{Final} and $T_{Initial}$ denotes the final and initial temperature of the biomass, respectively. c denotes the specific heat capacity of water and HV^{coal} denotes the heating value of coal.

$$Q^{coal} = \left(MC_t^{IN} - MC^{OUT}\right) \times Biomass_t^{IN} \times c \times (T_{Final} - T_{Initial}) \quad (8)$$

$$F_t^{COAL} = \frac{Q^{coal}}{HV^{coal}} \quad (9)$$

There is also a noteworthy constraint shown in Eq. (10) whereby the $Biomass_t^{IN}$ must be less than or equal to the $Biomass_t^{SUPPLY}$. If $Biomass_t^{IN}$ is greater than $Biomass_t^{SUPPLY}$, an alternative solution will have to be considered to ensure the demand is met.

$$Biomass_t^{IN} \leq Biomass_t^{SUPPLY} \quad (10)$$

Subsequently, a total of 10,000 randomized NPV samples are generated *via* the *Monte Carlo* simulation, where the values of $Biomass_{EFB}^{QUALITY}$, $Biomass_{PKS}^{QUALITY}$, $Biomass_{MF}^{QUALITY}$, $Biomass_t^{SUPPLY}$, $C_t^{BIOMASS}$ and C_t^{SYNGAS} for each sample is randomized based on the historical statistical data collected. With this, the supply uncertainty, biomass quality uncertainty, price variation of biomass, and syngas can be incorporated into the model. The NPV and payback period (PBP) are the expected outcome from the model whereby the PBP can be determined through the subtraction method.

2.3 *Illustrative Example*

A palm-based biomass gasification plant is used as an illustrative case study to demonstrate the utilization of the Monte Carlo model to perform decision-making for biomass selection. The palm-based biomasses considered are EFB, PKS, and MF. The randomized parameters considered in this case study include syngas price, biomass supply, cost of biomass, and biomass quality whereby their respective value is tabulated in Table 3.

The mean, μ, and standard deviation, σ are obtained from the calculation of the minimum and maximum values for each of the uncertainty. The minimum and maximum values can be extracted from the historical statistical data for the uncertainties. On the other hand, other non-randomized parameters that were used in developing the case study model are shown in Table 4. It is worthy to note that the current case study does not consider the transportation cost for the materials (i.e., biomass and syngas).

The validation of the simulated process flowsheet (Fig. 2) has been tabulated in Table 5. The process is simulated, modified, and validated based on the works of Han et al. (2017). It is observable that the percentage difference calculated is relatively low with the highest being nitrogen content with a 5.80% difference. This is partially due to the under-production of one of the components that is methane in the equilibrium modeling environment (Song et al., 2013).

After validation of the simulated process, the next step was to extract the design matrix from DoE software and perform the simulation based on the design matrix. Subsequently, the generic correlation equation relating biomass quality to specific syngas yield was extracted (see Eq. (11)). The equation extracted met with a few criteria that ensure the applicability of the equation. One of the criteria is the equation's Prob > F value for lack of fit was 0.3668 that implies the equation provides high probability for good fitting. On top of that, the equation has an R^2 value of 0.9020 whereas the adjusted R^2 value, and the predicted R^2 value are 0.8917, and 0.8686, respectively. If the predicted R^2 value and adjusted R^2 value have a difference smaller than 0.20, it signifies that the equation is used which is the case reflected here (StatEase, 2020).

$$\begin{aligned}\text{specific syngas yield} = {} & 2.09922 + 1.80801 \times 10^{-3}\left(Q^{\text{MC}}\right) \\ & - 5.26,803 \times 10^{-3}\left(Q^{A}\right) + 9.59237 \times 10^{-3}\left(Q^{C}\right) \\ & - 0.033116\left(Q^{H}\right) - 4.39309 \times 10^{-3}\left(Q^{O}\right) \\ & - 4.31267 \times 10^{-3}\left(Q^{S}\right)\end{aligned} \tag{11}$$

Table 3 Randomised parameters used for this case study

Parameter	Remark	Value	Reference
$\text{Biomass}_t^{\text{SUPPLY}}$ (kg/hour)	EFB PKS MF	$\mu = 82{,}206.75$; $\sigma = 3675.36$ $\mu = 26{,}156.69$; $\sigma = 1169.43$ $\mu = 48{,}576.72$; $\sigma = 2171.81$	DQS Certification (2018)
$\text{Biomass}_{\text{EFB}}^{\text{QUALITY}}$ (wt %)	Moisture content Ash content Carbon content Hydrogen content Oxygen content Sulfur content	$\mu = 7.79$; $\sigma = 2.60$ $\mu = 6.75$; $\sigma = 1.14$ $\mu = 42.28$; $\sigma = 9.23$ $\mu = 6.70$; $\sigma = 1.37$ $\mu = 42.88$; $\sigma = 8.74$ $\mu = 0.31$; $\sigma = 0.15$	Sohni et al., (2018) Mahlia et al., (2001) Yoo et al., (2019)
$\text{Biomass}_{\text{PKS}}^{\text{QUALITY}}$ (wt %)	Moisture content Ash content Carbon content Hydrogen content Oxygen content Sulfur content	$\mu = 5.87$; $\sigma = 4.03$ $\mu = 7.48$; $\sigma = 6.05$ $\mu = 47.8$; $\sigma = 6.51$ $\mu = 7.66$; $\sigma = 1.92$ $\mu = 40.22$; $\sigma = 4.12$ $\mu = 0.7$; $\sigma = 0.71$	Sohni et al., (2018) Mahlia et al., (2001) Ahmad et al., (2014) Aziz et al., (2011)
$\text{Biomass}_{\text{MF}}^{\text{QUALITY}}$ (wt %)	Moisture content Ash content Carbon content Hydrogen content Oxygen content Sulfur content	$\mu = 5.06$; $\sigma = 0.43$ $\mu = 4.9$; $\sigma = 4.95$ $\mu = 46.29$; $\sigma = 1.29$ $\mu = 8.30$; $\sigma = 3.25$ $\mu = 39.37$; $\sigma = 3.78$ $\mu = 0.49$; $\sigma = 0.26$	Mahlia et al., (2001) Aziz et al., (2011) Garba et al., (2017)
C_t^{BIOMASS} (USD/t)	EFB PKS MF	$\mu = 120$; $\sigma = 33.6$ $\mu = 60$; $\sigma = 31.2$ $\mu = 6$; $\sigma = 5.04$	Abas et al. (2011) Lam et al., (2013) Reduan (2017) Lo et al., (2021) Agensi Inovasi Malaysia (2013)
C_t^{SYNGAS} (USD/t)		$\mu = 470.4$; $\sigma = 100.8$	Zuldian et al., (2017)

Following the formulation of the generic correlation equation, the next step would be to evaluate the economic performance *via* Monte Carlo simulation. Figure 3 illustrates the probability density result for the NPV value. It can be observed that MF has a greater probability to achieve a higher NPV value that is 16% to achieve a probability of approximately USD 24 million. On the other hand, PKS has a lower NPV achievable that is approximately 16.5% to achieve an NPV of approximately USD 22 million. EFB then poses the least favorable NPV outcome among the three types of palm-based biomass with a 17% of achieving an NPV value of approximately USD 18 million. Three factors contributed to the resulting outcome, i.e., (i) cost of purchasing biomass, (ii) carbon content of biomass, and (iii) standard deviation of uncertainties.

Table 4 Fixed parameters used for this case study

Parameter	Remark	Value	Reference
MC^{OUT}(wt %)	Desired quality	4	–
in(%)	Common assumption	10	–
t	Expected minimum lifespan	20	–
C^{CAPEX}(M USD)	–	5.3	Susanto et al. (2018) Aghabararnejad et al. (2015)
C_t^{OPEX}(M USD)	–	0.38	AlNouss et al. (2020) Spath et al. (2005)

Table 5 Validation of simulated process flowsheet

Syngas composition	Units	Simulated results	Reference results	Percentage difference (%)
CO	mol%	20.98	20.93	0.24
H_2	mol%	18.07	18.32	1.36
CO_2	mol%	13.55	12.87	5.28
N_2	mol%	47.39	44.79	5.80

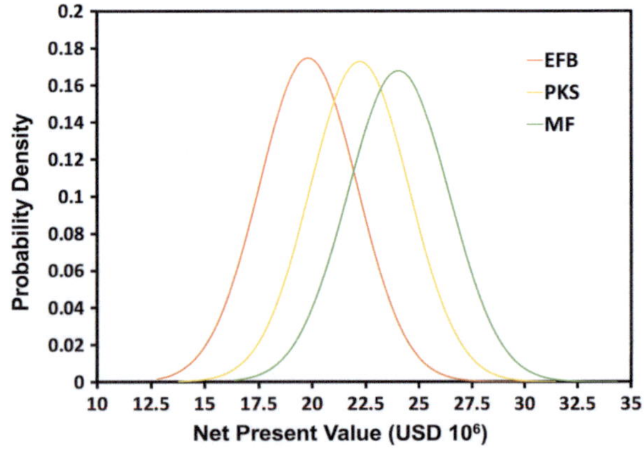

Fig. 3 Probability Density of NPV for biomass gasification plant

Firstly, it is observable from Table 3 that the mean acquisition price of MF is the lowest among the three types of palm-based biomass. Not to mention there is a significant difference between MF and EFB of approximately 95%. Thus, this will ultimately affect the total procurement cost, $C_t^{PROCURE}$ (see Eq. (11)). A lower

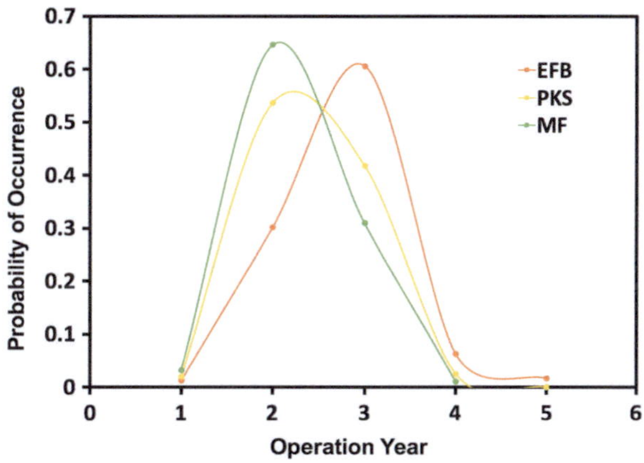

Fig. 4 Probability occurrence of PBP for biomass gasification plant

biomass acquisition cost will result in a lower procurement cost and operating cost that will ultimately increase the profit gained by the process. Secondly, the carbon content of biomass would also affect the syngas produce from the biomass gasification process. This is because there are two products formed during the biomass gasification process containing the carbon element that is carbon monoxide and carbon dioxide. On top of that, carbon monoxide is the critical component of syngas. Therefore, a higher amount of carbon content would be favorable. Table 3 shows that EFB offers the lowest average carbon content. Apart from that, it is noticeable that the standard deviation of the uncertainties for EFB is relatively greater in value as compared to MF and PKS. This is resulting from a larger difference in the minimum and maximum value extracted from the historical statistical data of the uncertainties. For instance, the mean and standard deviation of carbon content for EFB are 42.28 wt% and 9.23 wt%, respectively, whereas the mean and standard deviation of carbon content for PKS are 47.8 wt% and 6.51 wt%, respectively. On the other hand, Fig. 4 shows the payback period of the gasification process for the three types of palm-based biomass. In agreement with the obtained NPV results, MF has a higher probability to offset the capital cost expenditure earlier as compared to EFB and PKS. Specifically, MF has a probability of 68% to offset the capital cost expenditure in the second year of plant operation. Hence, the selected biomass that will be more feasible is MF.

3 Facilities Location Decision

Apart from the aforementioned biomass selection, Monte Carlo simulation can also be utilized in location selection for biomass conversion plants. Previously in Sect. 2, the biomass that showed a more favorable outcome was MF. Therefore, this section will target biomass conversion process using specifically MF as the feedstock. As the influence of biomass quality has been considered in the previous case study, biomass quality uncertainty will not be incorporated into the model developed in this section. This section will highlight more on location selection. Different locations pose different effects on the total transportation cost due to several variables (i.e., number of trips, distance, etc.). Hence, this section will demonstrate the use of Monte Carlo to select a financially feasible location for an MF-based gasification plant.

3.1 Model Formulation for Location Selection

Similar to the former case study, NPV and PBP are used to evaluate the economic feasibility of the candidate locations for setting up the biomass conversion plant. The same formulations (Eqs. (3) and (4)) are used to compute the overall NPV for this case study, while PBP is determined using the same aforementioned subtraction method.

Note that in this case study, $\mathrm{CF}_t^{\mathrm{IN}}$ and $\mathrm{CF}_t^{\mathrm{OUT}}$ are determined using the Eqs. (12) and (13), respectively.

$$\mathrm{CF}_t^{\mathrm{IN}} = F_t^{\mathrm{SYNGAS}} \times C_t^{\mathrm{SYNGAS}} \tag{12}$$

$$\mathrm{CF}_t^{\mathrm{OUT}} = C^{\mathrm{CAPEX}} + C_t^{\mathrm{OPEX}} + C_t^{\mathrm{PROCURE}} + C_t^{TR} \tag{13}$$

where F_t^{SYNGAS} refers to the amount of syngas produced, while C_t^{TR} denotes the total transportation cost.

The procurement cost of raw material, C_t^{PROCURE} is determined using Eq. (7), whereas C_t^{TR} is calculated using Eq. (14), where d^S refers to the traveling distance between biomass supply and the MF-based gasification plant; d^D refers to the traveling distance between the MF-based gasification plant and the demand point (port); $\mathrm{Cap}^{\mathrm{TRUCK}}$ denotes the vehicle load limit; while the fuel consumption rate of the vehicle and the respective fuel price in year t is expressed as ψ^{FUEL} and C_t^{FUEL}, respectively. For a round trip, the distances traveled have to be multiplied by "2" as shown in Eq. (14).

$$C_t^{\mathrm{TR}} = 2 \times \frac{\left(\mathrm{Biomass}_t^{\mathrm{IN}} \times d^S + F_t^{\mathrm{SYNGAS}} \times d^D\right)}{\mathrm{Cap}^{\mathrm{TRUCK}}} \times \psi^{\mathrm{FUEL}} \times C_t^{\mathrm{FUEL}} \tag{14}$$

Thereafter, *Monte Carlo* simulation (10,000 samples) is conducted to determine the probability profile for NPV and PBP for each location with the consideration of the uncertainties for five input parameters (i.e., F_t^{SYNGAS}, $\mathrm{Biomass}_t^{\mathrm{SUPPLY}}$, C_t^{BIOMASS}, C_t^{FUEL} and C_t^{SYNGAS}). Then, the optimal location to set up the MF-based gasification plant.

3.2 *Illustrative Example*

In this illustrative example, the developed model was used to select the optimal location to set up an MF-based gasification plant. Three candidate locations (Plant A, Plant B, and Plant C) have been considered in this case study. The distance between each location and specifications (i.e., loading capacity and fuel consumption) for the transportation vehicle are the significant data required to be collected. The respective distance data is tabulated in Table 6. Additionally, two types of transportation modes, i.e., a dump truck (truck A) for MF delivery and a fuel tanker (Truck B) for syngas delivery, are considered in this example (note: the specifications of the transportation modes are tabulated in Table 7).

On the other hand, the randomized input parameters for the developed *Monte Carlo model* are listed in Table 8, whereas Table 9 shows the fixed parameters used in the developed model. Note that the mean, μ, and standard deviation, σ shown in Table 8 are obtained from the historical statistical data.

Figure 5 showed the NPV results for the three plant locations investigated. Based on Fig. 5a, it shows that the plant locations seem to have minimal impact on the NPV result. It is worthy to note that the amount of biomass required to meet the demand is less than the biomass availability. However, it is assumed that all the biomass will be transported to the gasification plant, while the excessive biomass will be stored for future usage. As the biomass feed is excess, the number of trips taken to complete the transfer of biomass to the MF-based gasification plant will be more than the number of trips taken to transfer syngas from the plant to the port. Another noteworthy statement is that all mean and standard deviation of the uncertainty listed in Table 8 are the same for all three plant locations. The only variation would be the total transportation cost that is influenced by distance and number of trips. With a zoom-in view (Fig. 5b), it is found that Plant C has a slightly higher probability (approximately 0.1%) than that of the other two plants in obtaining an NPV value of USD 41.5 million. As a result, Plant C poses the lowest mean and standard deviation among the three plants (USD 41.58 million ± 4.94 million). Despite having the lowest mean, the lowest standard deviation implies that there would be less risk

Table 6 Fixed parameters used for this case study (Lo et al., 2020)

Plant	d^S(km)	d^D(km)
Plant A	18.6	78.5
Plant B	42.8	41
Plant C	80.4	24.2

Table 7 Specification for transportation modes

Truck	Truck type	Loading capacity (t)	Fuel consumption (l/100 km)
Truck A	Dump truck	30	47
Truck B	Fuel tanker	30	43

Table 8 Randomized parameters used for this case study

Parameter	Value	Reference
$Biomass_t^{SUPPLY}$(t/year)	$\mu = 48{,}576.72; \sigma = 2171.81$	DQS Certification (2018)
$C_t^{BIOMASS}$ (USD/t)	$\mu = 6; \sigma = 5.04$	Lo et al. (2021) Agensi Inovasi Malaysia (2013)
C_t^{SYNGAS} (USD/t)	$\mu = 470.4; \sigma = 100.8$	Zuldian et al. (2017)
C_t^{FUEL} (USD/L)	$\mu = 0.44; \sigma = 0.21$	Trading Economics (2020)
F_t^{SYNGAS} (MWth)	$\mu = 3262.09; \sigma = 1140.58$	MarketsandMarkets Research Private Ltd. (2020)

Table 9 Fixed parameters used for this case study

Parameter	Value	Reference
in(%)	10	–
t	20	–
C^{CAPEX}(M USD)	5.3	Susanto et al. (2018) Aghabararnejad et al. (2015)
C_t^{OPEX}(M USD)	0.38	AlNouss et al. (2020) Spath et al. (2005)

of fluctuation of the NPV for Plant C. To note, the lower mean observed in Plant C is partially due to the greater distance between the biomass source and Plant C that is d^S. As previously mentioned, the biomass will be in excess, thus, requiring more trips taken to transport the biomass to the gasification plant. Additionally, the fuel consumption of the dump truck is greater than the fuel consumption of the fuel tanker (see Table 7). The combinatory effect of a greater d^S, a greater number of trips and a higher fuel consumption rate for the dump truck directly increase total transportation cost. On the other hand, the mean and standard deviation for NPV for Plant A and Plant B is USD 41.66 million $\pm$ 4.98 million and USD 41.65 million $\pm$ 4.99 million, respectively. Looking at the probability of occurrence for PBP in Fig. 6, it is observable that Plant B has a higher probability to offset the capital cost invested when it reaches 1.75 years of operations. However, looking into Fig. 6(b), it is observable that Plant A has a higher probability of approximately 0.4% to offset the capital investment earlier than is into 1 year of operation. Hence, after careful consideration of the results obtained, Plant A is selected as the more feasible plant location.

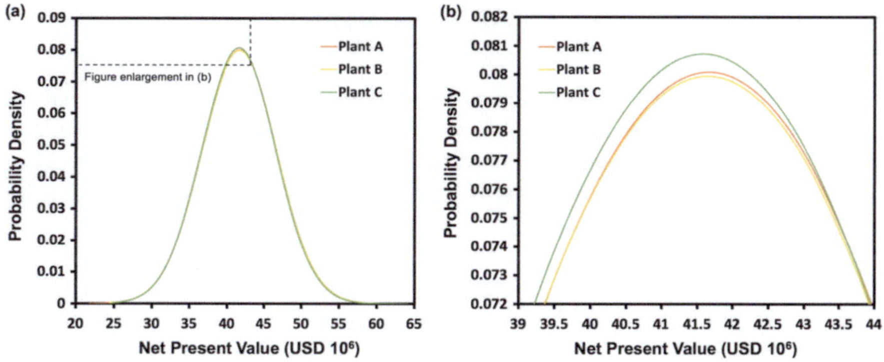

Fig. 5 **a** NPV for Plant A, Plant B and Plant C, **b** enlargement of a section of (**a**)

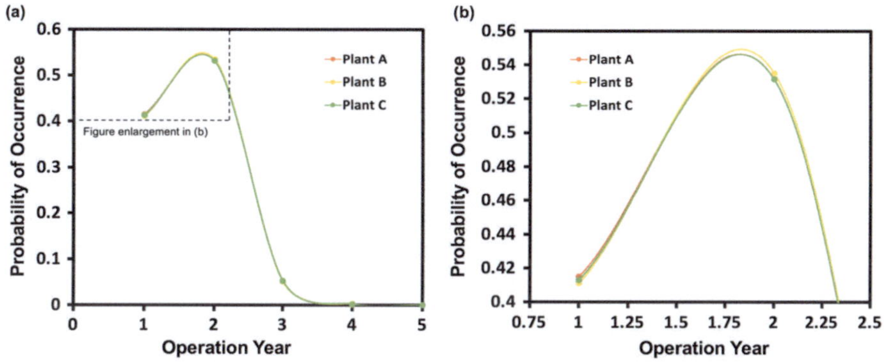

Fig. 6 **a** PBP for Plant A, Plant B and Plant C, **b** enlargement of a section of (**a**)

4 Policy Selection Decisions

Aside from the aforementioned process pathway selection, the utility of *Monte Carlo* simulation can be extended to policy selection. To date, it has been used as a supporting tool to help decision-maker in determining the optimal policy for various problems, including but not limited to pricing policy for parking management (Wang et al., 2021); E-hailing platform policy (Shou et al., 2020); replacement policy for equipment composed multiple non-identical components (Barde et al., 2019); and renewable energy schemes (Chou & Ongkowijoyo, 2014). This section, on the other hand, demonstrates its utility in selecting a policy (or action plan) to mitigate the overall business risk of a biomass-based polygeneration system.

4.1 Model for Biomass-Based Polygeneration System

Again, both NPV and PBP are used to evaluate the effectiveness of the available action plans. Note that since the monthly variations of the parameters are considered in this case study, the formulations used to determine NPV have now been revised to incorporate index m (for month). For instance, NPV is now determined using Eq. (15), where $\text{NCF}_{m,t}$ refers to the net cash flow at month m in year t.

$$\text{NPV} = \sum_{t}\sum_{m} \frac{NCF_{m,t}}{(1+in)^{t}} \tag{15}$$

Equation (16) is used to determine $\text{NCF}_{m,t}$, where the input and output cash flows are denoted as $\text{CF}^{\text{IN}}_{m,t}$ and $\text{CF}^{\text{OUT}}_{m,t}$ respectively, TAX refers to the corporate tax rate, while ITA indicates the tax exemption indicator. It is worth noting that the qualifying rate and limit for the exemption are set at 80% and 85%, respectively (Lembaga Hasil Dalam Negeri Malaysia (LHDN), 2018).

$$\text{NCF}_{m,t} = \left(\text{CF}^{\text{IN}}_{m,t} - \text{CF}^{\text{OUT}}_{m,t}\right) \times (1 - \text{TAX}) + \text{ITA} \times \text{TAX} \forall t \in T, \quad \forall m \in M \tag{16}$$

Generally, polygeneration plant yields multiple profitable products. Taking a polygeneration plant which consists of pyrolysis process and co-generation as an example, the $\text{CF}^{\text{IN}}_{m,t}$ can be determined using Eq. (17):

$$\text{CF}^{\text{IN}}_{m,t} = F^{\text{OIL}}_{m,t} \times C^{OIL}_{m,t} + \text{Elec}^{\text{EXP}}_{m,t} \times C^{\text{FIT}} \quad \forall t \in T, \forall m \in M \tag{17}$$

where the first term refers to the sales obtained from selling bio-oil ($F^{\text{OIL}}_{m,t}$) with bio-oil price of $C^{\text{OIL}}_{m,t}$; while the second term shows the profit obtained by exporting the generated power ($\text{Elec}^{\text{EXP}}_{m,t}$) with a feed-in-tariff (FiT) rate of C^{FIT}.

Equations (18) to (20) show the mass conversion equations for the three pyrolysis products, which include $F^{\text{OIL}}_{m,t}$, syngas ($F^{\text{GAS}}_{m,t}$) and biochar ($F^{\text{CHAR}}_{m,t}$):

$$F^{\text{OIL}}_{m,t} = \text{Biomass}^{\text{DRY}}_{m,t} \times y^{\text{OIL}} \quad \forall t \in T, \forall m \in M \tag{18}$$

$$F^{\text{GAS}}_{m,t} = \text{Biomass}^{\text{DRY}}_{m,t} \times y^{\text{GAS}} \quad \forall t \in T, \forall m \in M \tag{19}$$

$$F^{\text{CHAR}}_{m,t} = \text{Biomass}^{\text{DRY}}_{m,t} \times y^{\text{CHAR}} \quad \forall t \in T, \forall m \in M \tag{20}$$

where $\text{Biomass}^{\text{DRY}}_{m,t}$ refers to the mass flow rate of dried biomass fed into the pyrolyser; while y^{OIL}, y^{GAS} and y^{CHAR} denote the product yield of bio-oil, syngas, and biochar, respectively.

It is worth noting that $F_{m,t}^{\mathrm{GAS}}$ contains various high energy content gases g (i.e., CO, H_2, and CH_4). Therefore, it can be used to produce energy (power ($\mathrm{Elec}_{m,t}^{\mathrm{GEN}}$) and heat ($\mathrm{Thermal}_{m,t}^{\mathrm{GEN}}$)) *via* co-generation unit (see Eqs. (21) and (22)). Based on commercial gas engine performance data, the amount of thermal energy recovered from a co-gen process is 1.2 times the amount of electricity being generated (GE, 2018).

$$\mathrm{Elec}_{m,t}^{\mathrm{GEN}} = \sum_{g} \left(F_{m,t}^{\mathrm{GAS}} \times y_g^{PY} \times \mathrm{LHV}_g \right) \times \xi^{\mathrm{COGEN}} \quad \forall t \in T, \ \forall m \in M \tag{21}$$

$$\mathrm{Thermal}_{m,t}^{\mathrm{GEN}} = 1.2 \times \mathrm{Elec}_{m,t}^{\mathrm{GEN}} \quad \forall t \in T, \ \forall m \in M \tag{22}$$

where y_g^{PY} and LHV_g refer to the molar composition and low heating value of gas g, respectively, while ξ^{COGEN} denotes the conversion efficiency of the co-generation unit.

$\mathrm{Elec}_{m,t}^{\mathrm{GEN}}$ can be used to fulfill the power demand for the pyrolyser ($\mathrm{Elec}_{m,t}^{\mathrm{REQ}}$; computed through Eq. (23)). If $Elec_{m,t}^{\mathrm{GEN}}$ is less than $\mathrm{Elec}_{m,t}^{\mathrm{REQ}}$, the balance will be covered by importing external power ($\mathrm{Elec}_{m,t}^{\mathrm{IMP}}$). In contrast, the excess power ($\mathrm{Elec}_{m,t}^{\mathrm{EXP}}$) will be exported back to the grid. This can be defined as Eq. (24), where ψ^{PY} refers to the thermal energy required to try per unit mass of biomass:

$$\mathrm{Elec}_{m,t}^{\mathrm{REQ}} = \mathrm{Biomass}_{m,t}^{\mathrm{DRY}} \times \psi^{\mathrm{PY}} \quad \forall t \in T, \ \forall m \in M \tag{23}$$

$$\mathrm{Elec}_{m,t}^{\mathrm{GEN}} + \mathrm{Elec}_{m,t}^{\mathrm{IMP}} = \mathrm{Elec}_{m,t}^{\mathrm{REQ}} + \mathrm{Elec}_{m,t}^{\mathrm{EXP}} \quad \forall t \in T, \ \forall m \in M \tag{24}$$

whereas $\mathrm{Thermal}_{m,t}^{\mathrm{GEN}}$ can be used to compensate for the thermal energy required during the biomass drying ($\mathrm{Thermal}_{m,t}^{\mathrm{REQ}}$; determined *via* Eq. (25)). Besides, the generated $F_{m,t}^{\mathrm{CHAR}}$ can also be used as a solid fuel to generate thermal energy. Coal ($F_{m,t}^{COAL}$) will be imported as additional solid fuel if the generated thermal energy from co-generation and char burning is insufficient to meet the energy consumption (see Eq. (26)):

$$\mathrm{Thermal}_{m,t}^{\mathrm{REQ}} = \mathrm{Biomass}_{m,t}^{\mathrm{IN}} \times \frac{\left(MC_{m,t}^{\mathrm{IN}} - MC^{\mathrm{OUT}} \right)}{100} \times \psi^{\mathrm{THERMAL}} \quad \forall t \in T, \ \forall m \in M \tag{25}$$

$$\mathrm{Thermal}_{m,t}^{\mathrm{REQ}} = \left(F_{m,t}^{\mathrm{COAL}} \times \mathrm{LHV}^{\mathrm{COAL}} + F_{m,t}^{\mathrm{CHAR}} \times \mathrm{LHV}^{\mathrm{CHAR}} \right) \times \xi^{\mathrm{DRY}} + \mathrm{Thermal}_{m,t}^{\mathrm{GEN}} \quad \forall t \in T, \ \forall m \in M \tag{26}$$

where $\mathrm{Biomass}_{m,t}^{\mathrm{IN}}$ refers to the amount of raw biomass sent to the polygeneration plant; $\mathrm{MC}_{m,t}^{\mathrm{IN}}$ and $\mathrm{MC}^{\mathrm{OUT}}$ present the moisture content before and after the drying process, respectively, the low heating values of coal and char are denoted as LHV^{COAL} and LHV^{CHAR}, respectively; ψ^{THERMAL} indicate the thermal energy required to remove per mass unit of water content, while the drying efficiency is expressed as ξ^{DRY}.

$\text{CF}_{m,t}^{\text{OUT}}$, on the other hand, is contributed by the capital expenditure (CAPEX) and operating expenditure (OPEX) of unit u (C_u^{CAPEX} and $C_{u,m,t}^{\text{OPEX}}$ respectively); transportation cost ($C_{m,t}^{TR}$); procurement cost for imported electricity, biomass, and coal ($C_{m,t}^{\text{PROCURE}}$); and carbon penalty ($C_{m,t}^{\text{PENALTY}}$).

$$\text{CF}_{m,t}^{\text{OUT}} = \begin{cases} \sum_u C_u^{\text{CAPEX}} \Big|_{t=0} \\ \sum_u C_{u,m,t}^{\text{OPEX}} + C_{m,t}^{\text{TR}} + C_{m,t}^{\text{PROCURE}} + C_{m,t}^{\text{PENALTY}} \Big|_{t>0} \end{cases} \quad \forall t \in T,\ \forall m \in M \tag{27}$$

$C_{m,t}^{\text{TR}}$ considers the cost associated with the transportation of the materials (including biomass and bio-oil). To note, the conventional 10-t truck are used as the transportation mode in this section. It is expressed as follow:

$$C_{m,t}^{\text{TR}} = 2 \times \frac{\left(\text{Biomass}_{m,t}^{\text{IN}} \times d^S + F_{m,t}^{\text{OIL}} \times d^D\right)}{\text{Cap}^{\text{TRUCK}}} \times \psi^{\text{FUEL}} \times C_{m,t}^{\text{FUEL}} \quad \forall t \in T,\ \forall m \in M \tag{28}$$

where d^S refers to the traveling distance between biomass supply and the polygeneration plant; d^D refers to the traveling distance between the polygeneration plant and the demand point; $\text{Cap}^{\text{TRUCK}}$ denotes the vehicle load limit; while the fuel consumption rate of the vehicle and the respective fuel price at month m in year t is expressed as ψ^{FUEL} and $C_{m,t}^{\text{FUEL}}$ respectively. For a round trip, the distance traveled has to be multiplied by "2" as shown in Eq. (28).

On the other hand, $C_{m,t}^{\text{PROCURE}}$ can be determined by multiplying the capacity of the imported material to their respective unit cost, where $C_{m,t}^{\text{BIOMASS}}$, $C_{m,t}^{\text{COAL}}$ and C^{ELEC} refer to the respective cost of biomass, coal, and imported electricity, respectively:

$$C_{m,t}^{\text{PROCURE}} = \text{Biomass}_{m,t}^{\text{IN}} \times C_{m,t}^{\text{BIOMASS}} + F_{m,t}^{\text{COAL}} \times C_{m,t}^{\text{COAL}} + \text{Elec}_{m,t}^{\text{IMP}} \times C^{\text{ELEC}} \quad \forall t \in T,\ \forall m \in M \tag{29}$$

In terms of $C_{m,t}^{\text{PENALTY}}$, it is expressed as the compensation cost required to recover the environmental damage caused by the carbon emission. It can be determined using Eq. (30):

$$C_{m,t}^{\text{PENALTY}} = C^{\text{CO}_2} \times \left(2 \times \frac{\left(\text{Biomass}_{m,t}^{\text{IN}} \times d^S + F_{m,t}^{\text{OIL}} \times d^D\right)}{\text{Cap}^{\text{TRUCK}}} \times \psi^{\text{FUEL}} \times y^{\text{CO}_2\text{\_TR}}\right.$$

$$+\text{Elec}_{m,t}^{\text{GEN}} \times y^{\text{CO}_2\text{\_COGEN}} + F_{m,t}^{\text{GAS}} \times y^{\text{CO}_2\text{\_PY}}\Big) \quad \forall t \in T,\ \forall m \in M \tag{30}$$

where $y^{\text{CO}_2\text{\_TR}}$, $y^{\text{CO}_2\text{\_COGEN}}$, and $y^{\text{CO}_2\text{\_PY}}$ refer to the estimated CO_2 emitted during transportation, co-generation unit, and pyrolysis process, while C^{CO_2} refers to the compensation cost required per unit of carbon emission.

The following two equations are applied to reflect the biomass availability constraint and bio-oil demand constraint, where $\text{Biomass}_{m,t}^{\text{AVAILABLE}}$ and $F_{m,t}^{OIL\_DEMAND}$ reflect the biomass availability and the bio-oil demand at month m in year t:

$$+\text{Biomass}_{m,t}^{\text{AVAILABLE}} \geq \text{Biomass}_{m,t}^{\text{IN}} \quad \forall t \in T,\ \forall m \in M \tag{31}$$

$$F_{m,t}^{\text{OIL}} \leq F_{m,t}^{\text{OIL\_DEMAND}} \quad \forall t \in T,\ \forall m \in M \tag{32}$$

It is worth mentioning that, a total of 10,000 samples are generated through the *Monte Carlo* simulation, where the values of $\text{Biomass}_{m,t}^{\text{AVAILABLE}}$, $F_{m,t}^{\text{OIL\_DEMAND}}$, $C_{m,t}^{\text{BIOMASS}}$, $C_{m,t}^{\text{COAL}}$, $C_{m,t}^{\text{FUEL}}$, $C_{m,t}^{\text{OIL}}$ and $MC_{m,t}^{\text{IN}}$ for each sample is randomized based on the statistical data. With this, the supply uncertainty, demand variation, price fluctuation, and seasonal biomass quality can, therefore, be incorporated into the model. The NPV and PBP (determined *via* subtraction method) of these samples are determined. Subsequently, the effectiveness of the proposed action plans is evaluated based on the improvement of these two economic indicators.

4.2 *Illustrative Example*

An oil palm biomass-based polygeneration plant is used as an illustrative case study to demonstrate how *Monte Carlo* simulation can be used to aid the policy selection decision. Empty fruit bunch (EFB) is collected from a nearby palm oil mill (located 10 km away from the plant). They are subsequently dried before feeding into the pyrolyser in the polygeneration plant. The produced bio-oil can be sold to a demand point which is located 15 km away from the plant, while the syngas and biochar are used as energy sources for power and thermal energy. The energy can be self-consumed so that the requirement of an external energy source can be mitigated. As mentioned in Sect. 4.2, the excess generated electricity can be exported to grid for additional revenue. The visual illustration of this case study is presented in Fig. 7, while all the important parameters used to develop the case study model are summarised in Tables 10 and 11. As mentioned, the fluctuations and uncertainties in operations, transportation, market demand, and price can be rigorously modeled by *Monte Carlo* simulations based on the actual statistical distributions. Figure 8 shows the 10,000 sample points generated based on the statistical data stated in Table 10 (i.e., the mean (μ) and standard deviation (σ)) for each randomized parameter. By judging on the frequency of the data points, one can estimate the occurrence probability of the parameters' magnitude. Taking biomass moisture content (Fig. 7b) as an example, the probability of getting a moisture content of less than 70% during the rainy season is significantly lower as compared to that of the dry season. In terms of EFB availability, 8500–11,000 t EFB/month, 11,000–12,400 t EFB/month, and 12,800–15,000 t EFB/month are available in low season, mid season, and the high season, respectively (Fig. 8a, b). Whereas the price fluctuations for EFB (24.55–73.65 USD/t), coal (24.55–147.30 USD/t), oil (0.74–0.88 USD/L), and diesel (0.44–0.60 USD/L) are shown in Fig. 8a, c. Aside from that Fig. 8d presents the oil demand variation, where the peak is found in the middle of the year (around June), and declines from December to February. Table 12, on the other hand, presents the description and the explanation of the seven action plans. To indicate the implementation of each action plan, the model settings stated in Table 12 were conducted (e.g., to represent the implementation of AP5, the TAX is omitted in the first five years.

Figure 9 shows the *Monte Carlo* simulation results obtained under different action plans. As shown in Fig. 9a, it is observed that introducing a new financing incentive (AP3-FI) attains the most attractive result, i.e., leading to a 58.51% (equivalent to ~ 1.6×10^6 USD) increment in the mean of NPV as compared to the base case (without

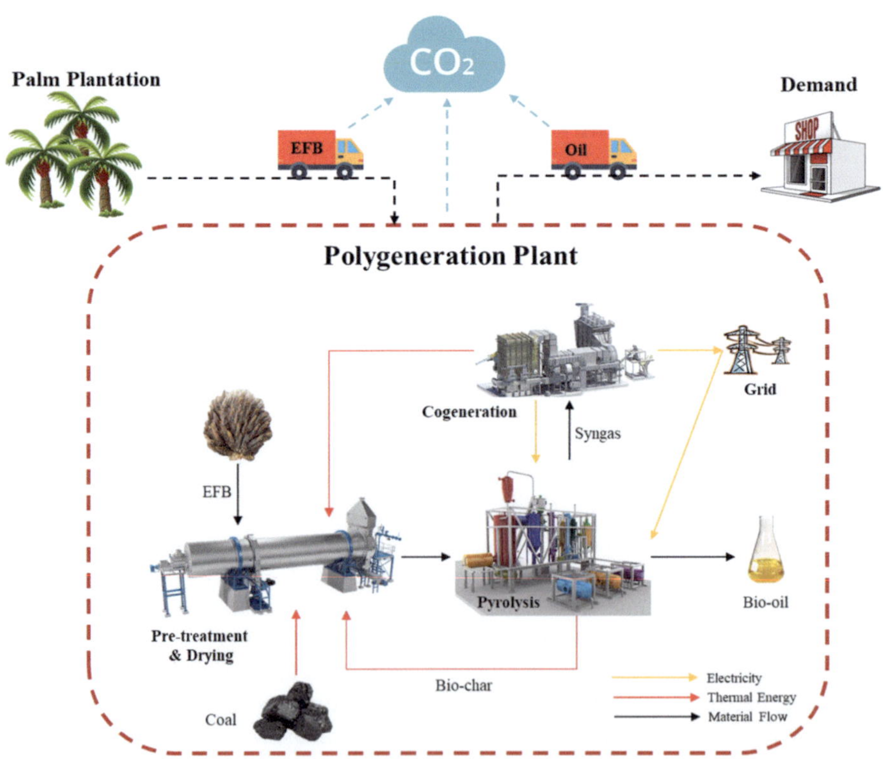

Fig. 7 Schematic diagram of the polygeneration plant case study (Ngan et al., 2020)

Table 10 Randomized parameters used for this case study (adapted from Ngan et al., 2020)

Parameter	Remark	Value	Reference
$\text{Biomass}_{m,t}^{\text{AVAILABLE}}$(t/month)	Low Season[a] Mid Season[a] High Season[a]	$\mu = 9890.82; \sigma = 323.69$ $\mu = 11{,}664.92; \sigma = 203.37$ $\mu = 13{,}853.1; \sigma = 350.30$	Malaysian Palm Oil Board (MPOB) (2018)
$F_{m,t}^{\text{OIL\_DEMAND}}$(t/month)	January[b] February[b] March[b] April[b] May[b] June[b] July[b] August[b] September October[b] November[b] December[b]	$\mu = 918.60; \sigma = 90.86$ $\mu = 915.60; \sigma = 65.80$ $\mu = 926.70; \sigma = 82.86$ $\mu = 927.75; \sigma = 69.03$ $\mu = 960.00; \sigma = 66.95$ $\mu = 957.45; \sigma = 48.72$ $\mu = 984.27; \sigma = 63.89$ $\mu = 978.95; \sigma = 59.60$ $\mu = 970.77; \sigma = 61.26$ $\mu = 936.82; \sigma = 68.21$ $\mu = 929.18; \sigma = 92.27$ $\mu = 924.95; \sigma = 89.10$	Trading Economics (2018)
$\boldsymbol{C}_{\boldsymbol{m,t}}^{\textbf{BIOMASS}}$(USD/t)	–	Max = 34.18; Min = 70.80	-
$\boldsymbol{C}_{\boldsymbol{m,t}}^{\text{COAL}}$(USD/t)	–	$\mu = 70.26; \sigma = 19.43$	Index Mundi (2018)
$\boldsymbol{C}_{\boldsymbol{m,t}}^{\text{FUEL}}$(USD/L)	–	$\mu = 0.52; \sigma = 0.03$	RinggitPlus (2018)
$\boldsymbol{C}_{\boldsymbol{m,t}}^{\text{OIL}}$(USD/L)	–	$\mu = 0.82; \sigma = 0.03$	–
$\text{MC}_{\boldsymbol{m,t}}^{\text{IN}}$(%)	Dry season Rainy season[c]	$\mu = 66.5; \sigma = 1.83$ $\mu = 76.5; \sigma = 1.83$	International Finance Corporation (IFC) (2017)

[a] Classified based on the monthly palm crude oil production (Andiappan et al., 2015)
[b] The local bio-oil demand is assumed similar to the pattern of the oil production in Malaysia
[c] Assumed to be 10% greater than that of during dry season

Table 11 Fixed parameters used for this case study (adapted from Ngan et al., 2020)

Parameter	Remark	Value	Reference
$MC_{m,t}^{OUT}(\%)$	–	10	–
$y^{OIL}(\%)$	–	27	Mohd (2017)
$y^{GAS}(\%)$	–	24	Mohd (2017)
$y^{CHAR}(\%)$	–	49	Mohd (2017)
$y_g^{PY}(\%)$	H_2 CO CH_4 CO_2	3.7 34.0 7.8 54.0	Mohd (2017)
$y^{CO_2\_TR}$(kg CO_2/L fuel)	–	2.68	Gu and Bergman (2015)
$y^{CO_2\_COGEN}$(kg CO_2/kWh)	–	0.525	Gu and Bergman (2015)
LHV_g(kJ/mol)	H_2 CO CH_4	240.2 283.5 801.4	PNAS (2018)
LHV^{COAL}(MJ/kg)	–	23	Othman et al. (2012)
LHV^{CHAR}(MJ/kg)	–	26	Mohamad et al. (2011)
$\xi^{COGEN}(\%)$	–	27[a]	–
$\xi^{DRY}(\%)$	–	85	–
ψ^{PY}(kWh/t EFB)	–	240	Rogers and Brammer (2012)
ψ^{FUEL}(L fuel/km)	–	0.213	How et al. (2016)
$\psi^{THERMAL}$(MJ/kg water removed)	–	4	Kovařík (2017)
TAX(%)	–	24	–
in(%)	–	10	–
C_u^{CAPEX}(M USD)	Pyrolyser Co-generation	1.54[b] 0.48	Wright et al. (2010), Energy Technology Systems Analysis Programme (ETSAP) (2010)
$C_{u,m,t}^{OPEX}$(USD/unit[c])	Pyrolyser Co-generation	50 60	How and Lam (2018) –
C^{FIT}(USD/kWh)	–	0.12	Sustainable Energy Development Authority Malaysia (SEDA) (2018)
C^{ELEC}(USD/kWh)	–	0.14	–
C^{CO_2}(USD/kg CO_2)	–	0.05	How et al. (2016)
Cap^{TRUCK}	–	10	–
d^S(km/trip)	–	10	–
d^D(km/trip)	–	15	–

[a] Assumed to be 60% of the typical gas engine efficiency
[b] Scaled by using six-tenths rule
[c] USD/t EFB for pyrolyser; USD/MWh for co-generation unit

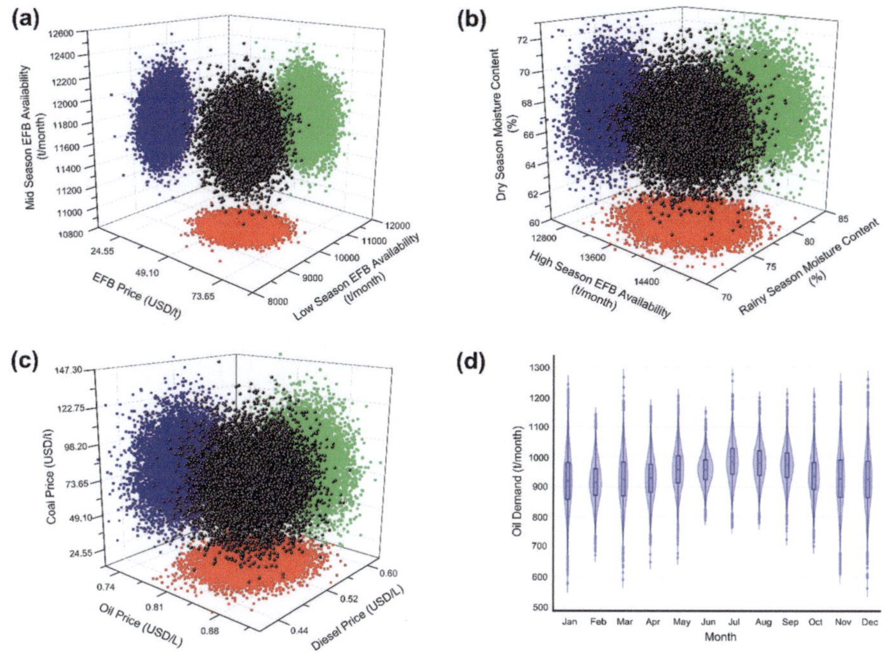

Fig. 8 Randomized input for: **a** Mid and high season EFB availability, EFB price; **b** dry and rainy season moisture content, low season EFB availability; **c** coal price, oil price, and diesel price, **d** oil demand (Ngan et al., 2020)

any action plan). On the other hand, as shown in Fig. 9c, PBP is also improved by 12.22% (equivalent to 7 months). Figure 9b shows that this action plan is capable to offer a maximum NPV of 6.14 million USD in 20 years, which is about 1.3 times greater than the next highest action plan (i.e., AP6-RF). Nevertheless, AP3-FI has demonstrated a mean PBP of 4 years which is the second shortest PBP among all action plans (Fig. 9c, d). Whereas the next most convincing action plan is to revise Fit-in-Tariff (AP6-RF) which has an expected NPV of between 2.9 and 3.4 million USD in 20 years and a mean PBP of 4 years. In other words, by adopting this action plan (increase the FiT rates to 0.132 USD/kWh), the overall NPV can be increased by 20.72% with a 14.04% reduction in PBP. In fact, multiple countries have achieved successful growth in renewable energy sectors through the implementation of FiT (e.g., Thailand has successfully increased the power generation capacity by renewable sources from 8% in 2015 to 17% by the end of 2017 with an attractive FiT rate (International Renewable Energy Agency (IRENA) 2017).

The action plan that ranked 3rd is AP5-TI (introducing tax incentives). Identical to financing incentives, the tax incentive is another type of financial instrument that is widely used by the government to spur up the growth of an industry. Some of the examples of tax incentives are tax returns, tax exemption, and tax reduction. The simulation result for implementing AP5-TI enhances the mean of the NPV by

Table 12 Description for each action plant (adapted from Ngan et al., 2020)

ID	Action plan	Description	Model setting
AP1-HD	Engage in demand contract	To hedge demand risk by committed into a supply contract with the consumer(s), to sell a fixed amount of product for a fixed duration, with a fixed price that is lower than that of the current market price	Set contracted demand as 1700 t/year $C_{m,t}^{OIL}$ is set 3% lesser than the current market price
AP2-HS	Engage in supply contract	To hedge supply risk by committed into a purchase contract with supplier(s), to buy a fixed amount of raw materials for a fixed duration, with a fixed price that greater than that of the current market price	St contracted supply as 10,000 t/year $C_{m,t}^{BIOMASS}$ is set 10% higher than the current market price
AP3-FI	Introduce new financing incentive	To reduce financing risk by providing financing incentives in the form of interest rate reduction to lower the debt obligation of industry players	Set in as 6%
AP4-SF	Substitute fossil fuel with biodiesel	To encourage the substitution of conventional fossil fuel which is less environmentally friendly with biodiesel to mitigate the overall CO_2 emissions	Set $C_{m,t}^{FUEL}$ as 0.69 USD/L and $y^{CO_2\_TR}$ as 2.1 kg CO_2/L
AP5-TI	Introduce new tax incentive	To reduce regulatory risk by showing favor in the form of tax exemption to encourage the utilization of biomass for wealth generation and development of green industry	Set TAX as 0% for the first 5 years
AP6-RF	Revise Feed-in-Tariff (FiT) rate	To promote higher utilization of renewable energy by revising the FiT rate to a higher rate to make it attractive for new entrants and investors to venture into the industry	Increase C^{FIT} by 10%

(continued)

Table 12 (continued)

ID	Action plan	Description	Model setting
AP7-CM	Introduce carbon management system	To promote sustainable development by introducing carbon management system upfront to avoid high carbon emission which could potentially result in carbon penalty	Set the removal efficiency as 80%[a]

[a] Assumed the total CAPEX to be 60% more expensive (Ooi et al., 2014) while the OPEX of the carbon management system is set as 0.037 USD/kg CO_2 removed (Rubin et al., 2015)

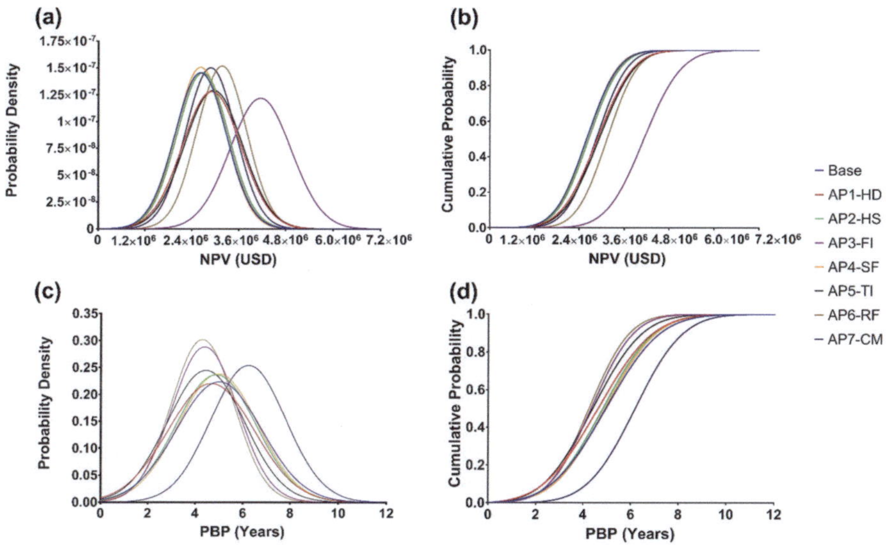

Fig. 9 The effect of the respective action plans on NPV (**a** probability density; **b** Cumulative probability) and PBP (**c** probability density; **d** cumulative probability) from 10,000 *Monte Carlo* simulations (Ngan et al., 2020)

12.62% (equivalent to ~ 0.34 × 10^6 USD), and reducing the payback period by 11.04% (equivalent to 6 months). It is followed by AP1-HD and AP2-HS which aim to mitigate the risks associated with the potential fluctuations in biomass supply and the product demand. It is a common strategy opted in the industry for production plants to engage in a long-term supply or/and demand contract to supply or purchase a fixed quantity of the product at a fixed price, for a fixed duration. The increase of mean of NPV for AP1-HD and AP2-HS is 10.84% (equivalent to ~ 0.3 × 10^6 USD) and 1.95% (equivalent to ~ 0.5 × 10^5 USD) respectively, while the mean PBP for each action plan is 6.68% (equivalent to 4 months) and 2.49% (equivalent to 1 month) respectively. It is worthfully to note that the combination of both action plans can yield better economic performance, i.e., 15.31% in terms of NPV improvement (equivalent to ~ 0.4×10^6 USD) and 9.46% in terms of PBP improvement equivalent to 5 months), which is comparable to AP5-TI.

Introduce carbon management system (AP7-CM) and substitute fossil fuel with biodiesel (AP4-SF) are ranked at the least. Both of these action plans aim to improve the overall economic performance *via* the mitigation of carbon emissions. Despite a significant increment (9.69% or equivalent to ~ 0.26×10^6 USD) of NPV are spotted for AP7-CM, the additional CAPEX required for the carbon capture system has prolonged the duration of PBP by 24.75% (equivalent to 2 years of additional project time). This shows that the uptake of CCS at this stage is categorized as a high-risk decision and not so favorable as compared to other action plans which could achieve a similar improvement in NPV, without the need to incur a higher upfront cost. Nevertheless, this finding is by no means an unchangeable fact as the incurred costs can be possibly reduced substantially as the technology matures (Mott MacDonald, 2012). Lastly, AP4-SF has shown a very minimal impact on both the NPV and PBP. This is due to the insignificant carbon penalty reduction (i.e., reduced about 0.03 USD/L) which is unable to compensate for the greater fuel cost (i.e., increased about 0.17 USD/L).

5 Further Reading

The following listed materials are recommended for further reading:

(i) Overview of sustainable biomass supply chain: from concept to modeling (Hong et al., 2016)

This paper provides readers with the fundamental knowledge and concept of sustainable biomass supply chain management, modeling, and optimization. In addition, key challenges and future prospects were elucidated comprehensively in this review article.

(ii) PCA method for debottlenecking of sustainability performance in integrated biomass supply chain (How & Lam, 2019)

This paper proposes a novel principal component analysis (PCA) aided debottlenecking approach to systematically remove the potential sustainability bottlenecks. Its effectiveness is benchmarked with another graph-theoretic based debottlenecking approach.

(iii) Techno-economic analysis for biomass supply chain: A state-of-the-art review (Lo et al., 2021)

This paper critically reviews the available approaches for the techno-economic evaluation of biomass supply chain (deterministic and stochastic). In addition, this review article outlines the key supply chain uncertainties and their corresponding impacts on the economic performance of the biomass business.

(iv) Debottlenecking of biomass element deficiency in a multiperiod supply chain system *via* element targeting approach (Lim et al., 2019)

This paper tackles the biomass availability fluctuation in a given regional biomass supply chain using element targeting approach and multiperiod analysis. This article demonstrates how the element targeting approach can be used to improve the overall biomass resources allocation and utilization.

(v) Synthesis of sustainable circular economy in palm oil industry using graph-theoretic method (Yeo et al., 2020)

This paper attempts to evaluate the feasibility of shifting the conventional palm oil industry toward circular economy using a powerful graph-theoretic approach, called process graph (P-graph). The potential bottlenecks that hinder the implementation of circular economy approach have also been discussed.

(vi) Bottleneck Tree Analysis (BOTA) with green and lean index for process capacity debottlenecking in industrial refineries (Teng et al., 2020)

This paper proposes a novel Bottleneck Tree Analysis (BOTA) to optimize the debottlenecking schedule in industrial refineries with the consideration of lean and green aspects. Such an effective scheduling approach can be easily adapted in palm oil refinery and biorefinery.

6 Conclusion

Palm biomass supply chain is deemed as a waste-to-wealth business. It contributes to economic development and serves as an effective strategy for waste management. Nevertheless, the complete shift to biomass as a feedstock is yet to be proven feasible and sustainable at a commercial scale. To fully explore its real potential and limits, sufficient knowledge and in-depth understanding of biomass supply chain modeling, particularly in stochastic modeling are essentially needed. This chapter, therefore, provides an overview of the key factors to be considered in stochastic modeling for biomass supply chain. It serves as a general guide to industry practitioners, academicians, and researchers who have interest in equipping themselves with knowledge and skills regarding stochastic modeling of the biomass supply chain. Three illustrative examples are used to demonstrate the utility of stochastic modeling in biomass supply chain research (i.e., biomass selection decision, facilities location decision, and policy selection decision). All these contents will help readers to understand stochastic biomass supply chain modeling from a wider perspective before venturing into this waste-to-wealth business.

Acknowledgements The author would like to acknowledge the financial support from the Ministry of Higher Education under the Fundamental Research Grant Scheme [grant number: FRGS/1/2020/TK0/SWIN/03/3] and Swinburne University of Technology Sarawak *via* Research Supervision Grant [grant number: 2-5545 RSG].

References

Abas, R., Kamarudin, M. F., Nordin, A. B. A., & Simeh, M. A. (2011). A study on the Malaysian oil palm biomass sector—supply and perception of palm oil millers. *Oil Palm Industry Economic Journal, 11*, 28–41.

Aboytes-Ojeda, M., Castillo-Villar, K. K., & Eksioglu, S. D. (2019). Modeling and optimization of biomass quality variability for decision support systems in biomass supply chains. *Annals of Operations Research, 314*, 319–346.

Agensi Inovasi Malaysia. National Biomass Strategy 2020: New wealth creation for Malaysia's biomass industry 2013.

Aghabararnejad, M, Patience, G. S., & Chaouki, J. (2015). Techno-economic comparison of a 7-mwth biomass chemical looping gasification unit with conventional systems. *Chemical Engineering & Technology, 38*(5), 867–878.

Ahmad, R., Hamidin, N., Md Ali, U., & Abidin, C. Z. A. (2014). Characterization of bio-oil from palm kernel shell pyrolysis. *Journal of Mechanical Engineering and Sciences., 7*, 1134–1140.

Ahmad, S., Ab Kadir, M. Z. A., & Shafie, A. S. (2011). Current perspective of the renewable energy development in Malaysia. *Renewable and Sustainable Energy Reviews, 15*(2), 897–904.

AlNouss, A., Parthasarathy, P., Shahbaz, M., Al-Ansari, T., Mackey, H., & McKay, G. (2020). Techno-economic and sensitivity analysis of coconut coir pith-biomass gasification using ASPEN PLUS. *Applied Energy, 261*, 114350.

Andiappan, V., Tan, R. R., Aviso, K. B., & Ng, D. K. S. (2015). Synthesis and optimisation of biomass-based tri-generation systems with reliability aspects. *Energy, 89*, 803–818.

Aziz, M.A., Uemura, Y., & Sabil, K.M. (2011). Characterization of oil palm biomass as feed for torrefaction process. In *2011 National Postgraduate Conference* (pp. 1–6).

Baral, N. R., Davis, R., & Bradley, T. H. (2019). Supply and value chain analysis of mixed biomass feedstock supply system for lignocellulosic sugar production. *Biofuel Bioprod Bior, 13*(3), 635–659.

Barde, S. R. A., Yacout, S., & Shin, H. (2019). Optimal preventive maintenance policy based on reinforcement learning of a fleet of military trucks. *Journal of Intelligent Manufacturing, 30*, 147–161.

Bussemaker, M. J., Day, K., Drage, G., & Cecelja, F. (2017). Supply chain optimisation for an ultrasound-organosolv lignocellulosic biorefinery: Impact of technology choices. *Waste Biomass Valori, 8*, 2247–2261.

Chou, J.-S., & Ongkowijoyo, C. S. (2014). Risk-based group decision making regarding renewable energy schemes using a stochastic graphical matrix model. *Automation in Construction, 37*, 98–109.

Department of Statistic Malaysia. (2020). Selected agricultural indicators, Malaysia, 2020. DOSM. [Online] Available at www.dosm.gov.my. Accessed March 1, 2021.

DQS Certification. MSPO Certification Summary Report 2018. https://www.dqs.com.my/wp-content/uploads/2019/07/SOPB_Galasah-Palm-Oil-Mill_2018-AA3_Report.pdf. Accessed April 7, 2021.

Energy Information Administration (EIA). 2021. Country Analysis Executive Summary: Malaysia. [Online] Available at https://www.eia.gov/international/content/analysis/countries_long/Malaysia/malaysia.pdf. Accessed March 26, 2021.

Energy Technology Systems Analysis Programme (ETSAP). 2010. Combined heat and power. Available at www.etsap.org. Accessed March 6, 2021.

Garba, K., Mohd Din, A.T., & Hameed, B. (2017). Pyrolysis of oil palm mesocarp fiber and palm frond in a slow-heating fixed-bed reactor: A comparative study. *Bioresource Technology, 241*.

GE. (2018). Jenbacher gas engines. [Online] Available at http://kts-eng.com/assets/files/J208.pdf. Accessed February 1, 2021.

Gu, H., & Bergman, R. (2015). Life-cycle GHG emissions of electricity from syngas produced by pyrolyzing woody biomass. In *Proceedings of the 58th International Convention of Society of Wood Science and Technology* June 7–12, 2015. Jackson Lake Lodge, Grand Teton National Park, Wyoming, U.S.A.

Han, J., Liang, Y., Hu, J., Qin, L., Street, J., Lu, Y., & Yu, F. (2017). Modeling downdraft biomass gasification process by restricting chemical reaction equilibrium with Aspen Plus. *Energy Conversion and Management, 153*, 641–648.

Hong, B. H., How, B. S., & Lam, H. L. (2016). Overview of sustainable biomass supply chain: From concept to modelling. *Clean Technologies and Environmental Policy, 18*, 2173–2194.

How, B. S., & Lam, H. L. (2018). Sustainability evaluation for biomass supply chain synthesis: Novel principal component analysis (PCA) aided optimisation approach. *Journal of Cleaner Production, 189*, 941–961.

How, B. S., & Lam, H. L. (2019). PCA method for debottlenecking of sustainability performance in integrated biomass supply chain. *Process Integration and Optimization for Sustainability, 3*, 43–64.

How, B. S., Ngan, S. L., Hong, B. H., Lam, H. L., Ng, W. P. Q., Yusup, S., et al. (2019). An outlook of Malaysian biomass industry commercialisation: Perspectives and challenges. *Renewable and Sustainable Energy Reviews, 113*, 109277.

How B.S., Tan K.Y., & Lam H.L. (2016). Transportation decision tool for optimisation of integrated biomass flow with vehicle capacity constraints. *Journal of Cleaner Production, 136* (Part B), 197–223.

International Energy Agency (IEA). (2019). Renewable energy. [Online] Available at https://www.iea.org/policiesandmeasures/renewableenergy/?country=Malaysia. Accessed March 20, 2021.

International Finance Corporation (IFC). (2017). *Converting biomass to energy: A guide developers and investors*. International Finance Corporation.

Index Mundi. (2018). Coal, Australian thermal coal monthly price—Malaysian Ringgit per Metric Ton. [Online] Available at www.indexmundi.com. Accessed January 10, 2021.

International Renewable Energy Agency (IRENA). (2017). Renewable energy outlook: Thailand, international renewable energy agency, Abu Dhabi. [Online] Available at www.irena.org. Accessed January 3, 2021.

Khatiwada, D., Leduc, S., Silveira, S., & McCallum, I. (2016). Optimizing ethanol and bioelectricity production in sugarcane biorefineries in Brazil. *Renewable Energy, 85*, 371–386.

Kieffer, M., Brown, T., & Brown, R. C. (2016). Flex fuel polygeneration: Integrating renewable natural gas into fischer-tropsch synthesis. *Applied Energy, 170*, 208–218.

Kovařík, P. (2017). *Drying of biomass with high water content (Master Thesis).* Czech Technical University, Prague.

Kristianto, Y., & Zhu, L. D. (2017). Techno-economic optimization of ethanol synthesis from rice-straw supply chain. *Energy, 141*, 2164–2176.

Lam, H. L., Ng, W. P. Q., Ng, R. T. L., Ng, E. H., Aziz, M. K. A., & Ng, D. K. S. (2013). Green strategy for sustainable waste-to-energy supply chain. *Energy, 57*, 4–16.

Lembaga Hasil Dalam Negeri Malaysia (LHDN). (2018). Investment tax allowance. LHDN. [Online] Available at www.hasil.gov.my. Accessed February 1, 2021.

Lim, C. H., How, B. S., Ng, W. P. Q., & Lam, H. L. (2019). Debottlenecking of biomass element deficiency in a multiperiod supply chain system via element targeting approach. *Journal of Cleaner Production, 230*, 751–766.

Lin, J. F., Gaustad, G., & Trabold, T. A. (2013). Profit and policy implications of producing biodiesel-ethanol-diesel fuel blends to specifications. *Applied Energy, 104*, 936–944.

Lo, S. L. Y., How, B. S., Leong, W. D., Teng, S. Y., Rhamdhani, M. A., & Sunarso, J. (2021). Techno-economic analysis for biomass supply chain: A state-of-the-art review. *Renewable and Sustainable Energy Reviews, 135*, 110164.

Lo, S. L. Y., Choo, J. J. L., Kong, K. G. H., How, B. S., Lam, H. L., Ngan, S. L., Lim, C. H., & Sunarso, J. (2020). Uncertainty study of empty fruit bunches-based bioethanol supply chain. *Chemical Engineering Transactions, 81*, 601–606.

Mahlia, T. M. I., Abdulmuin, M. Z., Alamsyah, T. M. I., & Mukhlishien, D. (2001). An alternative energy source from palm wastes industry for Malaysia and Indonesia. *Energy Conversion and Management, 42*, 2109–2118.

Malaysian Palm Oil Board (MPOB). (2018). Production of crude palm oil for the month of December 2017. [Online] Available at bepi.mpob.gov.my. Accessed January 10, 2021.

MarketsandMarkets Research Private Ltd. (2021). Syngas & derivatives market by production technology, gasifier type, feedstock (coal, natural gas, petroleum byproducts, biomass/waste), application (chemicals, fuel, and electricity), and region—global forecast to 2025. [Online] Available at https://www.marketsandmarkets.com/Market-Reports/syngas-market-1178.html. Accessed March 10, 2021.

Martinkus, N., Latta, G., Brandt, K., & Wolcott, M. (2018). A multi-criteria decision analysis approach to facility siting in a wood-based depot-and-biorefinery supply chain model. *Front Energy Res, 6*, 124.

Mohamad, A. S., Loh, S. K., Nasrin, A. B., & Choo, Y. M. (2011). Production and characterization of bio-char from the pyrolysis of empty fruit bunches. *American Journal of Applied Sciences, 8*(10), 984–988.

Mohd N.A. (2017). *Conventional and microwave pyrolysis of empty fruit bunch and rice husk pellets*. Ph.D. Thesis, The University of Sheffield, Sheffield, U.K.

Monthly Production of Oil Palm Products Summary 2019 & 2020. Malaysia Palm Oil Board. (2020). http://bepi.mpob.gov.my/index.php/en/production/production-2020/production-of-oil-palm-products-2020.html. Accessed March 26, 2021

MacDonald, M. (2012). Potential cost reductions in CCS in the power sector. Department of Energy and Climate Change, Mott Mac Donald, London, U.K.

Ngan, S. L., How, B. S., Teng, S. Y., Leong, W. D., Loy, A. C. M., Yatim, P., Promentilla, M. A. B., & Lam, H. L. (2020). A hybrid approach to prioritize risk mitigation strategies for biomass polygeneration systems. *Renewable and Sustainable Energy Reviews, 121*, 109679.

Ooi, R. E. H., Foo, D. C. Y., & Tan, R. R. (2014). Targeting for carbon sequestration retrofit planning in the power generation sector for multi-period problems. *Applied Energy, 113*, 477–487.

Othman N.F., Boosroh M.H., Hassan H., Mohan C., & Aziz W.A.L.W.A. (2012). Gasification of triple fuel blends using pilot scale fluidized bed gasification plant. In *Proceeding of International Conference on Science, Technology & Social Sciences (ICSTSS)* (pp. 471–477).

PNAS. (2018). SI appendix general information and assumption. [Online] Pnas.org. Available at http://www.pnas.org. Accessed March 6, 2021.

Reduan, H. (2017). FGv unit to export 60,000 tonnes of palm kernel shells to Japan. *New Straits Time*.

RinggitPlus. (2018). Petrol price Malaysia live updates (RON95, RON97 & Diesel). [Online] Available at ringgitplus.com. Accessed January 10, 2021.

Rogers, J. G., & Brammer, J. G. (2012). Estimation of the production cost of fast pyrolysis bio-oil. *Biomass and Bioenergy, 36*, 208–217.

Rubin, E. S., Davison, J. E., & Herzog, H. J. (2015). The cost of CO_2 capture and storage. *International Journal of Greenhouse Gas Control, 40*, 378–400.

Shou, Z., Di, X., Ye, J., Zhu, H., Zhang, H., & Hampshire, R. (2020). Optimal passenger-seeking policies on E-hailing platforms using Markov decision process and imitation learning. *Transportation Research Part c: Emerging Technologies, 111*, 91–113.

Sohni, S., Norulaini, N. A. N., Hashim, R., Khan, S. B., Fadhullah, W., & Mohd Omar, A. K. (2018). Physicochemical characterization of Malaysian crop and agro-industrial biomass residues as renewable energy resources. *Industrial Crops and Products, 111*, 642–650.

Song, G., Feng, F., Xiao, J., & Shen, L. (2013). Technical assessment of synthetic natural gas (SNG) production from agriculture residuals. *Journal of Thermal Science, 22*, 359–365.

Spath, P., Aden, A., Eggeman, T., Ringer, M., Wallace, B., & Jechura, J. (2005). Biomass to hydrogen production detailed design and economics utilizing the battelle columbus laboratory indirectly-heated gasifier. [Online] Available at https://www.nrel.gov/docs/fy05osti/37408.pdf. Accessed November 30, 2020.

StatEase. ANOVA Output. StatEase; 2020.

Susanto, H., Suria, T., & Pranolo, S. H. (2018). Economic analysis of biomass gasification for generating electricity in rural areas in Indonesia. *IOP Conference Series: Materials Science and Engineering , 334*, 012012.

Sustainable Energy Development Authority Malaysia (SEDA). (2018). FiT rates for biomass (solid waste) (16 years from FiT commencement date). Available at www.seda.gov.my/. Accessed March 6, 2021.

Tanzer, S. E., Posada, J., Geraedts, S., & Ramírez, A. (2019). Lignocellulosic marine biofuel: Technoeconomic and environmental assessment for production in Brazil and Sweden. *Journal of Cleaner Production, 239*, 117845.

Teng, S. Y., How, B. S., Leong, W. D., Teoh, J. H., Cheah, A. C. S., Motavasel, Z., & Lam, H. L. (2019). Principal component analysis-aided statistical process optimisation (PASPO) for process improvement in industrial refineries. *Journal of Cleaner Production, 225*, 359–375.

Teng, S. Y., How, B. S., Leong, W. D., Teoh, J. H., & Lam, H. L. (2020). Bottleneck Tree Analysis (BOTA) with green and lean index for process capacity debottlenecking in industrial refineries. *Chemical Engineering Science, 214*, 115429.

The Oil Palm. History of the Industry. (2021). [Online] Available at http://theoilpalm.org/about/#History_and_Origin. Accessed March 25, 2021

Trading Economics. (2018). Commodity crude oil. [Online] Available at www.tradingeconomics.com. Accessed January 10, 2021.

Trading Economics. (2020). Malaysia gasoline price. [Online] Available at https://tradingeconomics.com/malaysia/gasoline-prices. Accessed March 20, 2021.

Wang, S., Levin, M. W., & Caverly, R. J. (2021). Optimal parking management of connected autonomous vehicles: A control-theoretic approach. *Transportation Research Part c: Emerging Technologies, 124*, 102924.

Wright, M. M., Satrio, J. A., Brown, R. C., Daugaard, D. E., & Hsu, D. D. (2010). *Techno-economic analysis of biomass fast pyrolysis to transportation fuels*. National Renewable Energy Laboratory (NREL), Colorado, U.S.A.

Yatim, P., Mamat, M. N., Mohamad-Zailani, S. H., & Ramlee, S. (2016). Energy policy shifts towards sustainable energy future for Malaysia. *Clean Technology Environment Policy, 18*, 1685–1695.

Yeo, J. Y. J., How, B. S., Teng, S. Y., Leong, W. D., Ng, W. P. Q., Lim, C. H., Ngan, S. L., Sunarso, J., & Lam, H. L. (2020). Synthesis of sustainable circular economy in palm oil industry using graph-theoretic method. *Sustainability, 12*, 8081.

Yoo, H.-M., Park, S.-W., Seo, Y.-C., & Kim, K.-H. (2019). Applicability assessment of empty fruit bunches from palm oil mills for use as bio-solid refuse fuels. *Journal of Environmental Management, 234*, 1–7.

Zakaria, A., Ismail, F. B., Hossain Lipu, M. S., & Hannan, M. A. (2020). Uncertainty models for stochastic optimization in renewable energy applications. *Renewable Energy, 145*, 1543–1571.

Zuldian, P., Fukuda, S., & Bustan, M. (2017). Economic analysis of coal gasification plant for electricity and thermal energy supplies in Indonesia. *Journal of Clean Energy Technologies, 5*, 193–198.

Process Integration Tools for Optimal Allocation of Palm Biomass

Dominic C. Y. Foo

Abstract Biomass has attracted good attention in the commercial and scholarly world in past decade, due to its role as an important renewable resource. Various systematic methods and tools have been developed in recent years to explore the optimal use of these biomass resources. In this chapter, two important *process integration* tools, i.e. graphical pinch diagram and mathematical programming model, are presented to determine the optimal allocation of palm biomass for bioenergy production. The graphical pinch diagram is useful for macro-level planning of palm biomass allocation, while the mathematical programming model provides detailed planning of the biomass allocation. Several examples involving palm biomass are used to elucidate the usage of these process integration tools.

Keywords Bioenergy · Pinch analysis · Superstructure · Mathematical programming · Power generation · Optimisation

1 Introduction

As reported by the IEA (International Energy Agency, 2021), the renewable energy use increased by 3% in year 2020, while demand for all other fuels declined. The main reason for this increase was due to the strong growth in electricity generation from renewable sources. It has been reported that the share of renewables in global electricity generation has increased from 27% in year 2019 to 29% in 2020 (International Energy Agency, 2021). Along with wind, solar and hydropower, biomass is among the most important renewable energy sources for sustainable electricity generation.

In the past decade, various *process system engineering* methods have been developed for optimal planning of biomass resources. For instance, in the work of Ng

D. C. Y. Foo (✉)
Department of Chemical and Environmental Engineering/Centre of Excellence for Green Technologies, University of Nottingham Malaysia, Broga Road, 43500 Semenyih, Selangors, Malaysia
e-mail: dominic.foo@nottingham.edu.my

D. C. Y. Foo et al. (eds.), *Sustainable Technologies for the Oil Palm Industry*,
https://doi.org/10.1007/978-981-19-4847-3_7

et al. (2013) and Ng and Ng (2013), mathematical programming was used to synthesise integrated palm oil processing complex that explores the optimal use of palm biomass resources for value-added product and energy generation. Another work based on mathematical programming was reported by Foo et al. (2013) for palm-based bioenergy supply chain. The work was later extended to determine minimum cost and CO_2 emission during biomass allocation with road transportation (Foo, 2019). Apart from mathematical tool, graphical tools based of ternary diagram (Tay et al., 2011) and Piper diagram (Pang et al., 2021) have also been developed. In the work of Tey et al., (2021), Douglas's hierarchical approach was extended for the synthesis of palm-based biorefinery. All these process system engineering methods aim to explore the optimal use of biomass resources.

In this chapter, two important process integration tools are presented for the optimal allocation of palm biomass for bioenergy production. A recently developed graphical pinch diagram is useful in macro-level planning of palm biomass allocation, in order to fulfil the targeted power output of biomass power plants. A mathematical model on the other hand is useful in identifying detailed allocation of palm biomass to the power plants and has the advantage for solving large-scale problem and cost optimisation. Three examples of palm biomass allocation for energy generation are demonstrated for the use of these process integration tools.

2 Conceptual Understanding

The allocation of biomass to the power plant can be represented by a source–demand representation as in Fig. 1. As shown, different types of biomass (e.g. palm or paddy biomass) are available for use; they are termed as *sources*. On the other hand, the power plants that require specific amount of biomass are termed as *demand*. The problem is to determine the optimal allocation among the sources and demand, so that the latter can fulfil its power generation requirement.

In order to fulfil the power generation target, the biomass requirement by each power plant needs to be calculated, based on the model reported in Foo et al., (2013) and Foo (2022). Equation (1) determines the steam requirement of turbine (STM_j) for power plant j with output P_j.

$$STM_j = \frac{P_j}{\eta_{Turb} \hat{H}_{\text{Turb}}} \tag{1}$$

where η_{Turb} and $\hat{H}_{\text{Turb}}$ are efficiency (%) and enthalpy (kJ/kg) for turbine calculation.

The biomass requirement of the boiler (D_j) can then be calculated using Eq. (2).

$$D_j = \frac{STM_j \hat{H}_{\text{Boil}}}{CV_{\text{Biom}} SC_{\text{Biom}} \eta_{\text{Boil}}} \tag{2}$$

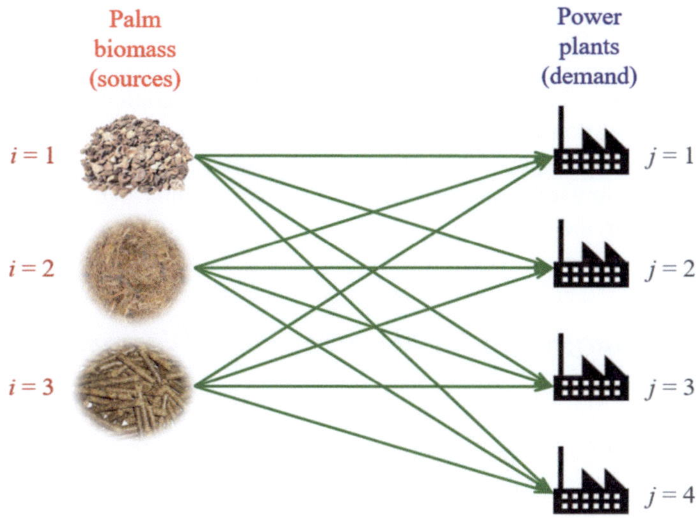

Fig. 1 Source–demand representation of biomass allocation problem

where η_{Boil} and $\hat{H}_{\mathrm{Boil}}$ are efficiency (%) and enthalpy (kJ/kg) for boiler calculation, and CV_{Biom} and SC_{Biom} are average values of calorific value (kJ/kg) and solid content (wt%) of biomass types that are fed to the boiler. Note that SC_{Biom} is the inverse of moisture content (MC_{Biom}, wt%) that is commonly used in the biomass sector.

Equations 1 and 2 may be combined to form Eqs. 3a and 3b.

$$P_j = C_j D_j \tag{3a}$$

$$C_j = \frac{\eta_{\mathrm{Turb}} \hat{H}_{\mathrm{Turb}} CV_{\mathrm{Biom}} SC_{\mathrm{Biom}} \eta_{\mathrm{Boil}}}{\hat{H}_{\mathrm{Boil}}} \tag{3b}$$

where C_j is characterised as the power generation factor for power plant j.

Similarly, Eqs. 4a and 4b may be used to calculate the power output (P_i) of biomass i with known calorific value (CV_i, kJ/kg) and solid content (SC_i, wt%).

$$P_i = C_i S_i \tag{4a}$$

$$C_i = \frac{\eta_{\mathrm{Turb}} \hat{H}_{\mathrm{Turb}} CV_i SC_i \eta_{\mathrm{Boil}}}{\hat{H}_{\mathrm{Boil}}} \tag{4b}$$

where C_i is power generation factor of biomass i. Note also that solid content of biomass can be calculated from its moisture content (MC_i, wt%).

3 Process Integration Methods

In this section, two simple methods based on process integration are presented for the optimum allocation of biomass resources. A graphical tool is first presented to determine high level targets for the biomass allocation problem, while the mathematical programming technique is next presented to determine detailed allocation of biomass sources to their demand.

3.1 Overall Targeting with Bioenergy Pinch Diagram (BEPD)

A novel graphical tool is presented here, known as the *bioenergy pinch diagram* (BEPD). Step for plotting the BEPD is given as follows (Foo, 2022).

1. Individual power plants are arranged according to the descending order of power generation factor C_j calculated using Eq. 3a, b.
2. Segment of the power plants is next plotted to form a *demand composite curve* on a power versus biomass capacity diagram (Fig. 2). Horizontal distance of the demand composite curve represents the maximum biomass capacity that can be handled by these plants ($\Sigma_j D_j$), while its vertical distance indicates their total power output ($\Sigma_j P_j$). Note that the slope of the individual segments corresponds to their power generation factor C_j.
3. A *source composite curve* is next plotted by combining one or more biomass sources on the same power versus biomass capacity diagram, which forms the BEPD. Note that the biomass sources are also arranged in descending order of their power generation factor C_i (determined using Eq. 4a,b). The horizontal distance of the demand composite curve represents the total available biomass ($\Sigma_i S_i$) for use, while its vertical distance indicates their maximum potential power output ($\Sigma_i P_i$). For such arrangement, the slope of the individual segments

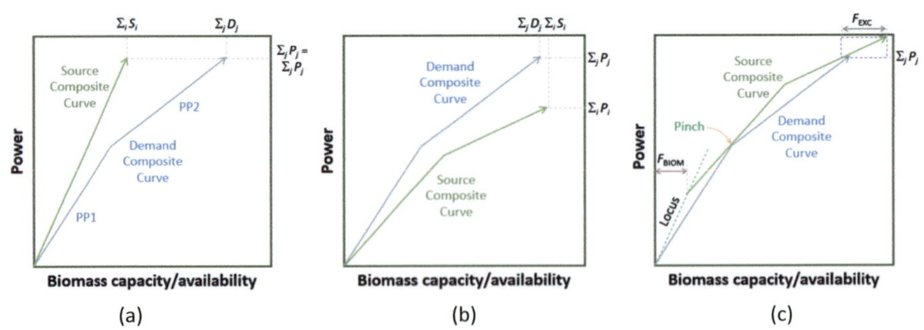

Fig. 2 **a** Feasible BEPD; **b** infeasible BEPD; **c** feasible BEPD with minimum biomass of higher power generation factor

corresponds to their power generation factor C_i. The BEPD is feasible when the source composite curve is located to the left of the demand composite curve and has at least the same vertical distance as the latter, such as that shown in Fig. 2a. For this case, the biomass will generate power that matches the total output of the power plants ($\Sigma_i P_i = \Sigma_j P_j$) and yet lower than their maximum handling capacity (i.e. $\Sigma_i S_i \leq \Sigma_j D_j$).

4. In cases where the source composite curve is found on the right and/or below the sink composite curve (such as that in Fig. 2b), the BEPD is considered infeasible. The feasibility can be restored by adding additional biomass with higher power generation factor; the latter is characterised by its locus of steeper slope. As shown in Fig. 2c, the source composite curve is then slid along this locus until it stay completely above and to the left of the demand composite curve and touches the former at the *pinch*. As biomass with higher power generation factor is usually more expensive, their usage is to be minimised. The exact amount of this additional biomass is indicated by the opening on the left of the BEPD (F_{BIOM}). On the other hand, dashed box at the right top corner of the BEPD indicates the excess amount of biomass beyond the power generation targets. The horizontal distance of the dashed box indicates its excess amount (F_{EXC}), which can be utilised for other commercial purposes.

The BEPD is an intuitive tool that is useful for the palm industrial practitioners for macro-level planning of biomass allocation. Its graphical interface fascinates discussion among team members of the biomass supply chain planners.

3.2 Optimal Allocation with Superstructural Model

The BEPD is a useful tool for overall planning of biomass allocation. For detailed allocation of biomass sources to the individual power plants and for large-scale case study (e.g. more than two power plants), the following superstructural model of Foo (2022) is useful for detailed planning.

Equation 5 described that the net output to be generated by power plant j is to be contributed by total of biomass source i ($f_{i,j}$), with individual power generation potential C_i. Each power plant j can only handle a maximum capacity of biomass (D_j), as described by Eq. 6. In Eq. 7, the unutilised biomass i (u_i) is given by the difference between its availability (S_i) and its total allocation to the power plants. All variables in this model must take non-negative values, as indicated by Eq. 8.

$$\sum_i f_{i,j} C_i \geq P_j \quad \forall j \tag{5}$$

$$\sum_i f_{i,j} \leq D_j \quad \forall j \tag{6}$$

$$u_i = S_i - \sum_j f_{i,j} \quad \forall i \tag{7}$$

$$f_{i,j} \geq 0;\ u_i \geq 0 \quad \forall i \forall j \tag{8}$$

The objective of the model can be set to minimise a specific type of biomass resource, due to its scarcity or higher cost; this is given in Eq. 9. Additionally, the main advantage of the superstructural model is its capability to minimise the overall cost of the biomass use; such objective is given in Eq. 10.

$$\min = \sum_i \sum_j f_{i,j} \tag{9}$$

$$\min = \sum_i \sum_j CT_i f_{i,j} \tag{10}$$

where CT_i is the unit cost of biomass i.

Since the above model is a linear programme (LP), any commercial software (e.g. MS Excel and MATLAB) may be used to obtain a global optimum solution, if it exists.

Several examples are used to illustrate the two process integration tools.

4 Biomass Allocation Examples

Three examples are used here to elucidate the two process integration methods. In these examples, various palm biomass resources are used for power generation. Important parameters for power generation of these examples are summarised in Table 1. The average calorific value of biomass used in these plants (CV_{Biom}) is estimated based on the calorific value of palm biomass, while the average moisture content (MC_{Biom}) of each power plants is given in the respective examples.

Table 1 Important parameters for power generation for both examples

Parameters	Values
Turbine	
Turbine efficiency, η_{Turb}	20%
Enthalpy of steam, $\hat{H}_{\text{Turb}}$	3140 kg/kg steam
Boiler	
Boiler efficiency, η_{Boil}	85%
Enthalpy of steam, $\hat{H}_{\text{Boil}}$	2669 kJ/kg steam
Average calorific value of biomass, CV_{Biom}	19,000 kJ/kg biomass

Table 2 Data for power plants in Example 1

Power plants	P_j (MW)	MC_{Biom} (%)	D_j (t/h)	C_j (MW/t/h)
PP1	14	55.8	30	0.47
PP2	26	59.0	60	0.43
Total	**40**		**90**	

Example 1 Two biomass power plants are to be built, with their power output (P_j) given in Table 2. The latter also shows the average moisture content (MC_{Biom}) that they can handle, which may be converted in to the solid content (i.e. $SC_{\text{Biom}} = 100\% - MC_{\text{Biom}}$) used in Eq. 3a, b. The biomass requirement of the individual power plants (D_j) is then calculated using Eq. 2, given as in column 4 of Table 2. Note that the power generation factor (C_j) in the last column of Table 2 can also be determined by dividing its power output with the biomass requirement (as an alternative to using Eq. 3b). Two types of palm biomass, i.e. PMF and EFB, are considered for used in these plants, with data given in Table 3. However, due to limited availability of PMF, its use is to be minimised. With the given calorific value (CV_i, kJ/kg) and moisture content (MC_i, wt%) of these biomass, their power output (P_i) can be calculated using Eq. 4 and summarised in Table 3.

Following the procedure (steps 1–3), a BEPD is plotted in Fig. 3, with power plants PP1 and PP2 form its demand composite curve, while its source composite curve consisted of EFB alone. The BEPD in Fig. 3 is infeasible, as the source composite curve is located on the right and below the demand composite curve. In other words, the power generation target of the power plants is not fulfilled.

Following step 4 of the procedure, a locus of PMF is added, with slope corresponds to its power generation factor. The source composite curve is then slid along the locus until it reaches the tip of the source composite curve and stays at its left side entirely; this forms a feasible BEPD in Fig. 4. The latter also shows the minimum PMF that should be used to restore the feasibility, i.e. 30 t/h (F_{PMF}).

We then move on to determine the exact allocation of PMF and EFB to the two power plants. The LP model is solved using MS Excel Solver, with setting shown in Fig. 5. The solution is also depicted in the source–demand diagram in Fig. 6a. As shown, the minimum PMF is determined as 30 t/h (cell C5), matching that obtained using the BEPD in Fig. 4.

Example 2 In this example, two biomass power plants (PP1 and PP2) are considered, with data given in Table 4 (Foo, 2022). Similar to Example 1, EFB of lower power generation factor (and cheaper) is prioritised for used in the biomass power plants.

Table 3 Data for biomass in Example 1

Biomass types	S_i (t/h)	CV_i (kJ/kg)	MC_i (%)	P_i (MW)	C_i (MW/t/h)
PMF	To be determined	19,000	36	To be determined	0.67
EFB	50	18,700	61	20	0.4

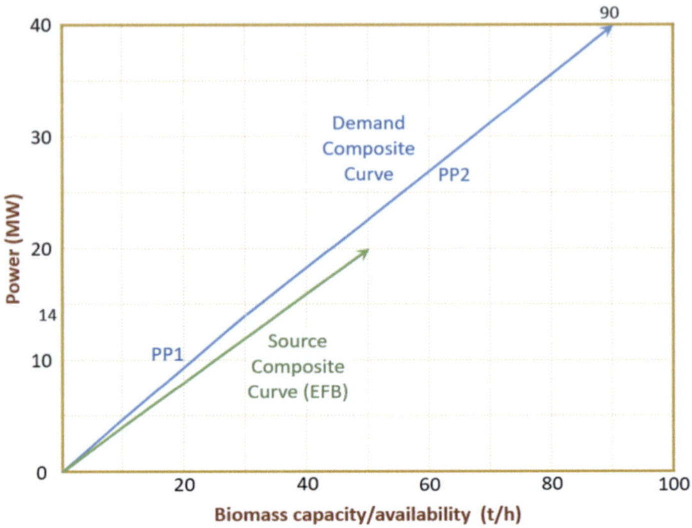

Fig. 3 Infeasible BEPD, where the source composite curve (EFB) is located on the right and below the demand composite curve

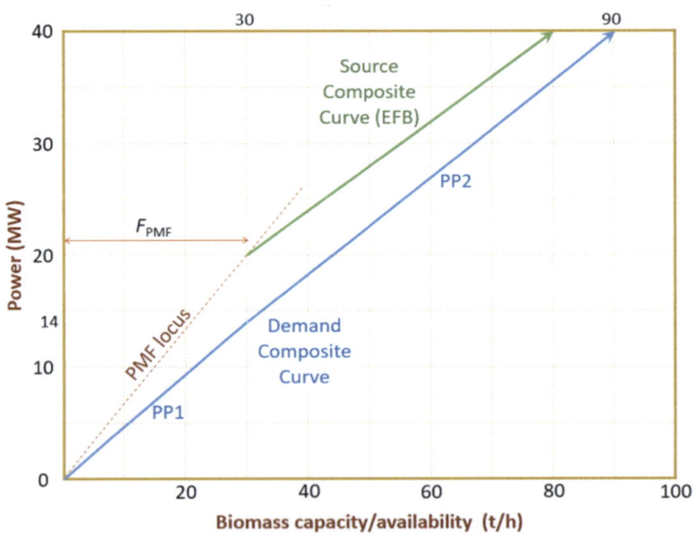

Fig. 4 A feasible BEPD with minimum usage of PMF

PKS that has higher calorific value and lower moisture content (and hence higher power generation factor and more expensive) may to be used to supplement the limited amount of EFB. Data for both biomasses are given in Table 5.

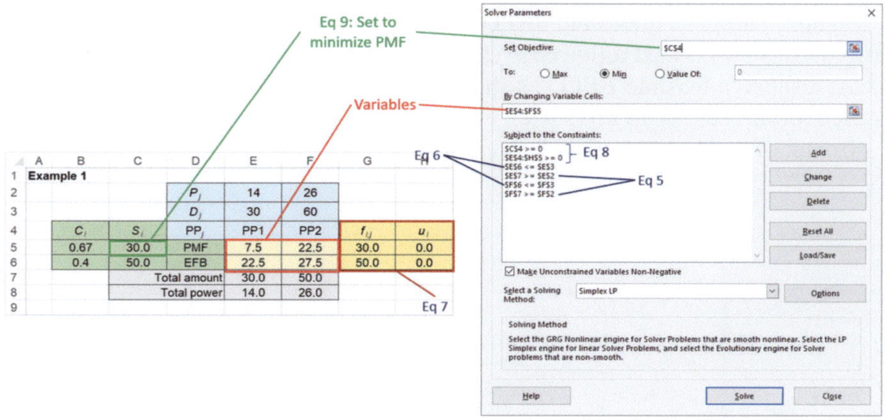

Fig. 5 LP model using MS Excel for Example 1 (all allocation in t/h; power in MW)

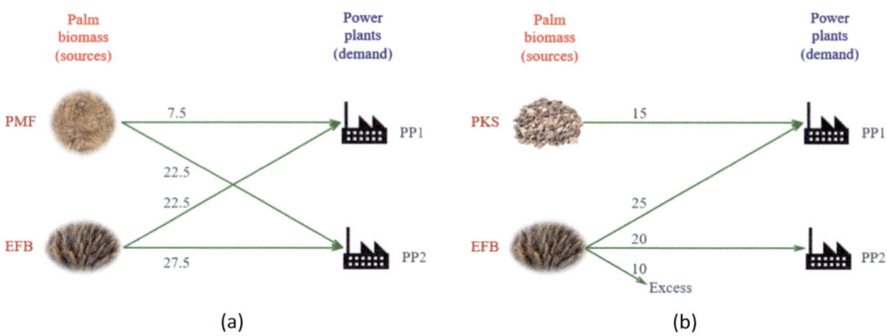

Fig. 6 Allocation of biomass resources for **a** Example 1; **b** Example 2 (all allocation in t/h)

Table 4 Data for power plants in Example 2

Power plants	P_j (MW)	MC_{Biom} (%)	D_j (t/h)	C_j (MW/t/h)
PP1	22	47.9	40	0.55
PP2	8	69.7	25	0.32
Total	**30**		**65**	

Table 5 Data for biomass in Example 2

Biomass types	S_i (t/h)	CV_i (kJ/kg)	MC_i (%)	P_i (MW)	C_i (MW/t/h)
PKS	To be determined	19,700	26	40	0.8
EFB	55	18,700	61	20	0.4

Figure 7 shows an infeasible BEPD is resulted following steps 1 – 3 of the BEPD procedure. Step 4 is next followed to slide the source composite curve along the PKS locus, resulted a feasible BEPD as shown in Fig. 8. As shown, the minimum PKS needed for this case is 15 t/h (F_{PKS}), while an excess of EFB is determined from the horizontal of the dashed box at the top right side of the BEPD, i.e. 10 t/h (F_{EXC}).

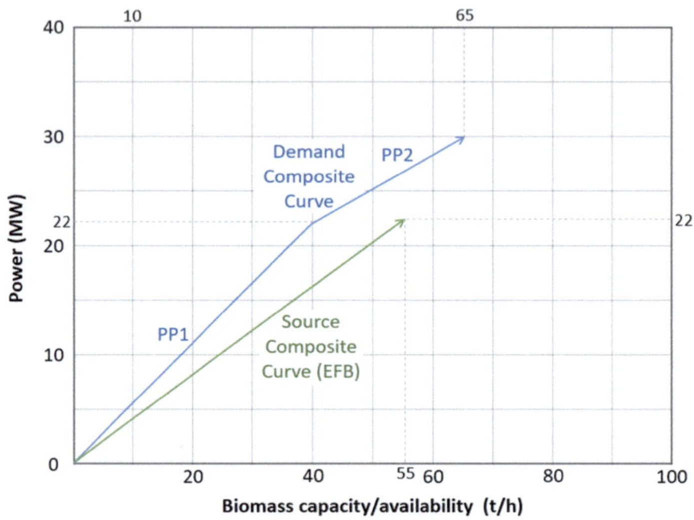

Fig. 7 Infeasible EBPD for Example 2

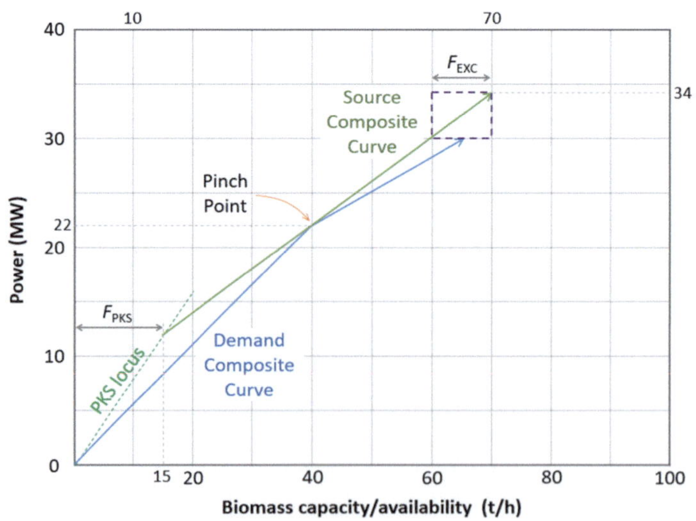

Fig. 8 Feasible EBPD for Example 2, with minimum PKS

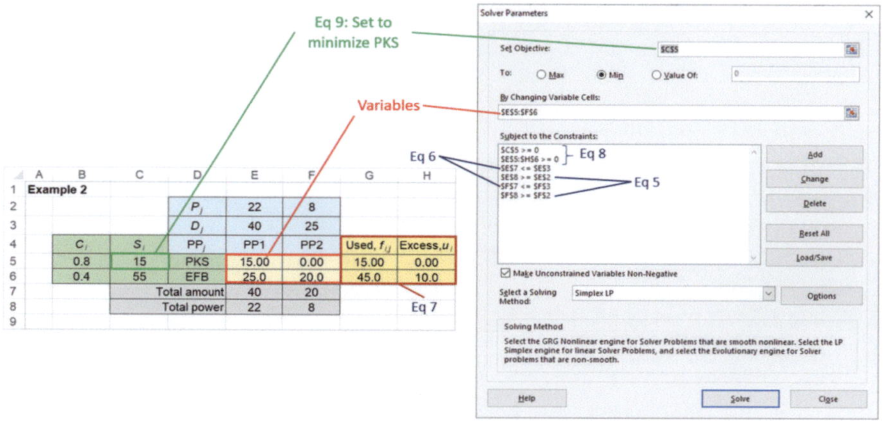

Fig. 9 LP model using MS Excel for Example 2 (all allocation in t/h; power in MW)

The LP model is next solved using MS Excel Solver, with setting shown in Fig. 9. As shown, the minimum PKS is determined as 15 t/h (cell C5), with 10 t/h of excess EFB (cell H6). The minimum PKS usage and excess EFB amount match those obtained using the BEPD (Fig. 9), with solution depicted in the source–demand diagram in Fig. 6b.

Example 3 This example considers biomass allocation for four power plants (Table 6), with four types of palm biomass being considered (Table 7). Apart from EFB, PMF and PKS, washed EFB pallet (WEP) is also used; the latter has low moisture content (9%) and hence high power generation factor (0.85 MW/t/h; see Table 7). In this example, it is desired to determine the minimum cost of allocation, while fulfilling the power generation target. The LP model is solved by minimising the total cost in Eq. (10), with setting shown in Fig. 10. With the unit cost of biomass in Table 7 (last column), the LP model determines the total cost as 1959 $/h (see MS Excel spreadsheet cell L9 in Fig. 11), with biomass allocation shown in Fig. 11a. The model determines that PKS should be used to supplement the use of PMF and EFB.

One may also explore the effect of biomass price as it is quite volatile. When the PKS price rises to 80 $/t, the LP model determines that it is cheaper to make use of WEP, with a total cost of 1986 $/t. The biomass allocation is shown in Fig. 11b. This can be seen that the LP model is useful in assisting oil palm industrial players in optimising their biomass usage for lower cost (which leads to higher profitability).

Finally, note that cost optimisation is the advantage of the LP model; it is not doable using the BEPD.

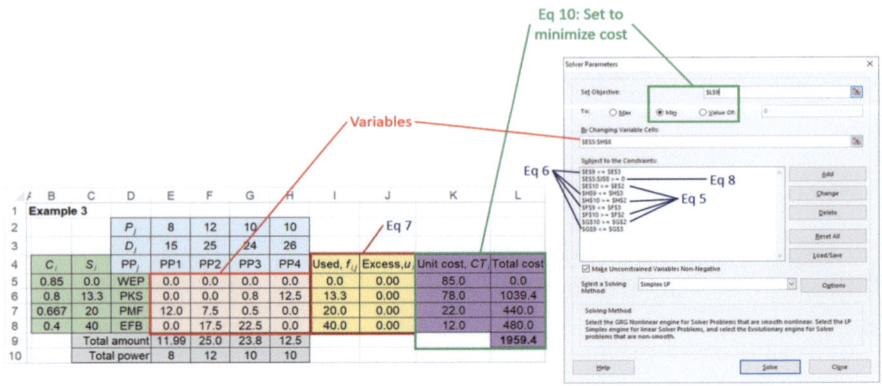

Fig. 10 Setting in MS Excel Solver for Example 3 (all allocation in t/h; unit cost in $/t; total cost in $/h)

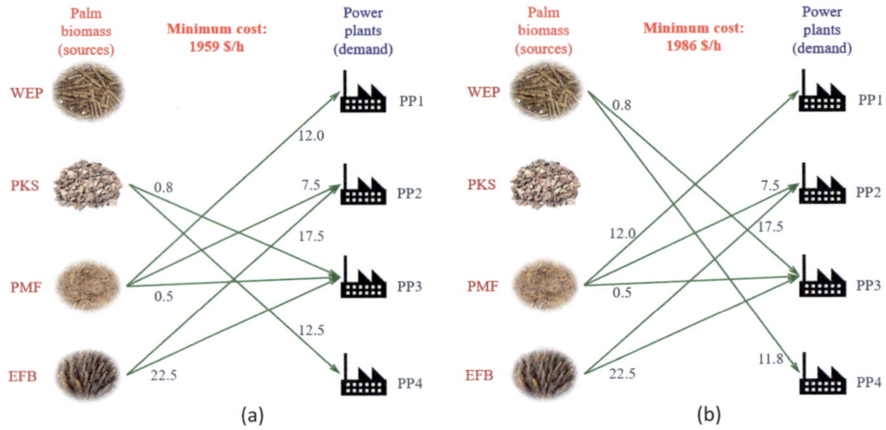

Fig. 11 Results for biomass allocation in Example 3: **a** for PKS price of $78/t; **b** for PKS price of $80/t (all allocation in t/h)

Table 6 Data for power plants in Example 3

Power plants	P_j (MW)	MC_{Biom} (%)	D_j (t/h)	C_j (MW/t/h)
PP1	8	49.0	15	0.53
PP2	12	54.0	25	0.48
PP3	10	60.0	24	0.42
PP4	10	63.2	26	0.38
Total	**40**		**90**	

Table 7 Data for biomass in Example 3

Biomass types	S_i (t/h)	CV_i (kJ/kg)	MC_i (%)	P_i (MW)	C_i (MW/t/h)	CT_i ($/t)
WEP	To be determined	17,000	9	To be determined	0.85	85.0
PKS	To be determined	19,700	26	To be determined	0.8	78.0
PMF	20	19,000	36	13.4	0.67	22.0
EFB	40	18,700	61	16	0.4	12.0

5 Conclusion

In this chapter, two process integration tools for biomass allocation are presented, i.e. bioenergy pinch diagram (BEPD) and superstructural model. The BEPD allows macro-level planning in determining the minimum biomass of higher power generation factor, by maximising the use of lower power generation biomass. The superstructural model, on the other hand, allows detailed planning and cost optimisation to be carried out. These tools complement each other in the optimum planning of biomass resources.

Acknowledgements The author gratefully acknowledged technical advice given by Albert Y. H. Law.

References

Foo, D. C. Y. (2022). A novel graphical targeting technique for optimal allocation of biomass resources. *Processes, 10*(5), 905. https://doi.org/10.3390/pr10050905

Foo, D. C. Y. (2019). A simple mathematical model for palm biomass supply chain. In D. Foo & M. Aziz (Eds.), *Green technologies for the oil palm industry* (pp. 115–130). Springer. https://doi.org/10.1007/978-981-13-2236-5_6

Foo, D. C. Y., Tan, R. R., Lam, H. L., Abdul Aziz, M. K., & Klemeš, J. J. (2013). Robust models for the synthesis of flexible palm oil-based regional bioenergy supply chain. *Energy, 55*, 68–73. https://doi.org/10.1016/j.energy.2013.01.045

International Energy Agency, I. (2021). Global energy review 2021 [WWW Document]. Global Energy Review 2021. https://www.iea.org/reports/global-energy-review-2021

Ng, T. L., & Ng, D. K. S. (2013). Systematic approach for synthesis of integrated palm oil processing complex. Part 1: Single owner. *Industrial & Engineering Chemistry Research, 52*, 10206–10220. https://doi.org/10.1021/ie302926q

Ng, T. L., Ng, D. K., & Tan, R. (2013). Systematic approach for synthesis of integrated palm oil processing complex. Part 2: Multiple owners. *Industrial & Engineering Chemistry Research, 52*, 10221–10235. https://doi.org/10.1021/ie400846g

Pang, Y. X., Foo, D. C. Y., Yan, Y., Sharmin, N., Lester, E., Wu, T., & Pang, C. H. (2021). Analysis of environmental impacts and energy derivation potential of biomass pyrolysis via Piper diagram. *Journal of Analytical and Applied Pyrolysis, 154*, 104995. https://doi.org/10.1016/J.JAAP.2020.104995

Tay, D. H. S., Ng, D. K. S., Kheireddine, H., & El-Halwagi, M. M. (2011). Synthesis of an integrated biorefinery via the C-H–O ternary diagram. *Clean Technologies and Environmental Policy, 13*, 567–579. https://doi.org/10.1007/s10098-011-0354-4

Tey, S.-Y., Wong, S. S., Lam, J. A., Ong, N. Q. X., Foo, D. C. Y., & Ng, D. K. S. (2021). Extended hierarchical decomposition approach for the synthesis of biorefinery processes. *Chemical Engineering Research and Design, 166*. https://doi.org/10.1016/j.cherd.2020.11.015

Innovations to a Palm Biomass-Fueled Power Plant

Shyam Lakshmanan, Yen Li Yung, Kalaiselvan Palanisamy, and How Kee Ling

Abstract The drive for sustainability has inspired palm oil refineries to use palm empty fruit bunch (EFB) fiber and palm kernel shell (PKS) as biomass fuels. Their usage, however, poses various challenges such as entanglement of fibers and blockage of fuel flow, as well as the tendency of PKS to form clinker in furnace. This chapter details some changes made for a water tube boiler to improve its fuel handling and combustion efficiency, which allowed handling of fiber with higher moisture content. It also shows how selection and location of multiple soot blowing systems have helped ash removal from the boiler tubes. Ash removal results in longer uninterrupted operation of this boiler, which is in excess of 60 days. The fuel feeding and furnace grate design, along with the heat recovery system has allowed higher EFB combustion at 95%; this has helped the boiler to achieve 30% lower fuel consumption. While many biomass boilers require an addition of 20–25% PKS, this new boiler can operate with less than 5% PKS. The new boiler is also able to handle EFB fiber with 55% moisture content due to the in-built drying zone within the furnace. The usage of heat pipe exchangers has improved heat recovery from the flue gas, resulting in increased boiler feed water temperature of 260° C. The chapter also shows how special surface coatings have been applied at various sections of the boiler to prolong its tube life.

Keywords Biomass boiler · Heat pipe exchangers · Palm kernel shell · Reciprocating grate · Soot blasters · Thermal spray coating

1 Introduction

IOI Bio-Energy Sdn Bhd operates biomass-fired boilers to generate power and steam. It is one of the few cogeneration power plants that uses palm biomass in Malaysia. The 20 acre site in Batu Sapi, Sandakan, and Sabah (Fig. 1) contains water catchment ponds, a raw water treatment plant, two biomass fueled boilers, a turbo-alternator capable of generating 15 MW, and other auxiliary plant and equipment. The plant

S. Lakshmanan · Y. L. Yung (✉) · K. Palanisamy · H. K. Ling
IOI Bio-Energy Sdn Bhd, Km12, Sungai Mowtas, Jalan Jaya Chip, Off Jalan Batu Sapi, 90000 Sandakan, Sabah, Malaysia
e-mail: yung@ioigroup.com

D. C. Y. Foo et al. (eds.), *Sustainable Technologies for the Oil Palm Industry*,
https://doi.org/10.1007/978-981-19-4847-3_8

obtains its water from the neighboring hills and processes this raw water to be suitable for industrial application. IOI Bio-Energy (IOIBE) treats raw water and supplies processed water, steam, and electrical energy to the IOIBE complex and to the neighboring refinery, IOI Edible Oils (IOIEO) complex. The boiler and equipment have experienced a fair share of problems and challenges which have since been rectified by the plant personnel. The company has implemented many initiatives over the years to show its commitment to operate in a sustainable manner.

The facilities available on the biomass-fueled power plant complex are given in Table 1.

Fig. 1 Aerial view of IOI Bio-Energy site with water catchment ponds in the foreground

Table 1 Facilities on-site (t/h = tons per hour; MWh/y = MegaWatt h per year)

No	Plants	Capacity
1	Enco biomass boiler	25 t/h
2	Water tube boiler	60 t/h
3	Raw water treatment	120 t/h
4	Power generation	60,000 MWh/y
5	Water catchment pond	160,000 m^3
6	Fuel storage (fiber)	30,000 t

2 Reducing Greenhouse Gas Emission

When the complex commenced operations in 1997, there were two fuel oil-fired boilers that were operated to supply steam to the IOIEO complex. In year 2004, the company installed and commissioned the Enco biomass boiler (http://enco.com.my) which demonstrates its commitments toward sustainability development. The Enco biomass boiler uses solid waste such as *empty fruit bunch* (EFB) fiber and *palm kernel shell* (PKS) as fuel. EFB and PKS are fed at a ratio of 4:1 and were obtained from palm oil mills located within a 50 km radius. These are both renewable fuel sources with net zero CO_2 emission. These fuel sources were used in substitution of fuel oil.

The normal practice in the oil palm estates is to allow the EFB fiber and PKS to mulch and decompose naturally and allow them to return to the estates as organic fertilizer. Note, however that during decomposition, they release methane (CH_4) and carbon dioxide (CO_2) (Brander, 2012). Their combustion in biomass boilers is considered to be carbon neutral, as there will be no additional release of CO_2 to the environment. It is hence a more environmental-friendly alternative as compared to mulching, as it prevents the formation of methane, which is 25 times more polluting compared to CO_2, in terms of global warming potential (Brander, 2012; Pachauri & Meyer, 2014). This was a major investment that propelled the company toward "greener" alternatives. After operating the Enco biomass boiler for five years, the company decided to install a cogeneration facility with boiler, which would generate both steam and power for site consumption.

Cogeneration is a term used to represent the simultaneous generation of steam and electrical power (Abbas et al., 2020). The cogeneration plant generates steam, and a portion of the steam is used to rotate a purpose-designed turbine that generates power. When the extraction turbine (Sinan Karakurt & Güneş, 2017) is suitably designed, steam can be extracted at the required pressure and flowrate to meet process steam as well as power generation requirements. Cogeneration is a more efficient method compared to only burning biomass for steam generation. The power generated by burning of biomass displaces power consumed from the grid, which is mainly generated from combustion of fossil fuels. Hence, utilization of biomass for power generation is a more environmental-friendly alternative.

This major investment further displays the company's aims of becoming a greener operator. Unfortunately, upon installing and commissioning the cogeneration facility boiler, many problems were encountered. These challenges prevented successful operation of the cogeneration facility. Despite many modifications and improvements undertaken over the years, there was only marginal improvement in the performance, with limited power generation. In 2017, the company decided to invest in a new water tube boiler with improved configuration. This water tube boiler which was fitted with an electro-static precipitator (ESP) was commissioned in 2019. A wet scrubber was installed for the older Enco biomass boiler, to comply with the Clean Air Act 2014 regulations (Legal Board Malaysia, 2018).

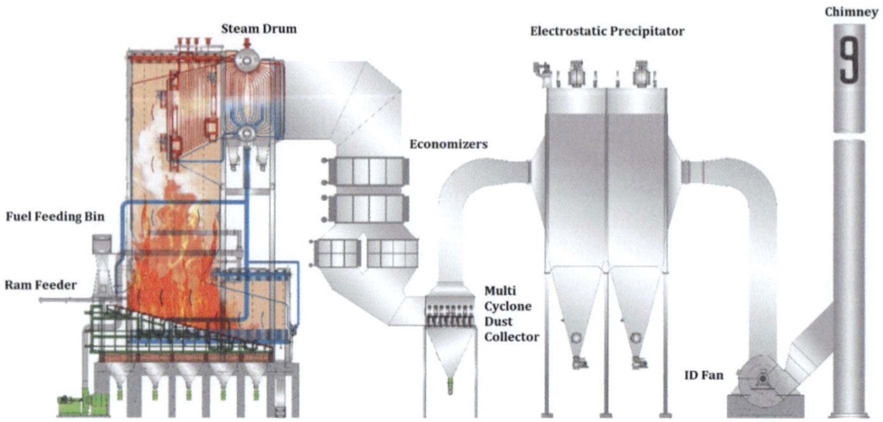

Fig. 2 Water tube boiler in single pass configuration (Lakshmanan et al., 2021)

Table 2 Emission in terms of particulate matter and opacity for each boiler

Improvement	Enco biomass boiler	Enco biomass boiler with scrubber	Cogeneration facility boiler	Water tube boiler
Particulate matter (g/Nm^3)	0.2–0.4	0.13–0.14	0.15–0.50	0.05–0.14
Opacity (%)	2.7–3.9	1.0–2.0	6.7–7.0	1.5

The boiler manufacturer's design team worked with IOIBE's steam engineers to incorporate changes to their standard water tube boiler design. Changes were made to the fuel feeding system, heat recovery, boiler feed water pumps, ash removal from the boiler, soot blasters, and surface coating protection to selected boiler surfaces to prolong the life span and robustness of the boiler. These changes are explained further in the following sections. Figure 2 shows the single pass water tube boiler from Mackenzie Industries (www.mackenzieind.com).

Table 2 gives the emission from each of the boiler chimneys. It can be seen that this water tube boiler with ESP recorded lower particulate matter and better opacity than the cogeneration facility boiler. The Enco biomass boiler operated with scrubber gives lower dust particulate and improved opacity, comparable to that obtained by the water tube boiler with ESP.

3 Single Pass Configuration of Water Tube Boiler

The Enco biomass boiler and cogeneration facility boiler had exhibited the tendency to have ash deposition at specific areas in the boiler. This deposition of ash affects heat

Table 3 Longer campaign runs before boiler shut down

Improvement	Enco biomass boiler	Cogeneration facility boiler	Water tube boiler
Mean time between cleaning	30 days	45 days	60–90 days

transfer in the boiler and necessitates regular shutdown for removal of accumulated ash. In order to minimize ash deposition, a single pass configuration was selected for this new water tube boiler. As shown in Fig. 2, this configuration resulted in lower pressure drop and consequently lower ash deposition within the boiler. As a result, this water tube boiler required less shutdown for ash removal. The frequent start and stop operations are undesirable as they lead to energy wastage. Less ash deposition also means that the boiler will have better heat transfer, which leads to lower fuel consumption.

Although the initial investment cost for a single pass boiler may be higher, it requires lower energy consumption for the induced draft fan; this converts to lower carbon footprint. Based on experience gained from the Enco and cogeneration facility boilers, several areas within the boiler were identified as areas prone to have ash deposition, which would affect heat transfer and flue gas flow. Hence, boiler shutdown is required for ash removal. This water tube boiler was designed to eliminate such ash-trapping zones by introducing new outlets for ash. In addition, soot blowers were installed to remove ash from such ash-trapping zones.

The longer campaign runs before shutdowns (Table 3) are evidence of the benefit of employing a single pass configuration for the boiler. Further advantage of the single pass and lower power requirements is captured in section of "Energy Conservation".

As given in Table 3, the water tube boiler had longer campaign runs with additional 15–60 days as compared to the cogeneration facility boiler and Enco biomass boiler.

4 Fuel Handling and Fuel Feeding

Many biomass boilers suffer with fuel feeding issues due to the complications involved in conveying and feeding EFB fibers to the furnace continuously. Many designers have assumed that EFB fibers could be easily conveyed. However, due to the fibrous nature of EFB fibers and its tendency to get entangled, it would result in bridging of fuel flow and is a problem frequently encountered by biomass boiler operators. Bridging of fuel results in interruption of fuel feed to the furnace, and if allowed to continue for even short spans of time could result in fluctuation in boiler steam pressure. The varying lengths of the EFB fibers (ranging from below 5 cm to above 20 cm) were also a challenge for boiler operators. The high and often widely varying moisture content (ranging between 40 and 60% of fuel) is another challenge faced by boiler operators. Another issue identified by boiler operators was that at

Table 4 Fiber moisture content handled by the boiler

Improvement	Enco biomass boiler	Cogeneration facility boiler	Water tube boiler
Moisture content in fuel (%)	45	45	55

high furnace temperatures, the ash tends to stick to the boiler tubes, which resulted in poor heat transfer.

This water tube boiler was able to handle fuel with higher moisture content. As given in Table 4, it was able to handle fuel with 10% higher moisture content, as compared to the other two boilers, mainly due to the use of biomass fuel feeding system onto the furnace grate and the drying zone on the grate.

For this new water tube boiler's fuel handling system, screw conveyors and holding bins were avoided, as they tended to get choked with fibers and bridging tended to occur within bins. This water tube boiler would need to be able to handle fibers of varying lengths. To get ideas on how best to address each of these issues, the plant engineers visited several biomass-fired boilers. These visits and their experiences have helped to distill out practical solutions for the various challenges. For this water tube boiler, conveying would be done using scraper conveyors (Barge & Shinde, 2017; Wankhade & Sharma, 2015; Mayer & Sussa, 2015), as shown in Fig. 3. This type of conveyor was chosen as it would not be affected by varying EFB fiber lengths or entanglement of fibers. The scraping action of the conveyor would still push the EFB fibers to their destination even if they were entangled.

To address the concern of tangled EFB fibers, large openings were placed below the scraper conveyors to permit the EFB fibers to be deposited at their destination. The above design improvements allow continuous conveying of the EFB fibers, with

Fig. 3 **a** Scraper conveyor and **b** Wide openings under the conveyor to avoid blockage and bridging

greater tolerance toward fiber entanglement. Such entanglement even if it did occur, would not affect the fuel feed to the furnace.

Feeding EFB fiber from the scrapper conveyor to the furnace was the next challenge. Since EFB fibers tended to bridge in conventional bins, an innovative approach was required to eliminate this problem. Furthermore, a suitable system would be required to inject the fiber into the furnace. To address these issues, the selected design system was one that used an upside down V-bin coupled with a ram pusher. This system as pictured at the fabrication yard is shown in Fig. 4.

With the upside down "V" bin, there is a wider mouth at the bottom compared to the top of the bin. This simple design effectively eliminated the possibility of bridging of EFB fibers occurring in the bin. Traditional V-shaped bins with a narrow throat could experience EFB fibers getting stuck within the bin. By merely inverting the "V", this problem was eliminated.

The bin shown in Fig. 4c is as wide as the furnace's grate. The scraper conveyor is placed directly above the bin, and its openings are strategically placed below the scrapper conveyor, which allows the deposition of EFB fiber into the four compartments of the bin. There are four ram pushers placed under this bin. Each ram pusher will extend and retract, in order to push EFB fiber from section of the bin above it, onto the corresponding section of the reciprocating grate, following preset frequency and sequences. Hence these four rams can be operated in tandem to ensure EFB fiber (fuel) is continuously fed into the furnace.

Fig. 4 Inverted "V" bin **a** Front view of the bin and service manhole, **b** Close up of the ram feeder located below the bin, and **c** EFB fiber filling the upside down V-bin

The ram pusher is neither affected by EFB fiber length nor entanglement. Entangled fiber and EFB fiber of any length will be pushed into the furnace. To preserve the life of the ram pushers which are exposed to the high heats in the furnace, they are continuously cooled with cooling water.

For the original power plant boiler, EFB fiber was fed into the furnace using a fuel feeding fan, as shown in Fig. 5. EFB fiber is conveyed using a screw conveyor and dropped into a bin. It then slides down a chute and a fuel feeding fan blows the EFB fiber onto the vibrating grate of the furnace.

In IOIBE's experience with its previous conveying system, the EFB fiber could get entangled at the screw conveyor and at the inlet of the bin. Any bridging that occurs here would immediately interrupt EFB fiber feed to the furnace, resulting in rapid decline in boiler steam pressure. The function of the fuel feeding fan is to blow the EFB fibers to the back of the vibrating grate so that it has sufficient time to dry on the grate prior to its combustion. However, two major factors make this rather challenging when EFB fiber is the fuel. With EFB fiber, there can be significant variation in its moisture content.

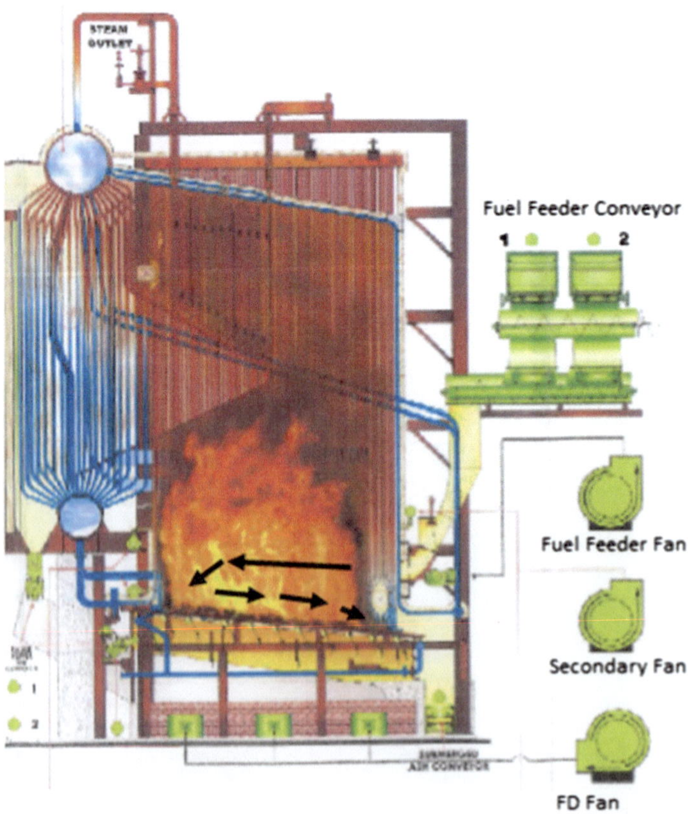

Fig. 5 Reverse feeding with fuel feeder fan (black arrows show direction of fuel flow)

When the moisture content of the fuel is higher, it tends to be heavier, which makes it more difficult for the fuel feeder fan to blow the EFB fiber to the back of the grate. When this occurs, the residence time of the EFB fiber in the furnace is reduced. This can result in incomplete combustion of the EFB fibers, as the EFB fibers do not get sufficient time in the furnace to first drive off its high moisture. When the EFB fibers get entangled and become a huge mass, it also becomes difficult for the fuel feeder fan to blow the EFB fibers to the back of the grate. The entangled EFB fibers also experience the same problem as wet EFB fibers and tend to get deposited toward the middle of the vibrating grate, affecting its complete combustion. The uneven distribution of EFB fiber on the grate affects the boilers flame and results in fluctuations in steam flow and pressure in the boiler.

A further complication results when there is improper distribution of EFB fiber on the grate. The primary combustion air is supplied from the section below the grate, so as to allow the air flow to assist in EFB fiber combustion. When there is poor EFB fiber distribution on the grate, the air stream will not push through the EFB fiber, and will take the path of least resistance. This would result in incomplete combustion occurring on the grate. The quality of the ash leaving the furnace will tell the boiler operator the degree of incomplete combustion of EFB fiber.

After experiencing these problems with the existing boilers, this new water tube boiler's EFB fiber feeding mechanism was designed to eliminate these problems. The screw conveyors were replaced with scraper conveyors; conventional bins were replaced with upside down V-bin (see Fig. 4); ram pushers were used to replace feed conveyors and reverse fuel feeding blowers. The ram pusher would push fuel into the furnace. A reciprocating grate with forward EFB fiber feeding (Fig. 6a) was selected for this furnace for further enhancement of EFB fiber conveying and feeding.

After installing these systems, less fuel supply interruptions was experienced for the furnace. The re-design of the fuel feeding and handling system have helped to eliminate the interruption to steam supply due to bridging of fuel, as given in Table 5.

A reciprocating grate was used for this furnace with a drying zone in addition to the combustion and ashing zones, as shown in Fig. 6. More explanation about this grate is given in the following section.

5 Fuel Combustion Grate

The EFB fiber is normally fiberized in the palm oil mills and contains more than 50% moisture. There may be a slight reduction in the moisture levels if the EFB fiber is properly stored in a covered shed. The Malaysian climate remains humid all year round, making it necessary for the EFB fibers to undergo drying operation for ease of combustion in the furnace. The general practice in most plants is to store the EFB fiber for several weeks with frequent turning to facilitate the drying of EFB fibers, in order to reduce its moisture content to 45%. This requires large storage space for the EFB fiber and additional costs for operating machinery for turning of the EFB fiber

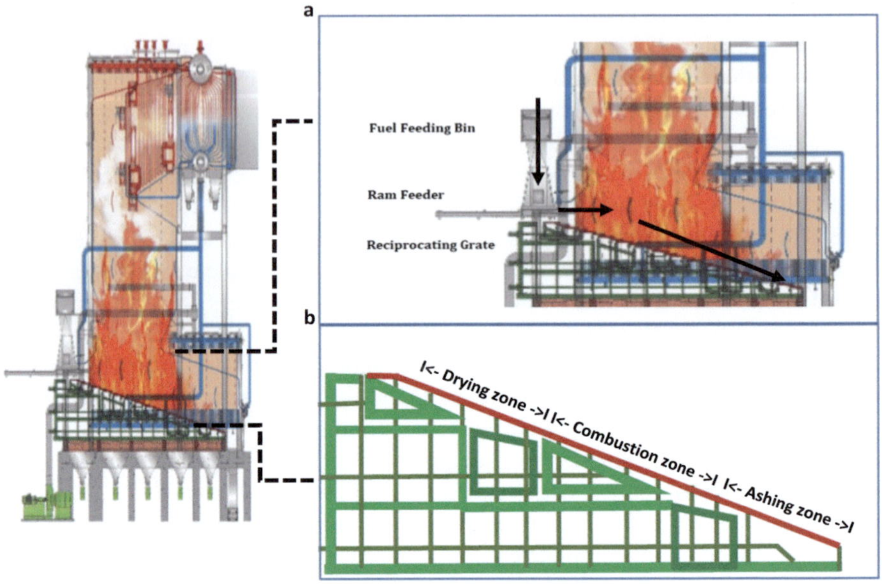

Fig. 6 **a** Feed forward system (Black arrows show direction of fuel flow) and **b** Different zones on the reciprocating grate

Table 5 Frequency of boiler interruption due to fiber feed problems

Improvement	Enco biomass boiler	Cogeneration facility boiler	Water tube boiler
Hours per week interruption	8	13	1

to aid in the drying process. Additionally, the annual rain seasons in Malaysia posed their own challenges too. Due to these reasons, the water tube boiler must be able to handle EFB fiber of 50% moisture content. To do so, the grate size of the water tube boiler was increased so as to provide sufficient drying zone within the furnace.

Hence, this water tube boiler was designed so that the EFB fiber could be dried at the furnace. This idea was implemented by installing a larger grate in this water tube boiler. This larger grate and the control logic for fuel feeding to this section was designed to cater for drying of the EFB fiber to be accomplished within the boiler. While some plants have opted to install an external dryer, IOIBE's design had it incorporated within the furnace of this water tube boiler. The combustion grate for this water tube boiler was separated into three sections, i.e., drying, combustion, and ashing zone, as shown in Fig. 6b. In the drying zone, recirculated flue gas is fed below the grate. The residual heat in the flue gas combined with the radiant heat within the furnace is used to dry these fibers in the drying zone.

The cogeneration facility boiler used vibrating grates, and cooling of the grates were accomplished using cooling tubes. However, due to the nature of the vibrating grate requiring it to be frequently vibrated to transport ash out of the furnace, frictional wear occasionally caused the cooling tubes to leak. Severe leaks to these tubes necessitated emergency shutdown of the cogeneration facility boiler to attend to these leaks. To minimize such interruptions for the water tube boiler, an air-cooled reciprocating grate was selected (Fig. 7). The air-cooled design for these grates eliminated the leaking problems of the cooling tubes. Since the grates are air-cooled, the air being supplied to the grate serves a dual purpose. While it is the primary air for combustion, it is also the cooling media for the grates. Due to this dual purpose of the primary air, it could not be heated to high temperature. In order to adequately cool down the flue gases leaving the boiler, additional economizers were installed immediately after the boiler to recover heat from flue gas (see heat recovery section) and heat up the boiler feed water.

The reciprocating grate is best described as a wide staircase, as shown in Fig. 7. Alternating steps of the grate are stationary, while the remaining grates move (see Fig. 7). The movement of the grate forces the EFB fiber to move along with the grate and onto the grate below it. This movement of the grate slowly pushes the EFB fiber to travel down the grate, one step at a time. The slow movement of the EFB fiber causes a tumbling effect and results in the exposure of the different surfaces of the EFB fiber; this enables more uniform heating and drying of the EFB fiber. The slow tumbling movement of the EFB fiber down the grates also provides more even distribution of the EFB fiber across the grate, which allows complete combustion of the fuel.

Fig. 7 Reciprocating grate inspection after six months operations

Table 6 Extent of combustion of fuel

Improvement	Enco biomass boiler	Cogeneration facility boiler	Water tube boiler
Extent of fuel combusted (%)	90	85	95

As mentioned earlier, a larger grate was used for this boiler. The point where EFB fiber is fed by the ram pusher onto the reciprocating grate is classified as the drying zone, which is the first and top most portion of the reciprocating grate. A portion of the flue gas exiting the boiler is recirculated to the drying zone. The hot flue gas is used to preheat the EFB fibers that have just entered the drying zone. The flue gas flowing through the EFB fibers also helps to vaporize some moisture contained in the EFB fibers. From the drying zone, the dried EFB fibers are pushed by the movement of the reciprocating grates, to the combustion zone. The reciprocating movement of the grates results in the EFB fibers tumbling down the steps of the grate and resulting in more complete combustion on the grate. From the combustion zone, the burning EFB fibers are next pushed to the ashing zone. In this zone, much of the EFB fibers have combusted, and mainly smoldering embers remain. The latter completely combusts in this zone before the ashes are dropped onto the ash conveyor. These changes were made for the grate of this water tube boiler to achieve better combustion of the fuel. The cogeneration facility boiler had problems to achieve complete combustion, with unburnt fuel clearly visible in the boiler ash. On the other hand, the water tube boiler has achieved combustion of 95% efficiency (Table 6), and hence resulted in lower fuel consumption of the boiler and lower greenhouse gas emissions.

Another improvement noticed with this water tube boiler was that it required less PKS as compared with the older boilers. This has resulted in significant cost reduction in operating the boiler. PKS prices have risen significantly to USD100/t in year 2021 (from USD60/t in year 2020). Hence, the reduced quantities of PKS required by this water tube boiler has helped reduce the cost of steam generation. The cost of EFB fiber has remained around USD15/t during this period. Table 7 shows that this water tube boiler was able to operate with lower shell usage and was around 50% better than the cogeneration facility boiler.

Table 7 Quantity of palm kernel shell required by each boiler

Improvement	Enco biomass boiler	Cogeneration facility boiler	Water tube boiler
Shell to fiber ratio	25:75	10:90	5:95

6 Boiler Tubes Protection

Failures have been observed on the surfaces of boiler tubes. These may be caused by corrosion and erosion at the flue gas side. Such failures could be traced to ash accumulation on these tubes which has led to condensation of flue gases on these ash deposits. Frequent start and stop of the biomass boilers have been observed to accelerate the tube failures. In some other parts of the biomass boilers, wear and tear on the tubes could be traced to erosion.

During annual shutdown of biomass boilers, it is a normal routine to check tube thicknesses at several locations in the boiler, using appropriate equipment. These are done as preventive measures to prepare for timely repair of tubes before a failure occurs. These checks help to identify any thinning of the tubes. Upon conducting, the thickness test on the cogeneration facility boilers, thinning was detected, and it was determined to be prevalent at locations above the grate. Further examination confirmed that the wear had occurred on the external surface of the tubes and not on the water side of these tubes. After careful examination of these tubes, it was determined that the thinning was due to erosive wear of these tubes. Studying the furnace design yielded an interesting observation; the passage above the grates was relatively narrow, causing higher flue gas velocity in this region. The high velocity in this region resulted in greater carryover of particles and erosion to the tubes located here. In order to extend the life of these tubes, thermal spray coating (TSC) was done to these tubes covering the areas where the tubes had thinned (Gonzalez et al., 2016; Metco, 2015, 2016).

Thermal spraying is a process by which the selected coating material such as metals, alloys, or non-metals are melted and immediately sprayed onto the surface to be coated using an inert gas, after proper initial surface preparation has been carried out (see Fig. 8). The tiny droplets of molten coating material which are sprayed using a specially designed spray gun impacts the surface, forming a protective coating on it. Proper selection of the coating material provides the desired properties for the protective layer, be it for corrosion resistance, erosion resistance, or other purposes.

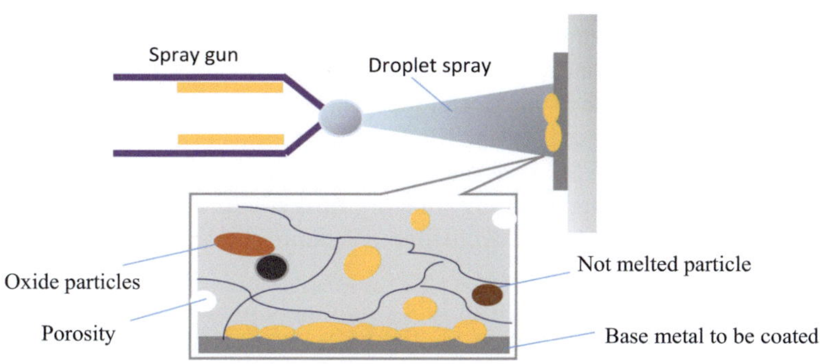

Fig. 8 Application of thermal spray coating

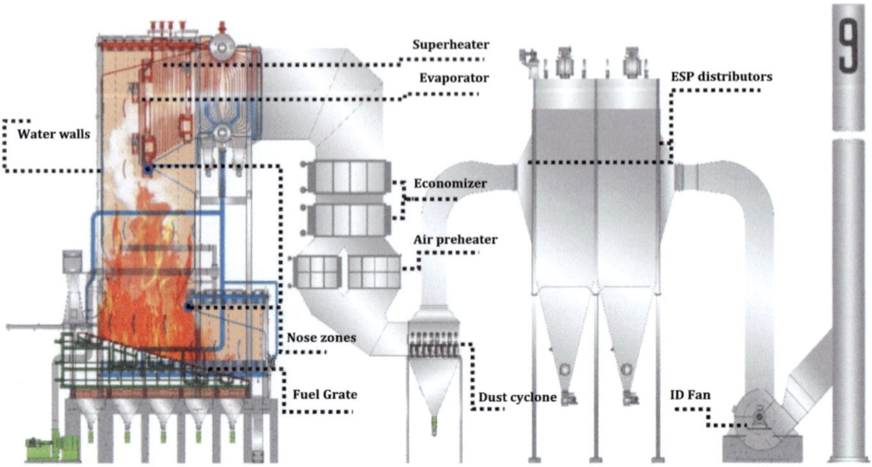

Fig. 9 Areas in the water tube boiler with TSC to protect the boiler tubes

In year 2016, TSC was conducted on the Enco biomass boilers to extend the operating life of its tubes. A literature review conducted showed that TSC is routinely carried out for boilers in advanced countries (Bala et al., 2017; Kumar et al., 2018; Prashar & Vasudev, 2021). One of the main advantages of TSC is its low heat impact on the base metal. Therefore, TSC does not affect the strength of the base metal and does not alter its metallurgical properties, compared to boiler tube repairs which are normally conducted by welding. Figure 9 shows areas in the boiler where TSC may be applied to extend the service life of these components.

The TSC conducted on the Enco biomass boiler had successfully extended the life of the boiler tubes. TSC was subsequently carried out on the cogeneration facility boiler to extend the service life of the boiler tubes. These tubes that underwent TSC have remained in service for more than 4 years. The successful use of TSC on the water walls of the cogeneration facility boiler was the main reason for implementing TSC at specific areas of the water tube boiler. The analysis of the flue gas flow through the water tube boiler identified several areas that would be exposed to erosive wear. These include the nose zones of the membrane walls, the evaporator tube bundle, and the first few rows of the superheater (see Fig. 9). Due to this observation, TSC was done to ensure prolonged service life for these tubes.

The thickness of the TSC coat can be checked using an ultrasonic thickness measuring gauge. By doing this, any thinning of the TSC can be determined and timely repairs can be carried out to avoid time-consuming and costly boiler shutdowns and tube replacements.

Table 8 shows that the economizer tubes were able to sustain in operation for a longer duration for the cogeneration facility boiler compared to the Enco biomass

Table 8 Life span of economizer tubes for each boiler

Improvement	Enco biomass boiler	Cogeneration facility boiler	Water tube boiler
Life of economizer tubes	1 year	5 years	>5 years

boiler. The Enco biomass boiler's economizer tubes failed due to the use of conventional economizers. The cogeneration facility boiler also used a conventional economizer with tubes. The low temperatures at the economizer tubes resulted in rapid corrosion of the tubes due to condensation of corrosive gases. The economizer for Enco biomass boiler has been replaced with a surface coated heat pipe exchanger in year 2016, and is still currently in operation. The heat pipe economizer is opened and inspected every 15 months and does not show signs of deterioration. For this water tube boiler, it is estimated that the economizer tubes will last longer than 5 years, based on the experience with the Enco biomass boiler's surface coated heat pipe economizer.

7 Energy Conservation

Inverters have been installed for all large motors of this water tube boiler for operating them at desired flowrates. This helps conserve energy by operating the motors at required flow conditions (Lakshmanan et al. 2020). The usage of dampers and valves to regulate air and water flows as done previously resulted in higher electricity consumption. For this water tube boiler, the speed of the motors is modulated by the inverters depending on process requirements. This results in reduced load for the motors and hence lower power is consumed by the blowers (see Table 9).

The inverters installed at this water tube boiler have reduced power consumption by 1,270,020 kWh per annum (April 2020–March 2021), when operating at around 70% of rated capacity, which translates to an annual saving of USD 94,869 (Table 9). This saving was determined as approximately 40% as compared to using dampers for controlling air flow. Note that the saving was close to the reported savings in the literature, i.e., 40–55% (Gozuk, 2021), where inverters were used (Global, 2011, Gozuk, 2015).

The steam generated by the cogeneration facility boiler was around 1.8 t steam per t of EFB fiber, as shown in Fig. 10. After commissioning this water tube boiler, steam generation has risen to 2.6 t steam per t fiber.

This translates to lesser fuel being consumed to generate steam and power for the complex. This will result in an even lower carbon footprint for the complex. Note also that the usage of biomass (EFB fiber and PKS shell) as boiler fuel is already considered as carbon neutral from the perspective of CO_2 emissions. With the increase of steam generation (per ton of fuel), lesser CO_2 is generated, and the environment protection is further enhanced.

Table 9 Power savings achieved by installing inverters for major drives

Mackenzie boiler	Motor rating (kW)	No inverter	With inverter		Savings	
		Freq (Hz)	Freq (Hz)	Running power (kW)	kWh/year	(USD)/year
Primary fan	132	50	35	125	51,240	3,828
Primary fan 2	37	50	35	35	14,640	1,094
Secondary fan	45	50	35	43	14,640	1,094
Flue gas return fan	60	50	35	43	124,440	9,296
Induced draft fan	560	50	35	460	732,000	54,680
Boiler feed pump	250	50	45	210	292,800	21,872
Fuel feeding conveyor	18.5	50	45	13	40,260	3,007
Total saving					1,270,020	94,869

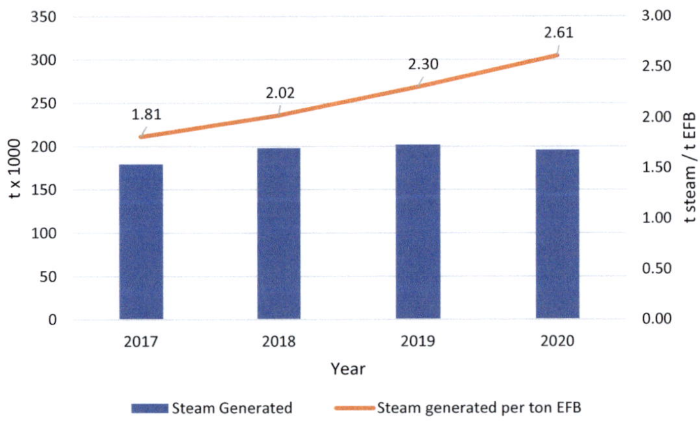

Fig. 10 Steam generation and unit fuel consumption by IOIBE complex

IOIBE's new water tube boiler supplies superheated steam to a turbine. The turbo-alternator generates 7 MW of power. Since this is more than IOIBE requires, it supplies the excess power to the neighboring IOIEO complex which then displaces an equivalent amount of electricity consumed from the grid. This green energy generated from palm biomass directly replaces energy generated from fossil fuels, significantly reducing the carbon footprint of products manufactured in the neighboring IOIEO complex.

8 Heat Recovery

To improve the heat recovery of the water tube boiler, heat pipe heat exchangers (Yang et al., 2003; Shelke & Gohel 2016) have been installed to serve as economizers. The heat pipe economizers used for the water tube boiler are provided with special surface coatings to prevent corrosion. The basic operating principles of these heat pipe heat exchangers (Vasiliev 2005; Chaudourne, 1992) are shown in Figs. 11 and 12. The heat pipe contains a heat transfer fluid within it. As shown in Fig. 11, the heat transfer fluid absorbs heat from the hot stream and vaporizes. The heat transfer fluid then moves to the cold stream. Upon contacting the cold stream, the fluid condenses and returns to the bottom of the tube where it again contacts the hot stream and vaporizes. This cycle is repeated, with the heat transfer fluid traveling between the hot stream and the cold stream, while transferring heat between them. This system which uses two-phases of the heat transfer fluid, gives it extremely high thermal conductance (Praful et al., 2020; Sadey & Kandula, 2013). Literature reports the thermal conductance to be many times higher than that from a copper rod of similar dimensions. (Chaudourne, 1992; Praful et al., 2020). For even greater heat recovery, a heat pipe economizer has even been installed to recover heat from the boiler's blowdown water as well. The heat pipes are basically specially manufactured tubes which are evacuated and filled with a suitable heat transfer fluid (Vasiliev, 2005; Chaudourne, 1992).

This heat pipe principle and design allow heat transfer to occur even though the hot and cold streams are kept separated from each other. By keeping the two streams apart, it avoids the condensation of corrosive gases present in the flue gases. This

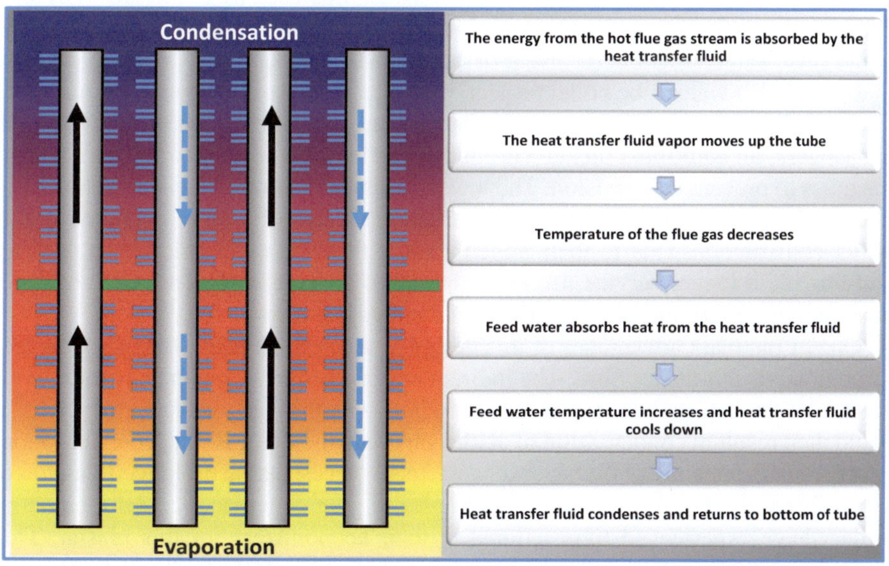

Fig. 11 Heat pipe heat exchanger operating principle

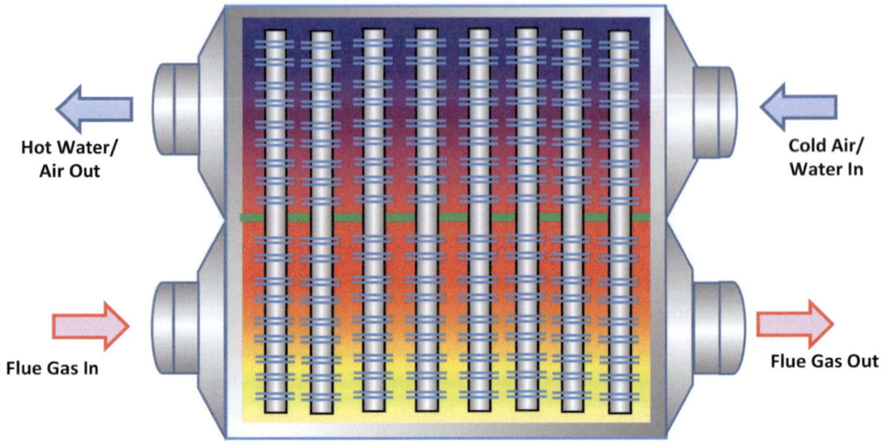

Fig. 12 Typical flow arrangement of heat pipe heat exchangers

prevents failure due to corrosion of the heat recovery unit (i.e., economizer in this case). As can be seen in Figs. 11 and 12, the hot and cold streams are kept separate from each other. Furthermore, with this design, it is also possible to place an insulation layer between these streams (Zhao Yong, 2014).

Various fluids may be used as heat transfer fluid in the evacuated tubes. The fluid used must have good heat transfer properties and will condense and vaporize in the temperature ranges that the heat pipe is to be utilized for. In the case of IOIBE, the heat transfer fluid utilized was pure water. Water was used as it has low viscosity and high thermal capacity. Furthermore, it does not pollute the flue gas stream or the boiler feed water stream in the event of heat pipe leakages.

For IOIBE's water tube boiler project, several heat pipe exchangers were installed for better heat recovery. In order to further enhance the service life of the heat pipe exchangers, an additional nickel–chromium coating was applied to the surfaces of the tubes to prevent its corrosion. The lower weight and compact design (Yang et al., 2003) of the heat pipe heat exchangers means they can be more economical for boiler construction. Due to their finned design, they are more efficient to operate and provide a longer service life as compared to conventional shell-and-tube type economizers commonly used for EFB fired biomass boilers.

One of the main advantages of this design is the indirect heat exchange effected through the heat transfer fluid. The indirect heat exchange arrangement avoids over cooling of the flue gas stream surfaces in contact with the cold stream surface, which could then result in condensation of acidic vapors in the flue gas. Such condensation could result in puncture of the heat exchange tubes, necessitating prolonged shutdown and tests to identify the leaked tube, in order to plug or replace it. Many biomass boilers have experienced tube failures in their economizers for this reason. The intermittent nature of the operation of palm oil mill boilers renders the situation even worse for any conventional economizers installed. However, heat pipe design heat exchangers may be safely used even for palm oil mill boilers to improve heat recovery.

Additionally, when operating a heat pipe heat exchanger, the failure of any single tube will not affect the heat exchange from the remaining tubes. Due to the inherent design of the heat pipe heat exchangers, there is no possibility of intermixing of the hot stream with the cold stream since the two streams are separated. If any heat pipe does leak, the heat transfer fluid within that tube would have escaped into the gas stream and it will only result in a very minor reduction in heat exchange. More importantly, the location of the leaked tube can be very quickly determined and the heat pipe may be replaced during the next scheduled shutdown.

As shown in Figs. 11 and 12, the heat pipes extend from the hot stream section to the cold stream section. The hot and cold sections are separated in order to avoid potential condensation of acidic vapors that are present in the flue gas stream. The latter flows through the lower section of the economizer and contacts the bottom section of the heat pipes. Upon contact, they heat up the heat transfer fluid within these tubes. The fluid vaporizes as a result of absorbing the heat and travels up the pipe similar to movement of convection currents. The upper portion of the heat pipes protrude into the section where the cold stream is flowing. When the cold stream comes into contact with these heat pipes, they are heated, while the vapors in the heat pipes are cooled and condense. The cooled heat transfer fluid then flows down the heat pipe. With the boiler in operation, this action is continuously repeated to provide heat recovery.

Four units of heat pipe economizers have been installed for this water tube boiler. This configuration heats up the boiler feed water from 108 °C to 260 °C before feeding it to the steam drum, with a corresponding reduction in the flue gas exit temperature. As has been mentioned earlier, this water tube boiler operates with an air-cooled reciprocating grate, hence the air is not preheated prior to supply to the furnace; thus more of the heat recovery is channeled to the boiler feed water.

Table 10 shows that the water tube boiler was able to achieve higher heat recovery with the use of heat pipe economizers on the flue gas stream. This resulted in higher boiler feed water temperature mainly through heat recovery from the flue gas leaving the boiler.

Table 10 Heat recovery achieved by each boiler

Improvement	Enco biomass boiler	Cogeneration facility boiler	Water tube boiler
Boiler feed water temperature (°C)	150	220	260
Superheated steam temperature	Saturated	480 °C	430 °C

Table 11 Boiler ash grade and quality produced from each boiler

Improvement	Enco biomass boiler	Cogeneration facility boiler	Water tube boiler
Ash grade	Regular II	Regular I	Premium
Ash K_2O	12–16%	9–15%	22–39%

9 Resource Utilization- Zero Waste

The plant minimizes water consumption and has implemented rainwater harvesting. All rainwater harvesting is done via gravity, to avoid additional pumping. The complex has its own raw water treatment plants, which also supplies treated water to the neighboring IOIEO complex. The new water tube boiler utilizes demineralized water, which is generated on-site with ultrafiltration (UF) and reverse osmosis (RO) processes. Reject water streams from the UF and RO systems are recycled to the fire-fighting system of the complexes, and for general washing. Additionally, it is used for quenching the water tube boiler's submerged ash conveyor and for the flue gas scrubber operation, in order to minimize fresh water consumption.

After combustion of the EFB fiber and PKS, the resulting boiler ash is collected and packed to be sent to estates as fertilizer. Boiler ash is known to be rich in nutrients and contains calcium and magnesium and has a sizeable potassium content (Zaj et al., 2018). Utilization of boiler ash as fertilizer can help the estates to reduce the cost of chemical fertilizers while it also helps in improving the texture and health of the soil. A further advantage of the boiler ash is its alkaline base which can assist to remediate peat soils. Additionally, the slow release of organic matter from boiler ash helps to enhance nutrient uptake efficiency (Khan & Qasim, 2008; Haron et al., 2008). The presence of some amount of un-combusted EFB fibers in boiler ash also helps to improve soil texture and moisture retention while it assists in the slow release of beneficial minerals. It is possible to further process palm boiler ash to be used in construction industries such as filler for cement and concrete (Lamers et al. 2018).

Table 11 shows that the water tube boiler is producing ash with a higher potassium (K_2O) content. This high K_2O content is a valuable parameter for fertilizers. This ash is currently sold at a premium to organic fertilizer producers, providing greater revenue to the plant.

10 Soot Removal—Detonation Pulse Technology

To extend the uptime of the water tube boiler, ash deposition on tubes and heat exchange surfaces must be minimized. This would require periodic removal of any ash build-up on these surfaces with the boiler in operation, so it does not affect its steam supply. This water tube boiler is equipped with steam soot blowers. The team identified suitable locations for placement of these soot blowers. Many retractable type soot blowers (Yeager, 2017) (Fig. 13) were installed in the superheater and

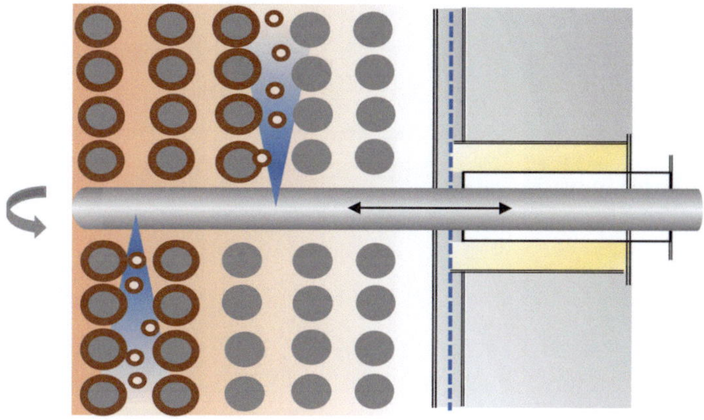

Fig. 13 Retractable steam soot blower

generation tube areas. Superheated steam was selected as the soot blowing media to prevent erosive wear from occurring on the tube surfaces. Saturated steam could have condensate which could cause erosive wear on the tube surfaces.

In addition to the steam soot blowers, shock wave soot blasters (Fig. 14) have also been placed in identified locations. The purpose of installing these devices is to maintain high uptime and boiler efficiency (while reducing fuel consumption). These soot blasters are mainly installed to avoid frequent boiler shutdown for removal of accumulated ash. Shock wave soot blasters utilize controlled mini explosions to remove ash and slag build-up on heating surfaces. In recent years, shock wave soot blasters have become more widely used for biomass-fired boilers. In addition to placing these shock wave soot blasters in the water tube boiler, the economizers have also been fitted with shock wave soot blasters to ensure that the heat exchange surfaces of these economizers remain clean to ensure good heat transfer.

The shock wave soot blasters utilize a mixture of air and a gas such as acetylene to create mini blasts in the furnace chamber. These blasts cause shock waves in the boiler that helps to dislodge ash build-up.

Table 12 shows the number of days of continuous operation of the boiler before it needs to be stopped for ash removal. The usage of the above-discussed strategies, i.e., shock wave soot blasters, steam soot blowers, single pass configuration, have contributed to clean boiler tubes for longer periods. The locations of these soot blowers and their ash collection chambers have been specifically designed for this boiler to maximize its operating time. These efforts have largely been successful in achieving longer uptime for the boiler which is highly desirable for power plant boilers.

Fig. 14 Shock wave soot blasters **a** Distributor header, **b** Mixing chamber and inlet nozzles (silver colored) placed horizontally, and **c** Mixing chamber and inlet nozzles placed around the economizers

Table 12 Comparison of boiler's operating parameters

No.	Descriptions	Enco biomass boiler	Cogeneration facility boiler	Water tube boiler
1	Boiler feed water temperature (°C)	108	220	260
2	Shell to fiber ratio	25:75	10:90	5:95
3	Extent of fuel combusted (%)	90	85	95
4	Moisture content in fuel (%)	45	45	55
5	Steam generated/t fiber	1.8	1.8	2.6
6	Life of economizer tubes	1 year	5 years	>5 years
7	Superheated steam temperature	Saturated	480 °C	430 °C
8	Mean time between cleanings	30 days	45 days	60–90 days
9	Hours per week interruption	8	13	1
10	Ash K_2O content (%)	12–16	9–15	22–39

11 Summary

While some attempt has been made to quantify the savings or improved performance data for the new water tube boiler in the previous sections, it must be appreciated that the improvements undertaken may have synergistic effects. Hence, it would be difficult to attribute a particular improved performance on one particular change that was implemented. Taken as a whole, the changes have resulted in enhanced overall performance of the boiler which have been summarized in Table 12.

This new water tube boiler and its turbo-alternator were successfully commissioned in October 2019, generating 60 t/h steam. When the boiler was supplying 25 t/h of steam to the neighboring IOIEO complex, the turbo-alternator was able to cogenerate 7.0 MW of power. The specific steam consumption of the turbine is 5.35 kg/kWh. The new power plant water tube boiler consumes less EFB fiber, and exhibits higher fuel combustion efficiency and can sustain with low PKS consumption as given in Table 12. The water tube boiler can also operate for longer periods without stopping for ash removal and produces boiler ash that has higher potassium (K) content.

12 Conclusion

As discussed in this chapter, the changes made to this water tube boiler have resulted in improved performance. The higher boiler feed water temperature as compared to earlier boilers is evidence of greater heat recovery from the flue gas. This has been accomplished through the usage of heat pipe exchangers. The larger fuel grate with a drying zone within the furnace has enabled the boiler to handle EFB fiber with higher moisture content. Recirculation of flue gas to this section of the grate has assisted in heat recovery while drying the EFB fiber. The drying of the EFB fibers within the furnace prior to combustion has resulted in more efficient EFB fiber combustion and lesser requirement of PKS addition as fuel, which also resulted in lower clinker formation in the furnace. The steam generation has thus improved to 2.6 t of steam generated per metric ton of EFB fiber (wet basis), i.e., 44% higher than the cogeneration facility boiler. Inspection of this water tube boiler during short shutdowns has shown no clinker build-up nor any major blockages on the refractory and grates. The tube surfaces do not show signs of ash build-up. The higher combustion efficiency achieved is observed in the quality of the ash generated, as compared to the previous cogeneration facility and Enco biomass boilers. The ash generated also contains higher K_2O, as a result of which it is sold at a premium to fertilizer producers.

The successful commissioning of this water tube boiler with various improvements implemented shows that an uninterrupted operation with palm-based biomass fuel is viable.

References

Abbas, T., Issa, M., Ilinca, A. (2020). Biomass cogeneration technologies: A review. *Journal of Sustainable Bioenergy Systems* (Scientific Research Publishing Inc), *10*(1), 1–15. Accessed February 10, 2021. https://doi.org/10.4236/jsbs.2020.101001

Bala, N., Singh, H., & Prakash, S. (2017). Performance of cold sprayed Ni based coatings in actual boiler environment. *Surface and Coatings Technology* (ISSN 0257-8972), *318*, 50–61. Accessed April 9, 2021. https://doi.org/10.1016/j.surfcoat.2016.11.075

Barge, N. G., Shinde, S. L. (2017). Design and analysis of drag chain to reduce failure modes for handling of fertilizer materials. *International Journal of Engineering Reasearch & Technology (IJERT)*, 705–710. Accessed April 23, 2021. https://doi.org/10.17577/IJERTV6IS060340

Brander, M. (2012). In G. Davis (Ed.). Accessed May 20, 2020. http://ecometrica.com/assets/GHGs-CO2-CO2e-and-Carbon-What-Do-These-Mean-v2.1.pdf

Chaudourne, S. (1992). The heat pipe heat exchangers: Design, technology and applications. In P. J. Heggs, Butterworth, & W. Roetzel (Eds.) *Design and Operation of Heat Exchangers. EUROTHERM Seminars* (pp. 1–19). Springer. Accessed February 20, 2021. https://doi.org/10.1007/978-3-642-84450-8_34

Global, Fuji Electric. (2011). *Energy savings with inverter.* Accessed April 27, 2021. https://www.fujielectric.com/products/ac_drives_lv/frenic-eco/products/in_eco/02_guatai.html

Gonzalez, R., Ashrafizadeh, H., Lopera, A., Mertiny, P., & Mcdonald, A. (2016). A review of thermal spray metallization of polymer-based structures. *Journal of Thermal Spray Technology, 25*(5), 897. Accessed March 12, 2021. https://doi.org/10.1007/s11666-016-0415-7

Gozuk. (2015). *Frequency inverter the best energy saving solution.* Accessed April 27, 2021. http://www.frequencyinverters.org/frequency-inverter-the-best-energy-saving-solution-334349.html

Gozuk. (2021). *Annual energy saving by frequency inverter on fans.* Accessed April 27, 2021. http://www.frequencyinverter.org/annual-energy-saving-by-frequency-inverter-on-fans-390874.html

Haron, K., Mohammed, A. T., Halim, R. M., & Din, A. K. (2008). *Palm-Based Bio-Fertilizer from Decanter Cake and Boiler Ash of Palm Oil Mill.* Selangor: MPOB. Accessed November 9, 2019.

Khan, M. J., & Qasim, M. (2008). Integrated use of boiler ash as organic fertilizer and soil conditoner with NPK in calcareous Soil. *Songklanakarin Journak of Science and Technology, 30*(3), 281–289. Accessed Nov 9, 2019.

Kumar, R., Kumar, R., & Kumar, S. (2018). Erosion corrosion study of HVOF sprayed thermal sprayed coatings on boiler tubes: A review. *International Journal of Science and Management Studies (IJSMS) E-ISSN, 1*(3), 2581–5946. Accessed April 9, 2021. http://www.ijsmsjournal.org/2018/volume-1%20issue-3/ijsms-v1i3p101.pdf

Lakshmanan, S., Yung, Y. L., Boon San, C., Tze Haw, C. (2020). Sustainable practice of and edible oils refining complex. *Journal of Oil Palm, Environment & Health, 11*, 42–56. https://doi.org/10.5366/jope.2020.05

Lakshmanan, S., Yung, Y. L., Palanisamy, K., & Ling, H. K. (2021). Lessons learnt from biomass-fueled power plant. *Journal of Oil Palm, Environment & Health, 11,* 6–10. https://doi.org/10.5366/jope.2020.02

Lamers, F., Cremers, M., Matschegg, D., Schmidl, C., Hannam, K., Hazlett, P., Madrali, S., Birgitte Primdal Dam, Roberto, R., Mager, R., Davidsson, K., Bech, N., Feuerborn, H. J., & Saraber A. (2018). *Options for Increased use of ash from biomass combustion and co-firing.* IEA Bioenergy. Accessed November 9, 2019.

Legal Board Malaysia. (2018). *Environmental Quality Act 1974 (Act127), Regulations, Rules & Orders.* Internationl Law Book Services.

Mayer, J., & Sussa, J. (2015). *Solidification Immobilization and Encapsulation of Waste and Contaminants.* PURATEK International. Accessed April 23, 2021. http://www.puratek.de/fileadmin/user_upload/030_Downloads/process/publication_SOLIDIFICATION.pdf

Metco, O. (2015). Surface Solutions Boilers_ Presentation to PLN_MH. April. Accessed November 9, 2019. https://www.oerlikon.com/ecomaXL/files/en/oerlikon_20150401_Oerlikon_Presentation_April_2015-2.pdf

Metco, O. (2016). An introduction to thermal spray. Oerlikon Metco. July 6. Accessed November 2019. BRO-0005.6_Thermal_Spray_Brochure_EN.pdf

Pachauri R. K., & Meyer, L. A. (2014). IPCC, 2014: *Climate change 2014: Synthesis report. Contribution of working groups I, II and III to the fifth assessment report of the Intergovernmental Panel on Climate Change.* 151. Accessed February 21, 2021. https://archive.ipcc.ch/pdf/assessment-report/ar5/syr/SYR_AR5_FINAL_full_wcover.pdf

Praful, S., Prajwal Rao, V., Vijeth, V., Bhagavath, S. V., Seetharamu, K. N., Rao, R. N. (2020). On the operating temperature of heat pipes. *International Conference on Thermo-fluids and Energy System (ICTES 2019).* IOP Publishing Ltd., *1473.* Accessed February 20, 2021. https://doi.org/10.1088/1742-6596/1473/1/012025

Prashar, G., & Vasudev, H. (2021). Application of thermal spraying techniques used for the surface protection of boiler tubes in power plants: Thermal spraying to combat hot corrosion. *Advanced Surface Coating Techniques for Modern Industrial applications, 23.* Accessed April 8, 2021. https://doi.org/10.4018/978-1-7998-4870-7.ch005

Sadey, R. R., & Kandula, J. (2013). Fabrication and analysis of heat pipe. *International Journal of Engineering Research & Technology (IJERT), 2*(04), 1740–1745. https://www.ijert.org/fabrication-and-analysis-of-heat-pipe

Shelke, S. A., & Gohel, N. S. (2016, June). Heat transfer perfroamnce of a vertical thermosiphon heat pipe heat exchanger using hybrid nanofluid for automobile engine exhaust heat. *International Journal of Current Engineering and Technology.*

Sinan Karakurt, A., & Güneş, Ü. (2017). Performance analysis of a steam turbine power plant at part load conditions. *Journal of Thermal Engineering, 3*(2), 1121–1128. Accessed February 21, 2021. https://doi.org/10.18186/thermal.298611

Vasiliev, L. L. (2005). Heat pipe in modern heat exchnagers. *Applied Thermal Engineering, 25*(1), 1–19. Accessed February 20, 2021. https://doi.org/10.1016/j.applthermaleng.2003.12.004

Wankhade, V., & Sharma, S. (2015). Design improvement for enhancing the perfromance of drag conveyor chain. *International Journal of Engineering research and General Science, 3*(2), 524–531. Accessed April 23, 2021.

Yang, F., Yuan, X., & Lin, G. (2003). Waste heat recovery using heat pipe heat exchanger for heating automobile using exhaust gas. *Applied Thermal Engineering, 23*, 367–372. Accessed February 20, 2021.

Yeager, M. (2017). *Increase the IQ of your intelligent sootblowing.* Accessed February 21, 2021. http://www.powermag.com/increase-the-iq-of-your-intelligent-sootblowing/

Zaj, G., Szyszlak-Bargłowicz, J., Gołebiowski, W., & Szczepanik, M. (2018). Chemical characteristics of biomass ashes. *Energies* (MDPI), *11,* 2885. Accessed November 9, 2019. https://doi.org/10.3390/en11112885

Zhao Yong, W. (2014). Heat pipe economizers-presentation given by Prof Wang Zhao Yong of Novoneng Singapore to IOI Edible Oils in March 2014.

Innovations in Palm Oil Processing

Mathematical Modelling and Optimisation for Fresh Fruit Bunch Harvesting and Evacuation

Chun Hsion Lim, Bing Shen How, Wendy Pei Qin Ng, Sue Lin Ngan, Steven Lim, and Hon Loong Lam

Abstract Sustainability development and practice in the oil palm industry have been a great concern among the global communities due to the increasing demand for palm-based products. The problems are mostly stressed on the oil palm cultivation and sustainable production of palm oil at the plantation site and mill, respectively. Labour shortage has always been a major issue in the oil palm industry. This chapter explores a potential solution to the labour-intensive fresh fruit bunch (FFB) harvesting and evacuation process using mathematical optimisation modelling approach, with the objective to determine the most efficient route for the processes. Demonstration

C. H. Lim (✉)
School of Engineering and Physical Sciences, Heriot-Watt University Malaysia, Jalan Venna P5/2, Precinct 5, 62200 Putrajaya, Malaysia
e-mail: l.chun_hsion@hw.ac.uk

B. S. How
Research Centre for Sustainable Technologies, Faculty of Engineering, Computing and Science, Swinburne University of Technology Sarawak, Jalan Simpang Tiga, 93350 Kuching, Sarawak, Malaysia
e-mail: bshow@swinburne.edu.my

W. P. Q. Ng
Petroleum and Chemical Engineering Programme Area, Universiti Teknologi Brunei, Jalan Tungku Link, Gadong BE1410, Brunei Darussalam
e-mail: peiqin.ng@utb.edu.bn

S. L. Ngan
UKM-Graduate School of Business, Universiti Kebangsaan Malaysia, 43600 Bangi, Selangor, Malaysia
e-mail: suelin.ngan@ukm.edu.my

S. Lim
Department of Chemical Engineering, Lee Kong Chian Faculty of Engineering and Science, Universiti Tunku Abdul Rahman, Jalan Sungai Long, Bandar Sungai Long, 43000 Kajang, Selangor, Malaysia
e-mail: stevenlim@utar.edu.my

H. L. Lam
Centre of Excellence for Green Technologies, University of Nottingham Malaysia, Broga Road, 43500 Semenyih, Selangor, Malaysia
e-mail: honloong.lam@nottingham.edu.my

D. C. Y. Foo et al. (eds.), *Sustainable Technologies for the Oil Palm Industry*,
https://doi.org/10.1007/978-981-19-4847-3_9

case studies were presented to illustrate the application of the model in various plantation site scenarios, including the variation in land topography, distribution of harvestable trees, and loading capacity of transporter. The study showed that computational assistance can enhance process efficiency and provide a platform for more systematic resource management. The implementation of computational effort in the agriculture industry is getting more feasible due to the increased accessibility to smart devices such as smart phone.

Keywords Optimisation · Harvest · Evacuation route · Mathematical modelling · Labour shortage · Harvesting and evacuation route optimisation (HERO) model

Nomenclature

Text	*Definition*
FFB	Fresh fruit bunch
GAMS	General algebraic modelling system
GDP	Gross domestic product
GPS	Global positioning system
HERO	Harvesting and evacuation route optimisation
MINLP	Mixed integer nonlinear programming
MPOCC	Malaysian palm oil certification council
MSPO	Malaysia sustainable palm oil
RSPO	Roundtable on sustainable palm oil
SMEs	Small and medium enterprises
Set	*Definition*
i	Departed point of location
j	Arriving point of location
Variables	*Definition*
l_j	Accumulative transporter load
r_j	Accumulative travelled distance, m
z	Total distance travelled, m
x_{ij}	Binary variables to govern the route selection
Parameters	*Definition*
d_{ij}	Distance between tree i and tree j,
p_{ij}	Binary parameter that denotes possible routes
t	Minimum number of trips required
W	Transporter capacity
y_i	Weight of harvestable FFB at location i

1 Sustainable Palm Oil

Palm oil (*Elaeis guineensis*) that is originated from West Africa was first industrialised in Malaysia in 1917 at Tennamaram Estate due to its versatility as a vegetable oil (Kushairi et al., 2017). Its cultivation is most suited in tropical regions such as Indonesia and Malaysia that have adequate rainfall, sunshine and optimum soil condition. Annually, the global consumption of palm oil was approximately 60% of the total vegetable oils in the market, which was equivalent to 45.3 million tonnes in 2009 (Ahmad et al., 2019; Begum et al., 2019). The production was increased to approximately 73.8 million tonnes in 2020 (Merrick, 2021). The palm oil industry is expected to bring significant impacts towards the society over the next four decades due to the increasing demand of palm-based products and energy, especially on palm oil exporters (Begum et al., 2019). Malaysia alone had exported up to 17.40 million tonnes of palm oil in 2020 (MPOB, 2021), contributing to an average of 5%–7% of the country's gross domestic product with an average total export revenue of USD 16.87 billion (MYR 70.14 billion) annually from oil palm products over the last 5 years (2016–2020)(MPOB, 2021).

The high demand for palm oil leads to huge requirements of natural resources for oil palm cultivation which creates environmental issues and challenges for sustainable production. For instance, the inevitable deforestation and replantating activities could lead to watershed degradation, increased fire risk, erosion and soil degradation, biodiversity loss, and greenhouse gas emission. Study had shown that 6% of the total annual global emission of greenhouse gas was contributed by carbon dioxide emission from wetland deforestation in Southeast Asia (Begum et al., 2019). Unethical oil palm stakeholders' practice of open burning for replantation in some regions have led to air pollution in the form of haze (Tan et al., 2018). The increase in land usage for oil palm farming was claimed to be leading to the extinction of orangutan and peatland destruction problems (Tan et al., 2009). The various environmental and sustainability concerns in the industry are causing the stakeholders to focus more on the development of cleaner practices in production to strike for a more environmentally friendly and sustainable production.

As an effort to promote sustainability in the palm oil industry, Roundtable on Sustainable Palm Oil (RSPO) was formed with representatives from the palm oil industry and civil society groups in year 2004 (Ruysschaert & Salles, 2014). RSPO aims to improve the sustainability of the palm oil industry in each layer of the supply chain and provides more engagement with its stakeholders in the process (Morgans et al., 2018). It adopts the concept of sustainable development that comprises economic viability, environmentally appropriate and socially beneficial to the management, and operation of palm oil production processes via certification scheme. The latter enables a systematic supply chain platform to trace the production of sustainable palm oil and to provide a good indicator for consumers to identify the sustainability of the purchased palm-based products. Malaysia Sustainable Palm Oil (MSPO) was officially launched at the Malaysia Palm Oil Board International Palm Oil Congress in 2013 (PIPOC 2013). MSPO is a Malaysian standard for oil palm management certification and supply chain that comprises the requirements set by the Malaysian Palm Oil Certification Council (MPOCC). The initiative of MSPO is similar to RSPO's, which aims to reduce the environmental impact and to ensure socio-economic well-being of the Malaysian oil palm industry. Other palm oil exporters like Indonesia and Thailand also took a similar initiative to uphold the concept of RSPO in the production of sustainable and traceable palm oil (Innocenti & Oosterveer, 2020).

Nonetheless, the huge number of small and medium enterprises (SMEs) in the palm oil industry, especially oil palm producers create another challenge in tracing the source of the sustainable palm oil and to differentiate them from those non-certified palm oil. End-users may concern about the origin and sustainability of the palm-based product; while the problem with poor tracing may lead to unfair pricing among the oil palm producers if the system fails to identify the certification status. The complex traceability problem could be addressed with the advancement of technology such as *block chain*. Besides, various studies have shown that development towards Industry 4.0 could improve the efficiency and sustainability of the palm oil industry. The adaptation potential of the smart technologies will be discussed in the next section.

2 Adaptation of Industry 4.0 Technology in Oil Palm Cultivation

Apart from the environmental concerns, abuse of human rights also could be a potential issue in a sustainable palm oil industry (Ayompe et al., 2021). Oil palm cultivation is well known for its extensive labour requirement in the harvesting and evacuation process, where farmers need to scout huge plantation site prior to manual harvesting and evacuation of oil palm fresh fruit bunch (FFB). As such, labour shortage is one of the main issues in the industry, where the ratio of land to labour is reported to be 10.9 ha: 1 labour (Alam et al., 2015). Shortage of labour could lead to sustainability issues when the ripe FFB is left unharvested or uncollected; or infeasible workload for farmers to cover large plantation area to maximise production yield. With the advancement of technology, majority of the industrial players are moving towards Industry 4.0 for process automation, digitalisation, and smart manufacturing. Automation and mechanisation of the process towards modern plantation operation are a promising improvement to tackle the labour shortage issue (Kushairi et al., 2017). For instance, advancement in autonomous vehicle technology with object detection capability via multi-camera system and radar has effectively helped in the improvement of the monitoring system in the oil palm plantation site (Chong et al., 2017). This can be incorporated onto small tractors and self-manoeuvre robots in oil palm plantation to detect the location of ripe fruits. The information could be used by farmers to plan for optimum harvesting pathway or potentially initiate auto-harvesting process in future. Smart spectrometer and remote sensor can be used to monitor oil palm trees in identifying poor yielding tree, which helps to improve the overall efficiency of the harvesting process (Kassim et al., 2014). Incorporation of smart technology could also enhance the concept of circular economy in the oil palm industry by reducing waste generation and optimising resource utilisation. Integrating the principles of circular economy in the oil palm industry helps to improve biodiversity, economic and social gains, and to minimise environmental pollution (Savolainen et al., 2020). Despite the huge potential in the global market, the adaptation of Industry 4.0 and circular economy in the oil palm industry remains challenging due to the unclear economic benefits of Industry 4.0 investment, lack of automation system virtualisation, lack of process design, unstable connectivity, and employment disruptions (Abdul-Hamid et al., 2020). A detailed analysis of the Industry 4.0 potential in the palm oil industry has been conducted by Lim et al. (2021b) to evaluate the possible implementation strategies. The study revealed that the harvesting process in the oil palm plantation site was one of the promising processes to incorporate Industry 4.0 technologies for improvement, such as introduction of smart spectrometer for palm fruit ripeness detection and cloud computing system to optimise the harvesting route. This would be able to address the critical issues in promoting sustainability and huge labour requirement as stated above.

3 Improving Oil Palm Fresh Fruit Bunch Harvesting and Evacuation Process

Various technologies and tools have been developed to improve the process efficiency of FFB harvesting. For instance, a four-wheel steering transporter with grabber was developed to assist farmers in the plantation site (Shuib et al., 2010); a six-wheel-drive transporter was designed to improve the accessibility, efficiency, and cost of FFB evacuation process (Shuib et al., 2020); and the use of solar energy to power electric vehicles in oil palm plantation field to reduce the carbon dioxide emission up to 8 kg CO_2/ha (Azwan et al., 2017). Apart from the establishment of new tools for FFB harvesting and evacuation, optimisation of the harvesting and evacuation route within the oil palm plantation is also part of the critical factor to effectively improve the efficiency and to reduce labour dependency. Similar logistic issues have been addressed in other industries to optimise the path selection within the supply chain. A route optimisation model for electric garbage truck fleet could enhance the sustainability and energy management by reducing approximately 31.7% energy consumption (Erdinç et al., 2019). Similar vehicle routing problem was tackled by researchers to reduce the field work time in corn field using fleet routing optimisation model (Seyyedhasani & Dvorak, 2018). Besides, researchers had also proposed a model to minimise the number of trucks needed for the transportation between agriculture fields and storage or processing facilities so that the cost can be minimised (Lamsal et al., 2016).

Despite the various research efforts to address the agriculture supply chain problems, none of the reported work is able to accurately describe the unique characteristics of oil palm plantation and compares to the existing harvesting and evacuation practise, especially with computational mathematical optimisation model. The lack of interest among oil palm farmers to invest into computational effort might be due to the fact that majority of the existing oil palm plantation sites are not equipped with adequate computational facility and expertise. Nonetheless, the global movement towards Industry 4.0 and improvement in quality of life has significantly enhanced the accessibility to smart devices such as smart phone, Internet, and computational power via cloud computing. This enables easier implementation of technological features in the plantation site, including SMEs (Shuib et al., 2010). In the context of FFB harvesting and evacuation, the conventional approach adopted by oil palm farmers is to visit each oil palm tree to observe the fruit ripeness prior to the arrangement of FFB harvesting process. The purpose of scouting is to estimate the quantity of ripe fruit and to plan for adequate labour and resources for the harvesting process. For some plantations, scouting is performed during the previous harvesting process to estimate the amount of ripe fruit for the next harvest cycle. Typically, the process of harvesting and evacuation would be conducted separately. During the harvesting process, the harvested FFB will be positioned near to the tree for evacuation at later stage using a transporter. Immediate collection of the FFB was found to be inefficient as it would impose longer wait-time during the harvesting process and leads to a higher carbon footprint when a mechanical vehicle is used (Wan et al., 2018).

4 Integration of Mathematical Optimisation Modelling in FFB Harvesting and Evacuation Process

The scouting, FFB harvesting, and evacuation processes are labour-intensive tasks, which the efficiency is highly dependent on farmer's experience and familiarity with the plantation site. Previous section has discussed the advantages of integrating Industry 4.0 technology in agriculture, such as using smart spectrometer and robotic to monitor FFB ripeness in order to reduce the dependency on labour. Nonetheless, it may not be a feasible solution due to its high investment and maintenance costs. Social issues may arise (e.g. reduced job opportunity) if the transition towards Industry 4.0 is too rapid. Despite that, the process could still be beneficial from the advancement in computational-aided technology by optimising the scouting, harvesting, and evaluation process via the minimisation of the travelled distance to improve the efficiency. The existing scout practice could provide the required information such as the amount of ripe FFB and its location for optimisation model to determine the shortest route. Optimisation model could also be used to obtain the optimum solution based on logistic cost, energy requirement, pollution, and carbon footprint. The software-based technology has the advantages of low capital and operating costs, and high accessibility via smart phone. If smart sensors are used across the plantation to detect fruits ripeness and are integrated with Internet of Things technology, this could eliminate the requirement of manual scouting in the near future.

As the preliminary step towards Industry 4.0, this chapter discusses the application of a mathematical optimisation model, i.e. *harvesting and evacuation route optimisation* (HERO) model, to minimise the travel distance requirement during FFB harvesting and evacuation process. Next section discusses the formulation of the model, and the following section demonstrates the application of the HERO model in various scenarios. The examples include (i) application of the HERO model on different plantation site floorplans, (ii) different densities of the harvestable FFB, and (iii) different transporter loading capacities.

5 Methodology

The ultimate goal of the HERO model (Lim et al., 2021a) is to determine the shortest route for oil palm farmers for FFB harvesting and evacuation process. The result provides a systematic guidance for the farmers to travel between the starting point, oil palm trees, and the ending point. The model requires to compute the optimum path that covers all oil palm trees with ripe FFB in the shortest distance. As such, the proposed model denotes two alias sets, where i refers to the departed point of location and j represents the arriving point of location. The context of location is referring to the location of the start point (collection point), oil palm trees, and/or end-point (another collection point).

Equation (1) shows the objective function of the model, where z is the total travelled distance, d_{ij} is the distance between location i and location j, and x_{ij} is binary variable to govern the route selection. Equation (2) calculates the minimum number of trips required, t, based on the total weight of harvestable FFB, y_i, and the transporter capacity, W. The minimum number of trips is the rounded-up integer/ceiling function of the calculated value.

$$\text{minimise } z = \sum_{i,j=1}^{I,J} d_{ij}x_{ij} \tag{1}$$

$$\frac{\sum_{i=1}^{I} y_i}{W} = t \tag{2}$$

To minimise the travelled distance, the proposed route should only visit each harvestable tree ($y_i, y_j > 0$) once (and departed from the same tree once). Both the starting point ($i = 0$) and end collection point ($j = N$) should be visited t times to achieve the minimum trip. Farmers are assumed to collect all harvestable FFBs each time they visited a particular tree. These conditions are governed by Eqs. (3) and (4). Equation (5) denotes the restriction of the possible route, p_{ij}, which is a binary parameter introduced in the model. For example, the value of p_{ij} is assigned to be zero when $i = j$ as the model should not propose the solution, where the farmers visit the same tree that they departed from. If farmers are free to travel from any point to any point within the region ($i \neq j$), the value of "one" will be assigned for the p_{ij}. This parameter improves the model flexibility to restrict any movement within the plantation site based on any known obstacles.

$$\sum_{j=1}^{N} x_{ij} = \begin{cases} t & i = 0, N \\ 0 \ \forall i \in I \wedge y_i < 0 \\ 1 \ \forall i \in I \wedge y_i > 0 \end{cases} \tag{3}$$

$$\sum_{i=1}^{N} x_{ij} = \begin{cases} t & j = 0, N \\ 0 \ \forall j \in J \wedge y_j < 0 \\ 1 \ \forall j \in J \wedge y_j > 0 \end{cases} \tag{4}$$

$$x_{ij} \leq p_{ij} \forall i, j \in I, J \tag{5}$$

Equations (6) and (7) are used to prohibit the occurring of the sub-tours. The accumulative travelled distance, r_j, is set to be at zero at the starting point, and the subsequent travelled distances are calculated accumulatively based on Eq. (7). Similarly, Eqs. (8) and (9) show the initial and accumulative transporter load, l_j, where the accumulated load should be less than or equals to the transporter's loading capacity, W, as shown in Eq. (10).

$$r_{j=1} = 0 \tag{6}$$

$$\sum_{i=1}^{N} x_{ij} \times \left(r_i + d_{ij}\right) = r_j \forall j \in N \wedge j \neq 1 \tag{7}$$

$$l_j = 0 \quad \text{For } j = 1 \tag{8}$$

$$\sum_{i=1}^{N} x_{ij} \times \left(l_i + y_j\right) = l_j \forall j \in N_t \wedge j \neq 1 \tag{9}$$

$$l_{j=N} \leq W \tag{10}$$

6 Case Studies and Discussions

6.1 Case Studies Descriptions

In this section, a demonstration case study will be discussed to show the application of the proposed HERO model in various scenarios for FFBs harvesting and evacuation process. Information was gathered from literature and oil palm producers to define the characteristics of the oil palm tree arrangement in the plantation site, and understand the current industry practice for the activity. For the case study, several assumptions were made based on generic oil palm plantation site characteristics.

1. The spacing between each palm tree is fixed at the optimum distance of 9 m in equilateral triangular patterns (Bonneau et al., 2018; Rafii et al., 2013).
2. Plantation site is separated into multiple small sections (based on the drainage system, obstacles, ownership boundary, and FFB collection point), where each section is optimised separately.
3. A maximum of 21 oil palm trees are considered in the proposed example.
4. Farmers and transporter are assumed to be able to travel freely from one tree to another within each section. Therefore, p_{ij} will be 1 when $i \neq j$.
5. Each palm tree is assumed to yield a single FFB with an average weight between 20 and 25 kg (Harun & Noor, 2002; Morel et al., 2016). The detailed FFB weight distribution of each tree is presented in Table 2.

Due to the presence of binary variables and nonlinear equations, the model was solved with BARON solver as a mixed-integer nonlinear programming (MINLP) problem using General Algebraic Modelling System (GAMS) version 23.4. Researches have shown that the branch and bound method is the most compromising algorithm among the heuristics and algorithms proposed for MINLP (Fernandes et al., 2009; Shahriari et al., 2016). BARON solver applies the branch and bound algorithm to diminish the domains of variables via bound tightening and reduction techniques to obtain global optimum (Khajavirad & Sahinidis, 2018; Kılınç & Sahinidis, 2018). The relative termination tolerance was specified at 1×10^{-9} to increase the accuracy of the global optimality. A total of three different scenarios were formulated to demonstrate the application of the HERO model, including (A) different floorplans, (B) different distributions of harvestable trees, and (C) different oil palm transporters loading capacities (see Table 1).

6.2 Scenario (A): Optimisation of Harvesting and Evacuation Route on Different Plantation Floorplans

Many oil palm farmers are SMEs, with different plantation sizes and shapes due to the segregation of land ownership, drainage system, and road access. Nonetheless, the typical plantation site floorplan can be described with simple geometry shapes

Table 1 Description of each case study presented

Scenario	Case	Floorplan	Distribution (%)	Palm transporter considered
Different floorplans (A)	i	Square	80	Unlimited
	ii	Vertical rectangle	80	Unlimited
	iii	Horizontal rectangle	80	Unlimited
	iv	Equilateral triangle	80	Unlimited
	v	Scalene triangle	80	Unlimited
Different distribution of harvestable trees (B)	i	Square	100	Unlimited
	ii[a]	Square	80	Unlimited
	iii	Square	60	Unlimited
Different oil palm transporters (C)	i	Square	80	Wheelbarrow (loading capacity, W = 150 kg)
	ii	Square	80	Mechanical buffalo (loading capacity, W = 700 kg)

[a]same case as A(i)

Table 2 Distribution of fresh fruit bunch weight, y_i at each location

Tree	Weight (kg)	Tree	Weight (kg)	Tree	Weight (kg)
1	21.2	8	22.4	15	22.4
2	22.4	9	21.9	16	21.2
3	21.8	10	20.8	17	23.5
4	20.9	11	23.6	18	21.5
5	24.7	12	22.9	19	24.2
6	21.3	13	24.1	20	22.0
7	23.9	14	22.7	21	23.6

or a combination of them, as shown in Fig. 1. The boundaries of the plantation site typically consisted of road access for transportation of FFB to processing mill. This example illustrates the application of the HERO model in handling different shapes of plantation sites to achieve the shortest route (Table 2).

For illustration, Fig. 2 shows the arrangement of oil palm trees within the different plantation floorplan in Cartesian coordinate system. There are two FFB collection points at the edge of plantation site which could be a start- or end-point of the harvesting and evacuation process. Current industry practice is to visit all trees for the harvesting process. In the ideal scenario, farmers should only visit each of the tree once between the collection points. The total required travelled distance to only visit all trees once was determined with the HERO model at 100% harvestable tree distribution. Noted that only experienced farmers could execute the shortest travelled distance

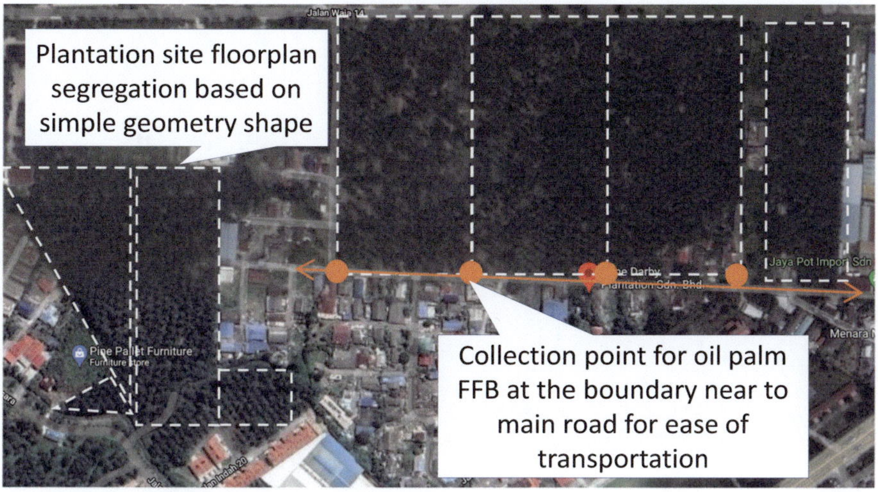

Fig. 1 Segregation of oil palm plantation site with simple geometry based on Google satellite image (Lim et al., 2021a)

to visit all trees. For a more sensible case study, the location of harvestable tree was located randomly up to 80% of the total number of trees within the system. The parameter to restrict the path selection, p_{ij}, was assigned to zero for $p_{\text{“tree without fruit”}\,j}$ and $p_{i\text{“tree without fruit”}}$. For example, $p_{\text{“5”}j}, p_{i\text{“5”}}, p_{\text{“7”}j}, p_{i\text{“7”}}, p_{\text{“14”}j}, p_{i\text{“14”}}, p_{\text{“16”}j}$, and $p_{i\text{“16”}}$ in Case A(i) are assigned with zero. All the cases in Scenario A were solved without consideration of transporter loading capacity limit to determine a single trip solution ($t = 1$). Figure 3 visualises the solution of the model by presenting the optimised binary variables, x_{ij}, into a harvesting and evacuation route for the plantation site. For all the sizes and shapes of the plantation floor plan, HERO model was able to provide a route with shorter travelled distance as compared to the existing practice of visiting all trees as given in Table 3. The reduction of travelled distances was in the range of 14.8% and 21.0%, which would be a significant improvement for huge plantation site to minimise labour and energy requirements, improve harvesting time, and reduce the carbon footprint from the motorised transporter.

6.3 *Scenario (B): Optimisation of Harvesting and Evacuation Route with Different Harvestable Fresh Fruit Bunch Distribution*

In this section, the difference in the ripe FFB distribution was discussed to understand the impact of the non-homogeneous pollination nature of oil palm which creates inconsistent amount of ripe FFB to the optimum route. Three cases with different amounts of harvestable FFBs at 60%, 80%, and 100% were discussed. The variation

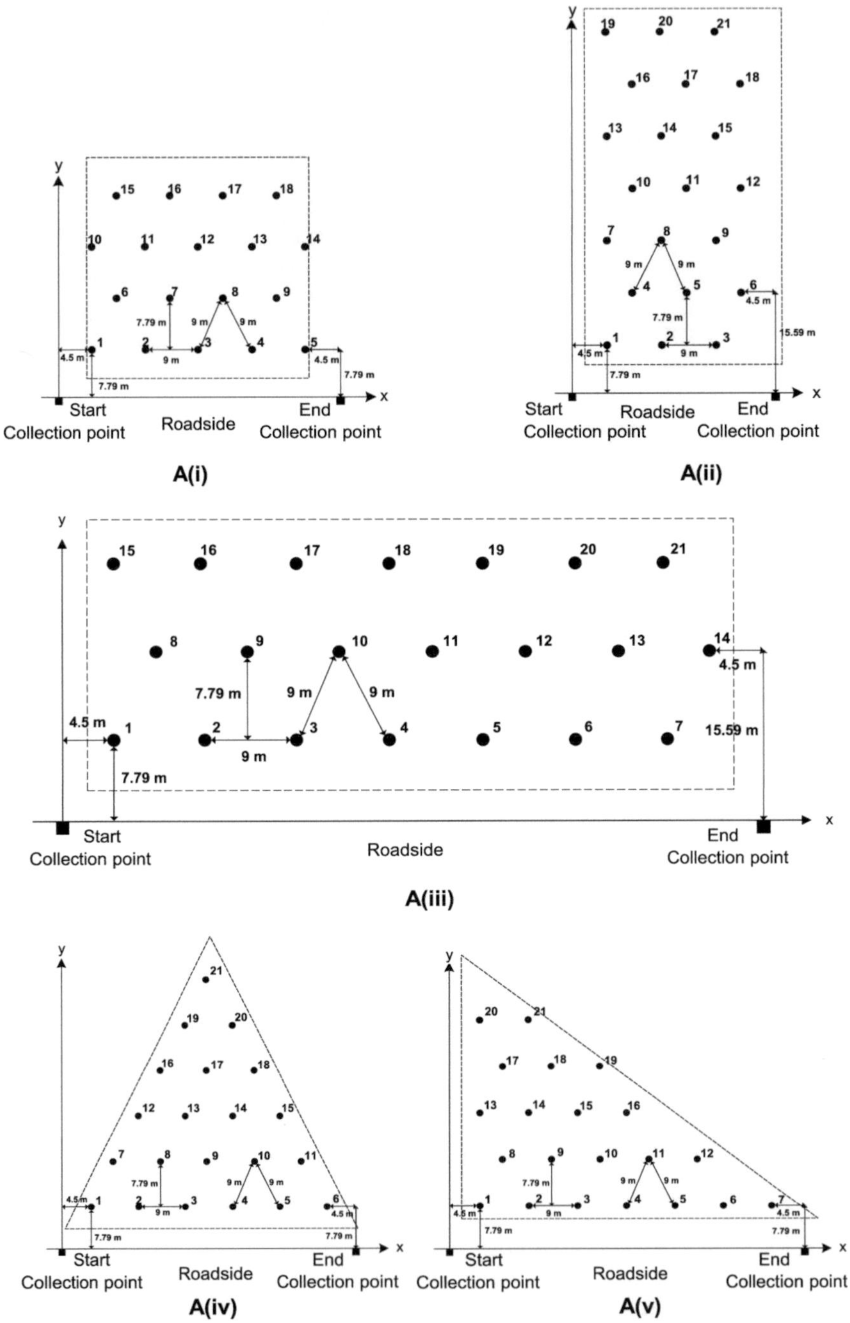

Fig. 2 Floorplan of different oil palm plantation sites (Lim et al., 2021a)

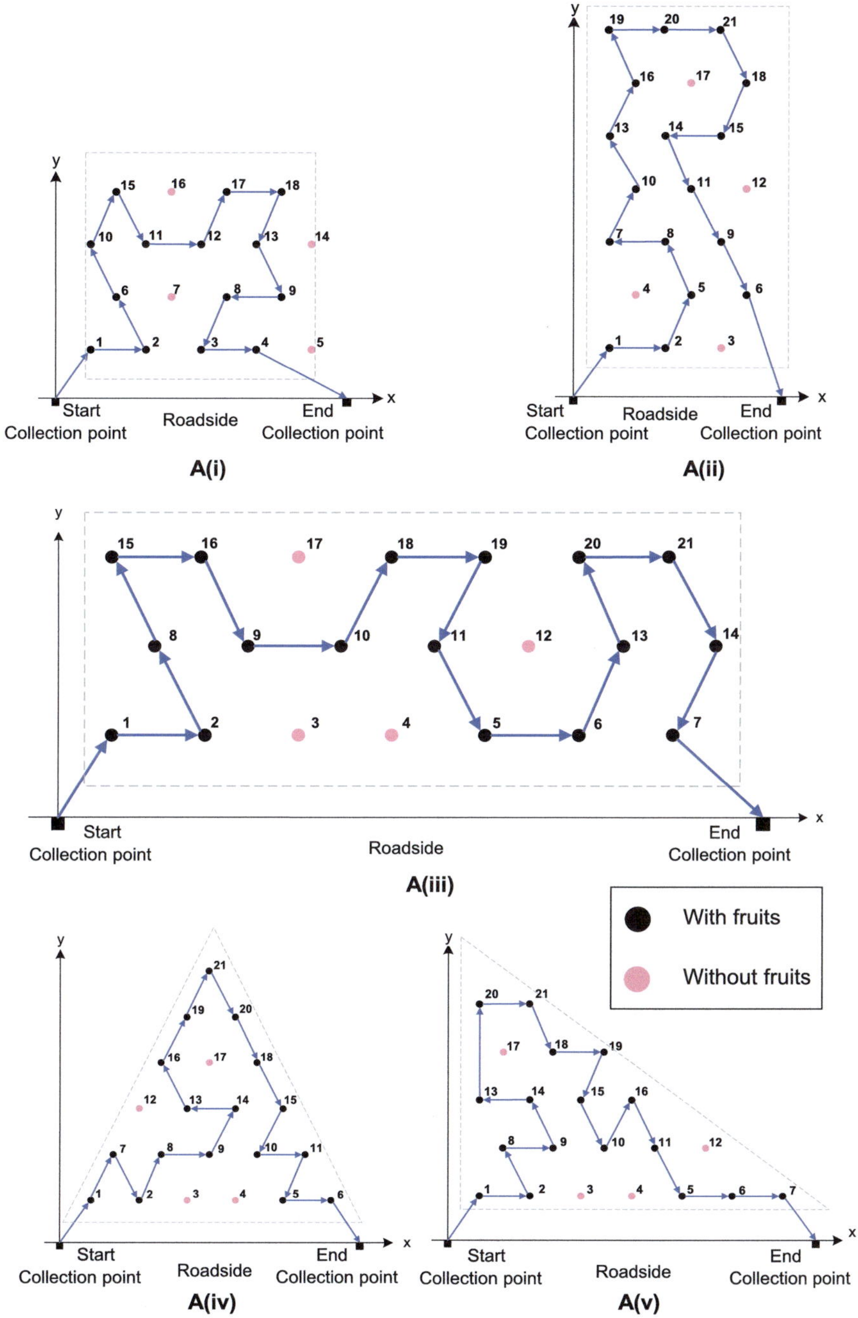

Fig. 3 Solution for shortest harvesting route for different floorplans (Lim et al., 2021a)

Table 3 Comparison of the proposed shortest harvesting distance with existing practice on different floorplans (Scenario A)

Case	Total distance proposed by HERO model (m)	Total distance required to visit all tree (m)	Reduction of distance travelled (%)
A(i)	141.6	171.0	17.2
A(ii)	169.2	214.2	21.0
A(iii)	164.9	198.0	16.7
A(iv)	162.0	198.0	18.2
A(v)	168.6	198.0	14.8

of FFBs distribution represents the low, average, and high yield of FFBs due to several factors, including different species, region, seasons, weather, and age. The loading capacity of the transporter was not taken into consideration for these cases, and all cases were standardised by using the square floorplan.

Figure 4 and Table 4 show the solutions of shortest route for the respective cases. In the maximum yield of FFB at 100% in Case B(i), farmers are required to travel 171.0 m to visit all the trees for harvesting and evacuation process. In Case B(ii) with 80% distribution (same as Case A(i)), the shortest distance required to complete the process was found to be at 141.6 m; while Case B(iii) (at 60% distribution) requires the shortest distance of 126.0 m due to the lowest number of harvestable trees. Compared to the conventional approach that requires farmers to visit all trees, Case B(iii) shows the highest improvement of 26.3% reduction in the travelled distance. Noted that in the solution for Case B(iii), the proposed optimum route passes through two non-harvestable trees, trees 4 and 17. The solution proposed by the HERO model will not deliberately avoid any non-harvestable tree if that is the shortest route, which is a logical solution. With the proposed HERO model, farmers could strategically avoid tree without fruits to improve the process efficiency.

The reason for harvesting FFB at low ripped fruit distribution is to prevent the build-up of free fatty acid within the ripped fruit (Basyuni et al., 2017). Besides, ripped fruit is more likely to fall onto the ground, resulting in the generation of more loose fruits; this will impose additional collection time or lower yield. (Zainon et al., 2017). Typically, farmers would visit the plantation sites for harvesting purpose once in 7–13 days for the optimum amount of matured FFB to be harvested (Castillo et al., 2017; Mohanaraj & Donough, 2016). The incorporation of optimisation model could improve the process efficiency and provide a better platform for resource planning and management.

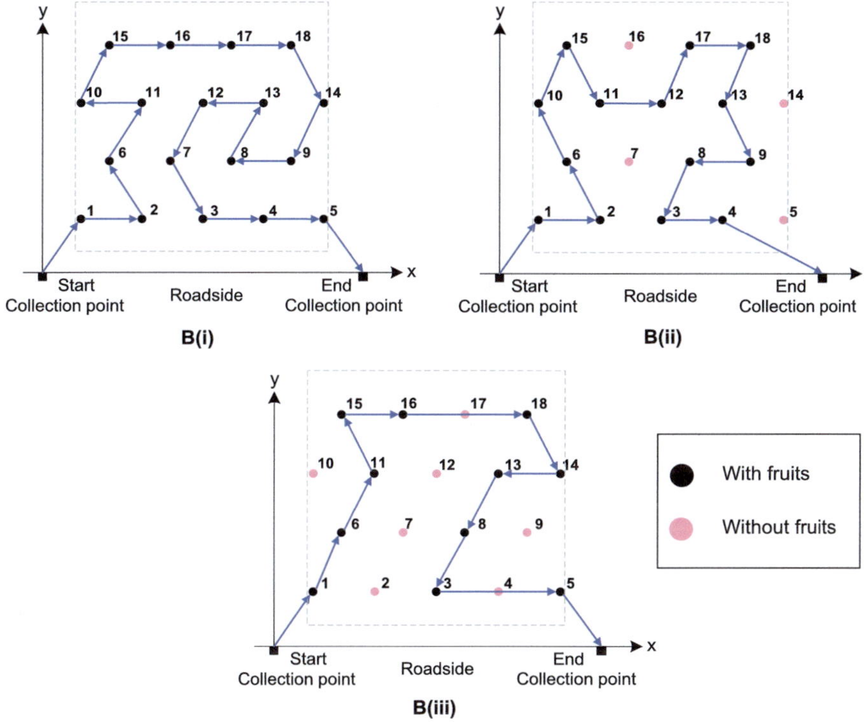

Fig. 4 Shortest harvesting route for different distributions of harvestable palm tree (Lim et al., 2021a)

Table 4 Comparison of the proposed shortest harvesting distance with existing practice at different harvestable fresh fruit bunch distributions (Scenario B)

Case	Total distance proposed by HERO model (m)	Total distance required to visit all tree (m)	Reduction of distance travelled (%)
B(i)	171.0	171.0	0
B(ii)	141.6	171.0	17.2
B(iii)	126.0	171.0	26.3

6.4 *Scenario (C): Optimisation of Harvesting and Evacuation Route with Different Transporter Loading Capacity*

Another factor to be considered in the FFB harvesting and evacuation process is the transporter loading capacity used by established enterprises and SMEs. Depending on the size of the plantation, terrain, facility, accessibility, and cost budget, each plantation may have different strategies and tools to conduct FFB evacuation process. For example, SMEs prefer wheelbarrow due to the lower capital and maintenance

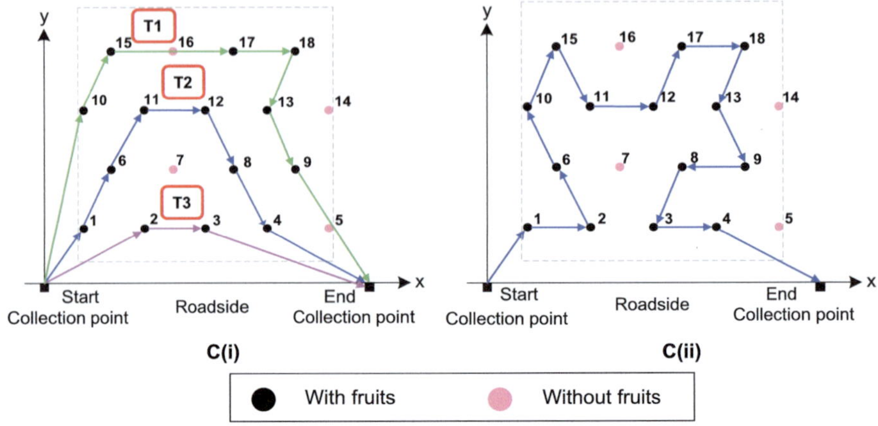

Fig. 5 Shortest harvesting route for different oil palm transporters loading capacities (Lim et al.. 2021a)

costs. Nonetheless, it is not ideal to cover large plantation site due to the limited loading capacity and the difficulty to operate on uneven ground (Shuib et al., 2020). Scenario C was conducted to investigate the impact of transporter loading capacity on the number of trips and the total travelled distance. Wheelbarrow and mechanical buffalo were considered in Case C(i) and C(ii) with the loading capacity of 150 kg and 700 kg, respectively (Muhamad & Aziz, 2018; Shuib et al., 2010). Both cases were conducted with a square floorplan and 80% of harvestable oil palm trees as shown in Fig. 5 with the optimised evacuation route. Due to the loading capacity limitation, Case C(i) requires three trips to complete the process ($t = 3$), while C(ii) requires single trip ($t = 1$).

Table 5 shows the computed result for Scenario C. For Case C(i), each trip of T1, T2, and T3 ended with the total load of the wheelbarrow at 133.2 kg, 132.3 kg, and 44.2 kg, respectively. The total travelled distance for the case was found to be 213.8 m to cover all harvestable trees. The result could be interpreted differently depending on the number of wheelbarrows available for the process. If only a single unit is available, farmers could follow the proposed route in series, such as T1 > T2 in reverse > T3 to transport the FFB to the collection points. Alternatively, three wheelbarrows can be deployed in parallel to each of the proposed paths to reduce the total evacuation time. Such strategy would require a higher capital cost on multiple units of transporter and farmers for the parallel operation. For Case C(ii), mechanical buffalo was able to complete the evacuation process in a single trip at a total load of 309.7 kg and a travelled distance of 141.6 m. The larger capacity enables a continuous collection process and reduces the required travel distance.

The optimum harvesting and evacuation route with different types of transporters could be utilised in the decision-making process. For instance, the optimum number of trip and travelled distance by each type of transporter also reflects the investment and operating cost, carbon footprint, and the process efficiency. For comparison,

Table 5 Comparison of the proposed shortest harvesting distance with different loading capacities (Scenario C)

Case	Transporter type	Loading capacity (kg)	Total load for each trip (kg)	Total distance proposed by HERO model (m)
C(i)	Wheelbarrow	150	T1 = 133.2	213.8
			T2 = 132.3	
			T3 = 44.2	
C(ii)	Mechanical buffalo	700	309.7	141.6

the cost of wheelbarrow and mechanical buffalo was reported to be approximately USD 36 (~MYR 150) and USD 1275 (~MYR 5,300) with an economic lifespan of 1 and 9 years, respectively (Syahrizan et al., 2017). In terms of environmental impact, mechanical buffalo had higher carbon footprint resulted from the motorised operation. The average productivity for both wheelbarrow and mechanical buffalo were reported to be approximately 1.75 and 8.5 t/d, respectively (Shuib et al., 2010). Hence, mechanical buffalo would be a more effective option for stakeholders' consideration if production is more than 7 t/d of FFB.

7 Conclusions

This chapter demonstrated the potential of integrating Industry 4.0 technology into the agriculture sector via computational-aided tools in the form of mathematical optimisation modelling approach, with the aim to improve the FFB harvesting and evacuation process. The HERO model utilises the data available from the existing processes to propose the shortest route for different scenarios, including different plantation floorplans, distribution of ripe FFB, and transporter loading capacity. The proposed solution can be used to achieve better resource management, route planning, and transporter investment strategy. The HERO model discussed in this chapter is part of the Industry 4.0 enabler in solving complex problems using computational power. The advancement of cloud computing and high-speed Internet eliminates the requirement of high investment cost in computational facilities in agriculture industry. This may motivate more oil palm stakeholders to invest in Industry 4.0 technologies, such as smart sensors and software to automate the process to reduce labour dependency and enhance process efficiency.

Acknowledgements Financial support from Universiti Tunku Abdul Rahman (research fund IPSR/RMC/UTARRF/2020-C2/L08) and Swinburne University of Technology Sarawak Campus (Research Supervision Grant 2-5545 RSG) is gratefully acknowledged.

References

Abdul-Hamid, A. Q., Ali, M. H., Tseng, M. L., Lan, S., Kumar, M., (2020). Impeding challenges on industry 4.0 in circular economy: Palm oil industry in Malaysia. Comput. *Operations Research, 123*, 105052. https://doi.org/10.1016/j.cor.2020.105052

Ahmad, F. B., Zhang, Z., Doherty, W. O. S., & O'Hara, I. M. (2019). The outlook of the production of advanced fuels and chemicals from integrated oil palm biomass biorefinery. *Renewable and Sustainable Energy Reviews, 109*, 386–411. https://doi.org/10.1016/j.rser.2019.04.009

Alam, A. S. A. F., Er, A. C., & Begum, H. (2015). Malaysian oil palm industry: Prospect and problem. *Journal Food Agriculture and Environment, 13*, 143–148.

Ayompe, L. M., Schaafsma, M., & Egoh, B. N. (2021). Towards sustainable palm oil production: The positive and negative impacts on ecosystem services and human wellbeing. *Journal of Cleaner Production, 278*, 123914. https://doi.org/10.1016/j.jclepro.2020.123914

Azwan, M. B., Norasikin, A. L., Sopian, K., Abd Rahim, S., Norman, K., Ramdhan, K., & Solah, D. (2017). Assessment of electric vehicle and photovoltaic integration for oil palm mechanisation practise. *Journal of Cleaner Production, 140*, 1365–1375. https://doi.org/10.1016/j.jclepro.2016.10.016

Basyuni, M., Amri, N., Putri, L. A. P., Syahputra, I., Arifiyanto, D. (2017). Characteristics of fresh fruit bunch yield and the physicochemical qualities of palm oil during storage in north sumatra, Indonesia. *Indonesian Journal of Chemistry, 17*, 182–190. https://doi.org/10.22146/ijc.24910

Begum, H., Alam, A. S. A. F., & Awang, A. H. (2019). Sustainability of Malaysian oil palm: A critical review. *International Journal Environment and Sustainable Development, 18*, 409–429. https://doi.org/10.1504/IJESD.2019.103467

Bonneau, X., Impens, R., Buabeng, M. (2018). Optimum oil palm planting density in West Africa. *OCL—Oilseeds fats, Crop. Lipids*, 25, A201. https://doi.org/10.1051/ocl/2017060

Castillo, E. G., Rodríguez C., L. F., Páez, A.F. (2017). Evaluación de dos procedimientos de cosecha de fruto de palma de aceite Elaeis guineensis Jacq. Un estudio de caso. *Agronomia Colombiana, 35*, 92–99. https://doi.org/10.15446/agron.colomb.v35n1.58524

Chong, K. L., Kanniah, K. D., Pohl, C., & Tan, K. P. (2017). A review of remote sensing applications for oil palm studies. *Geo-Spatial Information Science, 20*, 184–200. https://doi.org/10.1080/10095020.2017.1337317

Erdinç, O., Yetilmezsoy, K., Erenoğlu, A. K., & Erdinç, O. (2019). Route optimization of an electric garbage truck fleet for sustainable environmental and energy management. *Journal of Cleaner Production, 234*, 1275–1286. https://doi.org/10.1016/j.jclepro.2019.06.295

Fernandes, F. P., Costa, M. F. P., & Fernandas, E. M. G. P. (2009). Overview on mixed integer nonlinear programming problems. *AIP Conference Proceedings, 1168*, 1374–1377. https://doi.org/10.1063/1.3241342

Harun, M. H., & Noor, M. R. M. (2002). Fruit Set and Oil Palm Bunch Components. *Journal of Oil Palm Research, 14*, 24–33.

Innocenti, E. D., & Oosterveer, P. (2020). Opportunities and bottlenecks for upstream learning within RSPO certified palm oil value chains: A comparative analysis between Indonesia and Thailand. *Journal of Rural Studies, 78*, 426–437. https://doi.org/10.1016/j.jrurstud.2020.07.004

Kassim, M. S. M., Ismail, W. I. W., Ramli, A. R., & Bejo, S. K. (2014). Image clustering technique in oil palm fresh fruit bunch (FFB) growth modeling. *Agriculture and Agricultural Science Procedia, 2*, 337–344. https://doi.org/10.1016/j.aaspro.2014.11.047

Khajavirad, A., & Sahinidis, N. V. (2018). A hybrid LP/NLP paradigm for global optimization relaxations. *Mathematical Programming Computation, 10*, 383–421. https://doi.org/10.1007/s12532-018-0138-5

Kılınç, M. R., & Sahinidis, N. V. (2018). Exploiting integrality in the global optimization of mixed-integer nonlinear programming problems with BARON. *Optimization Methods Software, 33*, 540–562. https://doi.org/10.1080/10556788.2017.1350178

Kushairi, A., Singh, R., Ong-Abdullah, M., (2017). The oil palm industry in Malaysia: Thriving with transformative technologies. *Journal of Oil Palm Research, 29*, 431–439. https://doi.org/10.21894/jopr.2017.00017

Lamsal, K., Jones, P. C., & Thomas, B. W. (2016). Harvest logistics in agricultural systems with multiple, independent producers and no on-farm storage. *Computers & Industrial Engineering, 91*, 129–138. https://doi.org/10.1016/j.cie.2015.10.018

Lim, C. H., Cheah, Z. H., Lee, X. H., How, B. S., Ng, W. P. Q., Ngan, S. L., Lim, S., & Lam, H. L. (2021a). Harvesting and evacuation route optimisation model for fresh fruit bunch in the oil palm plantation site. *Journal of Cleaner Production, 307*, 127238. https://doi.org/10.1016/j.jclepro.2021.127238

Lim, C. H., Lim, S., How, B. S., Ng, W. P. Q., Ngan, S. L., Leong, W. D., Lam, H. L. (2021b). A review of industry 4.0 revolution potential in a sustainable and renewable palm oil industry: HAZOP approach. *Renewable and Sustainable Energy Reviews*, 135, 110223. https://doi.org/10.1016/j.rser.2020.110223

Merrick, K. (2021). Palm-oil market in 2020 [WWW Document]. M. P. Evans Gr. PL. https://www.mpevans.co.uk/palm-oil/market/palm-oil-market

Mohanaraj, S. N., & Donough, C. R. (2016). Harvesting practices for maximum yield in oil palm: Results from a re-assessment at IJM plantations Sabah. *Oil Palm Bull, 72*(72), i.

Morel, A., Friedman, R., Tulloch, D. J., & Caldecott, B. (2016). *Stranded assets in palm oil production: A case study of Indonesia*. University of Oxford.

Morgans, C. L., Meijaard, E., Santika, T., Law, E., Budiharta, S., Ancrenaz, M., & Wilson, K. A. (2018). Evaluating the effectiveness of palm oil certification in delivering multiple sustainability objectives. *Environmental Research Letters, 13*, 064032. https://doi.org/10.1088/1748-9326/aac6f4

MPOB. (2021). Economics and industry development division—statistics [WWW Document]. Malaysian Palm Oil Board. http://bepi.mpob.gov.my/

Muhamad, Z.-M., & Aziz, M. F. A. (2018). Mechanization in oil palm harvesting. *International Journal Academic Research in Business and Social Sciences, 8*, 247–256. https://doi.org/10.6007/ijarbss/v8-i5/4098

Rafii, M. Y., Isa, Z. A., Kushairi, A., Saleh, G. B., & Latif, M. A. (2013). Variation in yield components and vegetative traits in Malaysian oil palm (Elaeis guineensis jacq.) dura×pisifera hybrids under various planting densities. *Industrial Crops and Products, 46*, 147–157. https://doi.org/10.1016/j.indcrop.2012.12.054

Ruysschaert, D., & Salles, D. (2014). Towards global voluntary standards: Questioning the effectiveness in attaining conservation goals. The case of the roundtable on sustainable palm oil (RSPO). *Ecological Economics, 107*, 438–446. https://doi.org/10.1016/j.ecolecon.2014.09.016

Savolainen, V., Clottey, V. A., Doubi, B. T. S., Konan, J. L., Quain, M., Bezeng, B. S., Logah, V., Wireko-Kena, A., Osekre, E. A., Atuah, L., Angui, C. M. V., Ameka, G., Turkson, B., Boatemaa, A., Anankware, J. P., Boafo, H. A., Agyei-Dwarko, D., Collins, C. M. (Tilly). (2020). Systems thinking creates opportunities for a circular economy and sustainable palm agriculture in Africa. *Current Research in Environtal Sustainability, 1*, 31–34. https://doi.org/10.1016/j.crsust.2020.05.001

Seyyedhasani, H., & Dvorak, J. S. (2018). Reducing field work time using fleet routing optimization. *Biosystems Engineering, 169*, 1–10. https://doi.org/10.1016/j.biosystemseng.2018.01.006

Shahriari, B., Ravari, M. R. K., Yousefi, S., & Tajdari, M. (2016). A heuristic algorithm based on line-up competition and generalized pattern search for solving integer and mixed integer non-linear optimization problems. *Latin American Journal of Solids and Structures, 13*, 224–242. https://doi.org/10.1590/1679-78252293

Shuib, A. R., Khalid, M. R., Deraman, M. S. (2010). Enhancing field mechanization in oil palm management. *Oil Palm Bull*, 1–10.

Shuib, A. R., Radzi, M. K. F. M., Bakri, M. A. M., & Khalid, M. R. M. (2020). Development of a harvesting and transportation machine for oil palm plantations. *Journal of the Saudi Society of Agricultural Sciences, 19*, 365–373. https://doi.org/10.1016/j.jssas.2020.05.001

Syahrizan, S., Farahida, Z., Nizar, K. M., Aluiddin, Z. M. R. B. M., Adzmi, Y., Shafiq, S., Ismail, R. (2017). Buffalo assistance for in-field collection of fresh fruit bunches in oil palm plantation, In *The 2nd International Symposium on Sustainable Agriculture and Agro-Industry (ISSAA2017)* (pp. 1–10).

Tan, K. T., Lee, K. T., Mohamed, A. R., & Bhatia, S. (2009). Palm oil: Addressing issues and towards sustainable development. *Renewable and Sustainable Energy Reviews, 13*, 420–427. https://doi.org/10.1016/j.rser.2007.10.001

Tan, S. T., Hashim, H., Abdul Rashid, A. H., Lim, J. S., Ho, W. S., & Jaafar, A. B. (2018). Economic and spatial planning for sustainable oil palm biomass resources to mitigate transboundary haze issue. *Energy, 146*, 169–178. https://doi.org/10.1016/j.energy.2017.07.080

Wan, A. J., Samsudin, S. N., Pebrian, D. E. (2018). Farm machinery operating costs in oil palm plantations an evaluation of repairs and fuel costs. In *MSAE Conference* (pp. 398–409).

Zainon, M. S., Mohd Taib, N. B., Bin Yusof, M. S., Bin Rahim, M. Z., Bin Mohd Tobi, A. L., Bin Othman, M. S. (2017). Study on handing process and quality degradation of oil palm fresh fruit bunches (FFB). *IOP Conference Series Materials Science Engineering 203*, 012027. https://doi.org/10.1088/1757-899X/203/1/012027

Zero-Waste Technologies for the Sustainable Development of Oil Palm Mills

Yeit Haan Teow, Mohd Sobri Takriff, Mohd Shahbudin Masdar, Sahilah Abdul Mutalib, Peer Mohamed Abdul, Jamaliah Md. Jahim, Zahira Yaakob, Shuhaida Harun, and Mohammed Faisal Mohammed Yunus

Abstract Wastes generated from oil palm processing can be potentially utilised for the production of renewable energy and organic fertilisers, as well as the recovery of used water. This chapter describes a research programme that adopted a sustainable approach. The primary objective of this research programme was to develop an integrated technology for zero-waste oil palm processing. Seven strategic thrust areas were identified to address the sustainability challenges of oil palm processing, namely, discharge of highly polluting effluent, emission of greenhouse gases and the generation of solid waste. The programme allows the generation of renewable energy from palm oil mill effluent (POME) and pretreated empty fruit bunches (EFB) hydrolysate from biohydrogen production. POME from a biohydrogen and biomethane two-stage fermenter was subjected to integrated algae wastewater treatment, carbon dioxide sequestration and membrane filtration to recover treated water for boiler feed application. Meanwhile, EFB and POME, along with the residual solids and algae biomass produced from these zero-waste technologies were used to produce organic fertiliser. A successfully tested laboratory-scale technology was

Earlier version of this manuscript was published in Journal of Oil Palm, Environment and Health (https://doi.org/10.5366/jope.2021.04).

Y. H. Teow (✉) · M. S. Takriff · M. S. Masdar · P. M. Abdul · J. Md. Jahim · Z. Yaakob · S. Harun
Department of Chemical and Process Engineering and, Faculty of Engineering and Built Environment, Universiti Kebangsaan Malaysia, 43600 Bangi, Selangor Darul Ehsan, Malaysia
e-mail: yh_teow@ukm.edu.my

Research Centre for Sustainable Process Technology, Faculty of Engineering and Built Environment, Universiti Kebangsaan Malaysia, 43600 Bangi, Selangor Darul Ehsan, Malaysia

M. S. Takriff
Chemical and Water Desalination Engineering Programme, College of Engineering, University of Sharjah, Sharjah, UAE

S. Abdul Mutalib
Department of Food Sciences, Faculty of Science and Technology, Universiti Kebangsaan Malaysia, 43600 Bangi, Selangor Darul Ehsan, Malaysia

M. F. Mohammed Yunus
Sime Darby Plantation Berhad, Main Block, Level 3, Plantation Tower, No. 2, Jalan PJU 1A/7, Ara Damansara, 47301 Petaling Jaya, Selangor Darul Ehsan, Malaysia

D. C. Y. Foo et al. (eds.), *Sustainable Technologies for the Oil Palm Industry*,
https://doi.org/10.1007/978-981-19-4847-3_10

scaled up to pilot scale for demonstration. A pilot plant installed under the Universiti Kebangsaan Malaysia (UKM)-Yayasan Sime Darby (YSD) zero-waste technology project at Sime Darby's Tennamaram Oil Palm Mill features various technologies for sustainable development, transforming a regular mill into a green technology factory. Results of the pilot plant operation showed that POME and solid biomass can be used to generate additional 20% renewable energy than the technologies adopted in existing mills at 1020 t/d of POME discharge. Organic fertiliser was produced at a lower cost and shorter duration and the recovered water was used as boiler feedwater. The UKM-YSD pilot plant is a pioneering technology with the possibility of changing perceptions, particularly amongst foreign importers, that oil palm production can be environment-friendly.

Keywords Zero-waste technology · Biohydrogen · Organic biofertiliser · CO_2 sequestration · Water recovery · Circular economy

1 Introduction

The oil palm industry is one of the most important agroindustries for rural development in Malaysia. It provides direct employment to nearly 1 million of Malaysia's population and indirect employment to another 250,000 individuals (MPOC, 2020). Besides, it also accounts for 7.0% of the country's national gross domestic product (MPOC, 2020). However, the extraction and purification processes in oil palm processing are invariably accompanied by a considerable amount of generated waste. For every tonne of processed fresh fruit bunches (FFB), 0.5–0.7 tonnes of palm oil mill effluent (POME) and 0.37 tonnes of solid waste are generated (The Institution of Engineers, Malaysia (IEM), 2019). The latter is composed of 0.22 tonnes of empty fruit bunches (EFB), 0.06 tonnes of kernel shell and 0.09 tonnes of fibres (The Institution of Engineers, Malaysia (IEM), 2019). POME is the single largest source of industrial wastewater pollution in Malaysia. Considering its extremely high biochemical oxygen demand (BOD) and chemical oxygen demand (COD), POME poses considerable threat to aqueous environments due to oxygen depletion and serious pollution problems (Sajjad et al., 2018). Conventionally, POME is biologically treated in a series of open ponds. However, this treatment method suffers from the emission of greenhouse gases (GHGs), particularly methane (CH_4) and carbon dioxide (CO_2) (Arisht et al., 2020; Ghani, 2019). In accordance with the 2018 report of the Roundtable on Sustainable Palm Oil (RSPO) greenhouse gas working group, 62.3% of the greenhouse gases emitted from oil palm mills are contributed by POME (Roundtable on Sustainable Palm Oil, 2018). Thus, whilst the oil palm industry is benefitting from a highly profitable commodity, its commitment to the environment should remain unquestionable.

The Universiti Kebangsaan Malaysia (UKM)-Yayasan Sime Darby (YSD) Chair for Sustainable Development adopted a multidisciplinary and multi-institutional approach for developing an integrated zero-waste oil palm processing technology.

Collaboration was established in February 2021 with research institutions and universities from the Netherlands, Indonesia, Thailand and Taiwan to ensure that the zero-waste technology is relevant and techno-economically feasible. The zero-waste research programme consisted of following seven thrust areas:

i. Thrust Area 1: Hydrogen gas (H_2) for power generation,
ii. Thrust Area 2: Bioconversion of EFB and POME into organic fertiliser,
iii. Thrust Area 3A: Utilisation of POME and EFB hydrolysate for biohydrogen production,
iv. Thrust Area 3B: Catalysis of biogas into hydrogen,
v. Thrust Area 4: Biomass pretreatment for biohydrogen production,
vi. Thrust Area 5: Algae CO_2 sequestration and effluent treatment,
vii. Thrust Area 6: Water recycling and reuse.

The UKM-YSD zero-waste research programme aims to transform oil palm mills into green, carbon (C)-neutral factories with zero discharge into the air, ground and water. The integrated zero-waste oil palm processing strategy will simultaneously increase revenue and improve the sustainability of the oil palm industry. In the zero-waste oil palm processing strategy, the by-products generated during oil palm mill processing become resources for other processes and thus, waste is eliminated. By-products are utilised in the generation of renewable energy, the production of biofertilisers and the reclamation and reuse of treated effluent. The development timeline of the UKM-YSD zero-waste technology project is provided in Fig. 1.

The concept of the integrated zero-waste oil palm processing strategy is illustrated in Fig. 2. This strategy is based on the circular economy principle wherein the by-products of a given process unit are used as feed materials in other processes to retain the resources within the production cycle. Major oil palm processing by-products, namely, POME and solid biomass, are utilised as feedstock to produce renewable

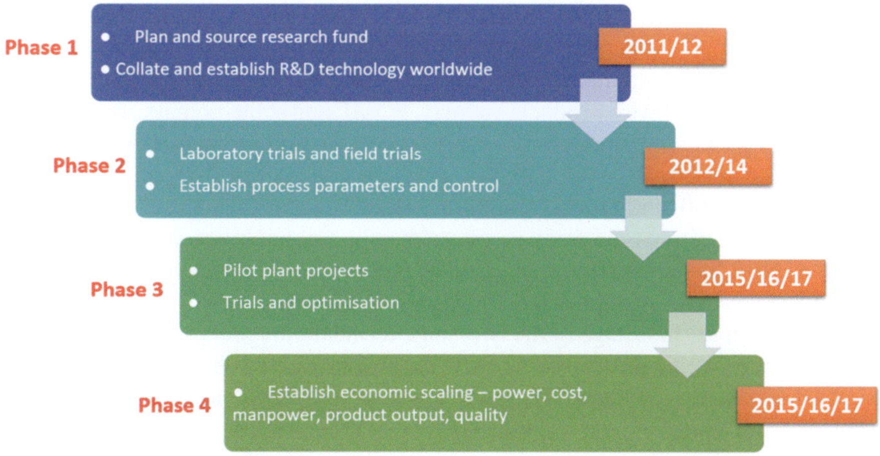

Fig. 1 Development timeline of the UKM-YSD zero-waste technology project

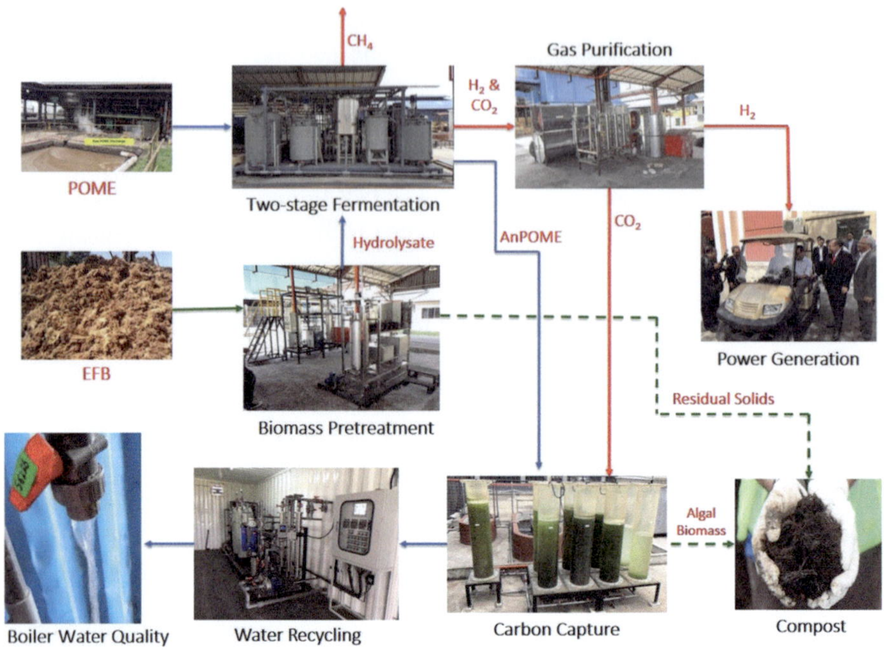

Fig. 2 Integrated zero-waste oil palm processing strategy (Teow & Takriff, 2021)

energy and organic fertilisers. They are also used to recover water. These resources are critical for the oil palm industry.

In a plantation, the cost of fertilisers is 30% of its operation (Harahap et al., 2019). Therefore the capability to convert palm biomass into fertiliser is not only economically sound but also in line with the RSPO certification. In accordance with this certification, biomass must be returned to the plantation to ensure that the land will not be stripped of the required nutrients. A typical oil palm mill with a capacity of 60 tonnes FFB/h requires 25 kWh of power for its operation (Kramannandita et al., 2014). Processing by-products, namely, POME and solid biomass, are adequate resources for generating renewable energy to power the mill's operation whilst excess power can be sold to the grid. The use of renewable energy from processing by-products will significantly reduce carbon footprint.

Fresh POME discharged from an oil palm mill and hydrolysate produced from pretreated EFB in Thrust Area 4 are used as feedstock to produce biohydrogen and biomethane in a two-stage fermentation process in Thrust Area 3A. The biohydrogen gas produced in the first stage of the two-stage fermentation process is sent to Thrust Area 1 for purification prior to being used for power generation. Meanwhile, anaerobic POME discharged from Thrust Area 3A (after biogas production) and CO_2 extracted from Thrust Area 1 are transferred to Thrust Area 5. Microalgae are used in Thrust Area 5 for further treating the POME discharged from Thrust Area 3B and sequestering CO_2 separated from biohydrogen gas in Thrust Area 1. An integrated membrane filtration system in Thrust Area 6 is used to treat POME further and recover water resources, which are used as boiler feedwater. EFB, POME and other residual solids produced from these zero-waste technologies are used to produce organic fertiliser in Thrust Area 2.

These thrust areas are illustrated in the following sub-sections of the chapter.

2 Thrust Area 1: Biohydrogen for Power Generation

Thrust Area 1 involves research in purifying biohydrogen for power generation. H_2 is clean renewable energy that reduces dependency on fossil fuels (Ramirez-Moreno et al., 2013). Biohydrogen produced from POME and EFB hydrolysate in fermentation (Thrust Area 3B) or a formulated gas mixture can be used in fuel cell systems to generate electricity and other industrial applications. The biogas produced in Thrust Area 3B is a mixture of H_2 and CO_2. In addition, it possibly contains certain amounts of carbon monoxide (CO), CH_4 and hydrogen sulphide (H_2S). However, biohydrogen must attain 99.9% purity to be applied to a polymer electrolyte membrane fuel cell (PEMFC) system. The presence of CO and H_2S may affect the function of fuel cell components, such as catalysts and membranes. They may be adsorbed onto the active sites of the catalysts, blocking catalyst activity. This condition leads to overpotentials at the electrodes, diminishing PEMFC performance due to a reduction in stack power (Wu et al., 2017).

The effect of CO_2 on PEMFC stack was tested using different CO_2 concentrations at constant H_2 concentration prior to entering the anode side of the fuel cell. The concentration of CO_2 varied from 0 to 60% in H_2 fuel. Figure 3 shows that the PEMFC stack operated at different CO_2 concentrations during its 2 h performance. As presented in the figure, the presence of CO_2 at the anode side of the fuel cell causes a huge decrease in fuel cell performance compared with that of pure H_2. Moreover, the continuous supply of CO_2 in the fuel stream causes a decline in fuel cell performance. A higher CO_2 concentration will result in a higher degree of decrease in stack power. Thus, purifying biohydrogen to more than 99% purity is necessary.

Several types of purification techniques, including membrane, absorption and adsorption techniques, were evaluated at laboratory scale. For the membrane technique, two types of membrane with different thickness values, namely, polysulphone (PSF) membrane (glassy polymer) and polydimethylsiloxane (PDMS) membrane (rubbery polymer), were tested to obtain a membrane with higher permeability and H_2 selectivity. The selectivity of H_2/CO_2 that used PDMS and PSF membranes at different thicknesses and operating pressure values is presented in Fig. 4. As shown in Fig. 4a, the selectivity of H_2/CO_2 decreased with increasing pressure regardless of the thickness of the membranes. The highest selectivity for the PDMS membranes was 1.19 and 1.06 for a membrane thickness of 75 mm and 200 mm, respectively (measured at 1 bar). This result indicates that the permeance of H_2 is more favourable

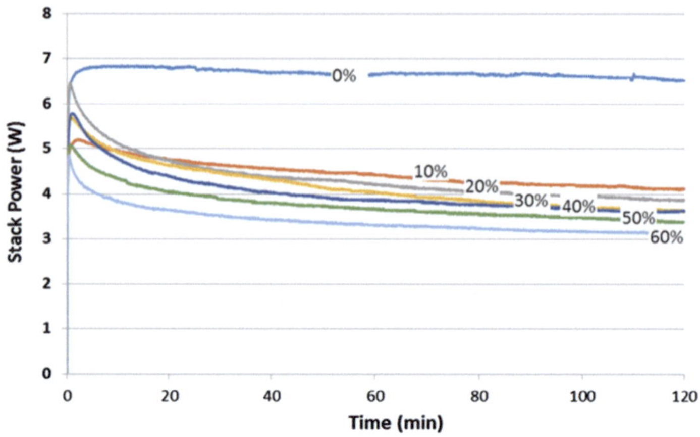

Fig. 3 Performance of PEMFC stack for 2 h of operation at different CO_2 concentrations (Chew et al., 2015)

than that of CO_2 at a pressure of 1 bar and vice versa for other pressure values (2–8 bar). The purification of H_2 requires higher H_2 permeance than CO_2 and thus, obtaining H_2/CO_2 selectivity of more than 1 is crucial. Subsequently, the selectivity of the PDMS (200 mm) and PSF (250 mm) membranes was measured and compared at 1–8 bar, as shown in Fig. 4b. The highest selectivity of 3.32 was recorded at 8 bar for the PSF membrane. The PSF membrane exhibited better H_2 permeance than the PDMS membrane. The selectivity of the PDMS membrane decreased with pressure, which is in contrast with the result of the PSF membrane. This finding validates that H_2 is less permeable than CO_2 at higher pressures for the PDMS membrane.

An absorption technique was also implemented to purify biohydrogen. The absorption of CO_2 was conducted at different concentrations of potassium hydroxide (KOH), ammonia and monoethanolamine (MEA) solutions under different operating conditions, such as feed flow rate (1–5 mL/s), temperature (ambient to 60 °C) and absorption time (2–60 min). Figure 5 shows the percentage of purified H_2 at solution concentrations of 0.01, 0.1 and 1 M at a feed flow rate of 1.0 mL/s. The highest percentage of purified H_2 was observed at the solution concentration of 1 M for all the solvents, wherein nearly 100% CO_2 was absorbed into the solution. For the solution concentration of 0.1 M, the KOH, ammonia and MEA solutions exhibited a drastic decrease within 15 min of absorption time with 89.9%, 88.9% and 88.3% of H_2 purification, respectively. At this concentration, the solutions became the limiting factor for CO_2 absorption because H_2 purity started to decrease with time. By contrast, the concentration of 0.01 M exhibited an insignificant change compared with the control for all the solvents because it started to decrease as early as 2 min of the absorption time.

In accordance with this early investigation, the membrane technique requires higher energy than the absorption technique to operate under high operating pressure.

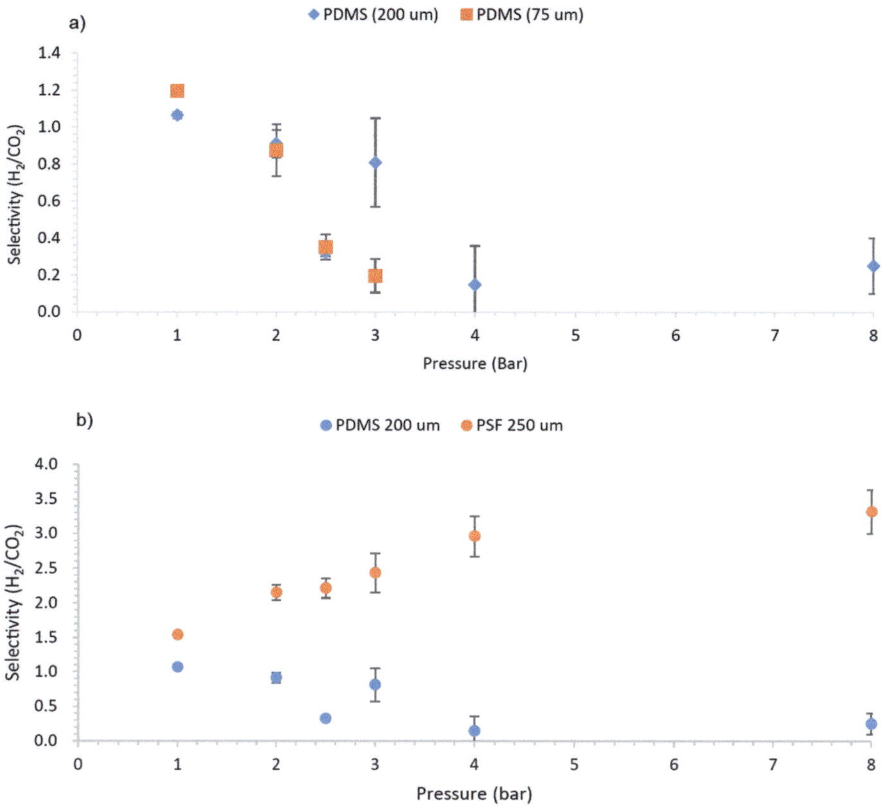

Fig. 4 Selectivity for H_2/CO_2 at different pressure values of **a** the PDMS membrane at different thickness values and **b** the PDMS and PSF membranes (Mohamad et al., 2016)

Hence, the adsorption technique was selected as an alternative technique for biohydrogen purification. A suitable adsorbent with a tendency to adsorb CO_2 was required. To improve CO_2 adsorption capacity, different types of adsorbents were synthesised, characterised and evaluated by impregnating commercial activated carbon (AC) with different types of chemical substances. Biohydrogen purification was conducted using an adsorption–desorption technology with a low cost and an efficient adsorbent, i.e. commercial AC (particle size: 3.0–4.2 mm) supplied by Effigen Carbon Sdn. Bhd (Sidek et al., 2019). Commercial AC is a suitable solid adsorbent for CO_2 and other contaminant gases due to its higher surface area, microporosity, thermal stability, removal capacity and lower cost per unit volume of adsorber compared with those of other mesoporous materials, such as zeolite, metal–organic compounds, porous silica and incandescent clay. This adsorbent can achieve high CO_2 adsorption capacity and is easy to regenerate. For example, Wu et al. obtained 56.12–77.17 mg/g adsorption capacity for CO_2 by using AC under ambient conditions (Wu et al., 2017). Moreover, a high CO_2 adsorption capacity was obtained at pressure values of 1.5–2.0 bar and

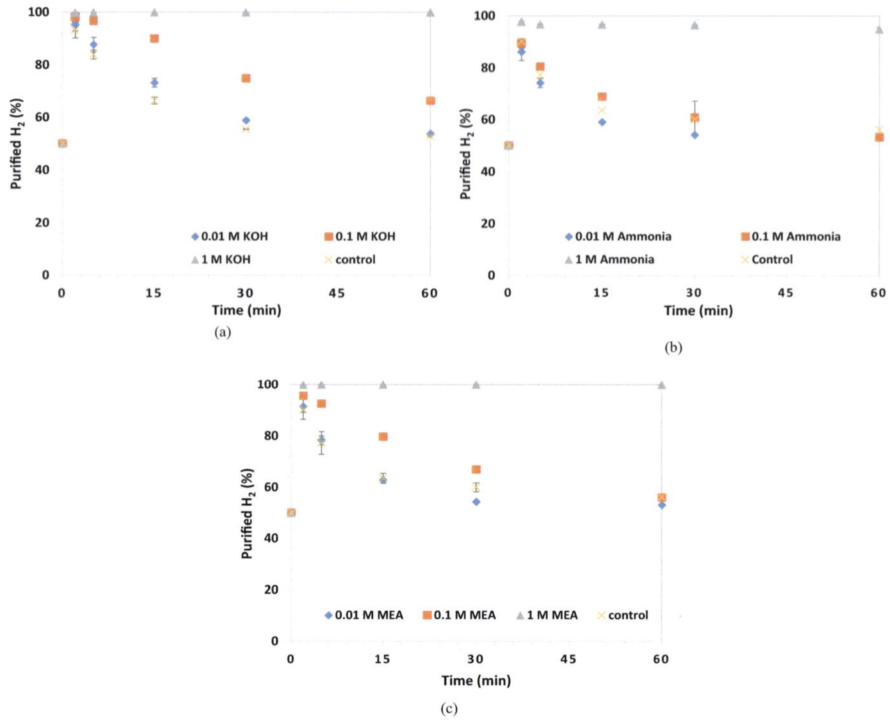

Fig. 5 Purified H_2 (%) at 1.0 mL/s of the inlet gas flow rate at 0.01, 0.1 and 1 concentrations of the **a** KOH, **b** ammonia and **c** MEA solutions within 60 min of absorption time (Mohamad, 2016)

a temperature of 60 °C. This statement is supported by the adsorption breakthrough curves of different types of AC used in the current study (Fig. 6). As shown in the figure, commercial AC took a shorter breakthrough time (12.3 min) in attaining CO_2 adsorption compared with modified AC (Sidek et al., 2019) by chemically capturing CO_2 through a chemisorption process (Chew et al., 2015).

After the completion of the adsorption cycle, spent AC was regenerated through desorption in an adsorber by using compressed air. Adsorption–desorption cycles were performed to determine the reusability (i.e. desorption efficiency) of AC. Commercial AC presented the lowest adsorption–desorption profile (500 min, as depicted in Fig. 7) due to its low adsorption capacity. Therefore, less time was required for the regeneration step (Sidek et al., 2019).

The biohydrogen produced after the adsorption–desorption process achieved more than 90% H_2 purity (Ru et al., 2018). The adsorbed CO_2 with a concentration range of 30–40 vol.% was sent to Thrust Area 5 for microalgae CO_2 sequestration. The purified H_2 supplied to the PEMFC system can generate electricity, which is then used as transportation fuel in a modified buggy car. This biohydrogen–fuel cell integration system is beneficial for creating a sustainable energy supply. It also minimises waste and pollution by utilising biohydrogen as a green energy resource.

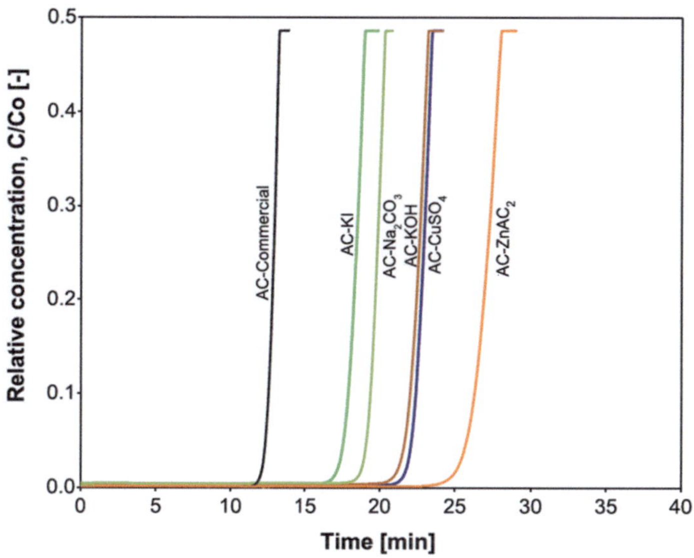

Fig. 6 Adsorption breakthrough curves of CO_2 onto fresh AC (Sidek et al., 2019)

3 Thrust Area 2: Bioconversion of EFB and POME into Organic Fertiliser

EFB and residual solids produced from the integrated zero-waste oil palm processing strategy are converted into organic fertiliser in Thrust Area 2. EFB and POME are used as potential raw materials for conversion into organic fertiliser because these materials contain high organic matter and nutrient contents that can be recycled into soil (Berhad, 2019). EFB and POME can improve soil structure. They can also increase water holding capacity and soil pH. Although EFB is used for natural composting in oil palm plantations, the degradation process is lengthy. Moreover, nitrogen fertiliser supplied to oil palm trees is not fully absorbed by the trees because part of the fertiliser is consumed by microbes for the degradation of EFB.

The conversion of EFB and POME into organic fertiliser will take at least 2–3 months. However, the composting period can be reduced with an appropriate strategy. The use of indigenous microorganisms for the conservation and utilisation of new resources is concomitant with the Third National Agricultural Policy to conserve and utilise natural resources on a sustainable basis (Ministry of Agriculture, 2021). Microorganism activities play important roles in the biogeochemical cycles of carbon, nitrogen, sulphur and other elements, enabling ecosystems to recycle these substances into a form that is utilisable for living organisms. Apart from in situ microorganisms, the introduction of microbial consortia is essential for further promoting this process.

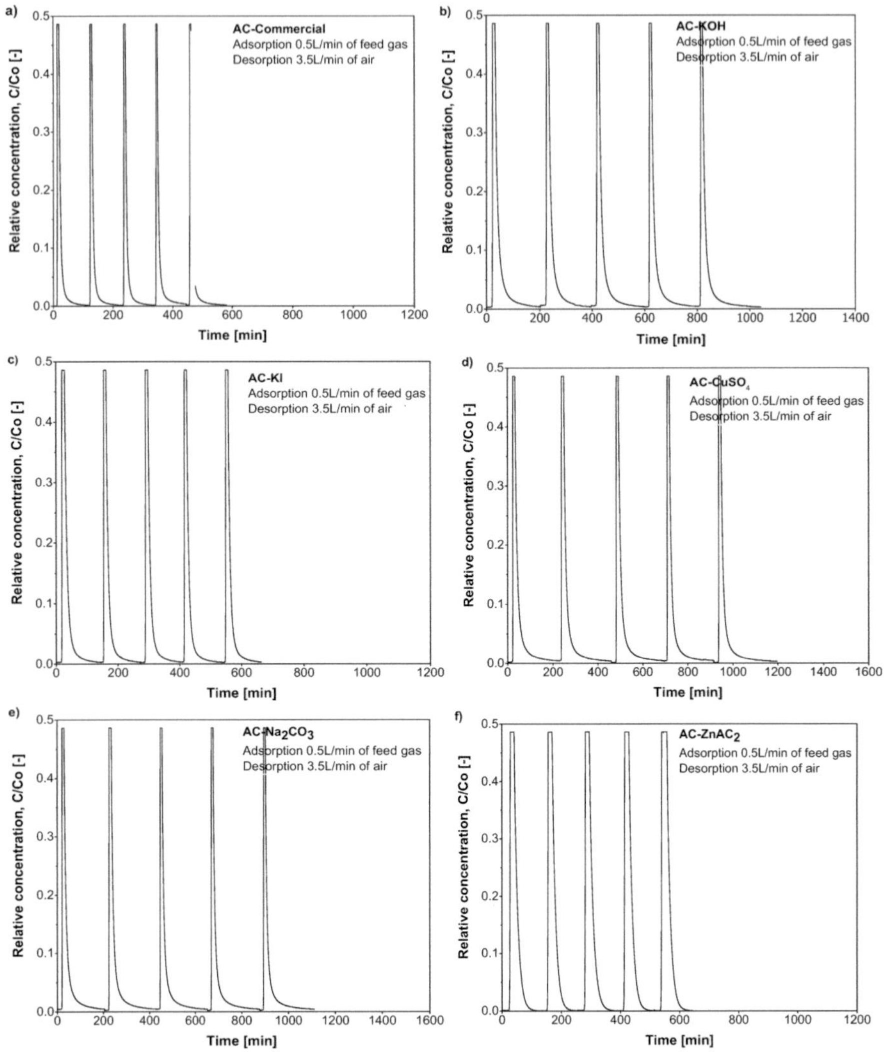

Fig. 7 CO_2 adsorption–desorption profiles of AC for five cycles: **a** commercial AC, **b** AC-KOH, **c** AC-KI, **d** AC-$CuSO_4$, **e** AC-Na_2CO_3 and **f** AC-$ZnAC_2$ (Sidek et al., 2019)

During the early stage of Thrust Area 2, thermophilic microorganisms were isolated and identified to improve the formulation of bacterial consortia. This newly developed formulation was then applied to EFB composting at 25 t per windrow capacity with the addition of POME. The introduced microbial consortia are important for degrading lignin and cellulolytic components during the decomposition of EFB and POME. The optimised formulation of microbial consortia was scaled up at

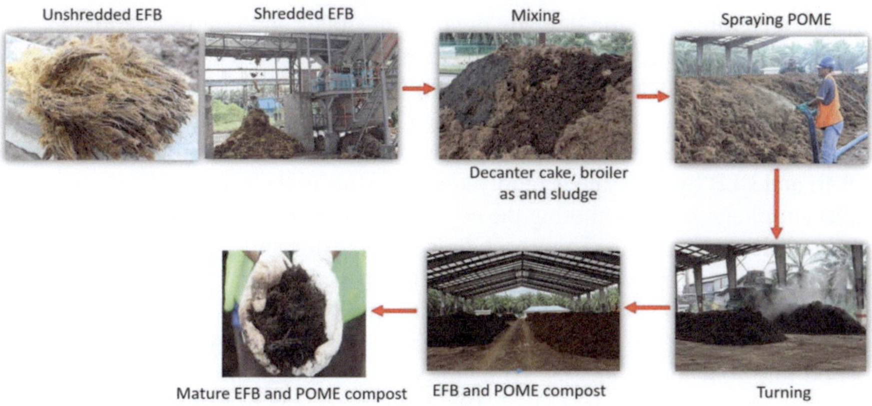

Fig. 8 Bioconversion of EFB and POME into organic fertiliser

operating capacities of 40, 80 and 400 tonnes. The bioconversion of EFB and POME into organic fertiliser is illustrated in Fig. 8.

The physical, chemical and microbiological parameters of the compost were analysed and monitored. A C/N ratio below 1:30 of the compost from shredded fresh EFB was achieved within 45 days of composting, with nitrogen >1.8%, phosphorus >0.5%, potassium >3.4% and moisture content <50%. The consistency and stability of the produced compost were validated using the Solvita test (https://solvita.com/).

4 Thrust Area 3A: Utilisation of POME and EFB Hydrolysate in Biohydrogen Production

The combustion of fossil fuels, such as coal, to generate energy produces high emissions of CO_2 into the atmosphere. CO_2 is a major contributor to global warming. Carbon sequestration is used to capture and store CO_2, allowing the continuous use of fossil fuels as the primary energy source in the world. However, the depleting supply of fossil fuels and stringent controls towards the release of greenhouse gases has accelerated the exploration of alternative energy sources with low/zero greenhouse gas emissions. Biohydrogen is actively promoted as a green viable replacement for fossil fuels. The energy evolved by H_2 is reported to be 2.75 times higher than that of conventional fossil fuel (Mahmod et al., 2021). In addition, combustion of H_2 has zero CO_2 emission and can be used in fuel cell technology.

POME and oil palm lignocellulosic biomass were utilised as substrates for biohydrogen production in Thrust Area 3A through dark fermentation. Fermentative biohydrogen production from organic waste materials exhibits high potential in meeting the demands of an ever-increasing world population without releasing any harmful substances into the environment (Maaroff et al., 2019). During the dark fermentation process, anaerobic bacteria degrade organic substrates and remove additional electrons to maintain ionic neutrality. Consequently, the redox reactions of ferredoxins yield molecular H_2 (Das & Veziroglu, 2008). The by-products of oil palm production have huge potential in the renewable energy field and the biorefinery industry, such as biohydrogen production. The concept of biohydrogen production with the use of POME and EFB hydrolysate is highly advantageous in the overall process economy whilst simultaneously promoting the concept of circular economy.

A traditional single-stage approach for anaerobic digestion (AD) only yields CH_4. AD is a multistep biological process that involves the transformation of complex organic compounds, such as polysaccharides in oil palm mill residues, into simple molecules, namely, CH_4, H_2 and CO_2. The biological steps in AD can be easily categorised into four major phases: (i) hydrolysis, (ii) acidogenesis, (iii) acetogenesis and (iv) methagenesis (Lay et al., 2019). The advantages of conventional AD were re-engineered into a two-stage anaerobic sequencing batch reactor (ASBR), leading to the accumulation of H_2. This process is typically disregarded in the single-stage system. The first reactor includes the first two phases of AD (hydrolysis and acidogenesis) and operates under thermophilic conditions (55 °C). The second reactor (i.e. the methanogenic digester) incorporates the remaining phases of AD and operates under mesophilic conditions (37 °C). Anaerobic bacteria in the hydrogenic reactor break down organic compounds into H_2, which is then harvested before CH_4 accumulation. This technique is essential for increasing H_2 yield by eliminating competitors, such as methanogens, during biohydrogen production.

This phenomenon results in the accumulation of H_2 via inexpensive dark fermentation (Das & Veziroglu, 2008). Figure 9 illustrates the schematic of the two-stage ASBR system. A thermophilic condition is optimum for biohydrogen production from POME with high yield (Zainal et al., 2018). The second mesophilic system is set close to ambient temperature by considering heating cost, particularly when the future upscaling of the process is considered.

The highest biohydrogen production of 1.67 ± 0.02 liquid H/liquid POME was attained under a thermophilic condition with an initial pH of 5.5, an initial sugar concentration of 13.57 g/L and a hydraulic retention time of 72 h. The fermentation of biohydrogen from locally available biomass, e.g. POME and EFB hydrolysate, is an attractive technology for bioenergy production.

5 Thrust Area 3B: Catalytic Conversion of Biogas into H

Thrust Area 3B involves synthesising catalysts for H production from biogas. Biogas that consists of CO_2 and CH_4 is a good renewable source for producing H. The latter is an alternative energy carrier that can be derived from partial oxidation (Wang et al., 2018) and dry and steam reforming (Álvarez et al., 2016). In general, CH_4 reforming with CO_2 or steam uses a catalytic system with a noble metal or a transition metal (Turchetti et al., 2016) as catalyst. Noble metal-based catalysts exhibit excellent catalytic performance in terms of feed conversion and H_2 yield. However, their high

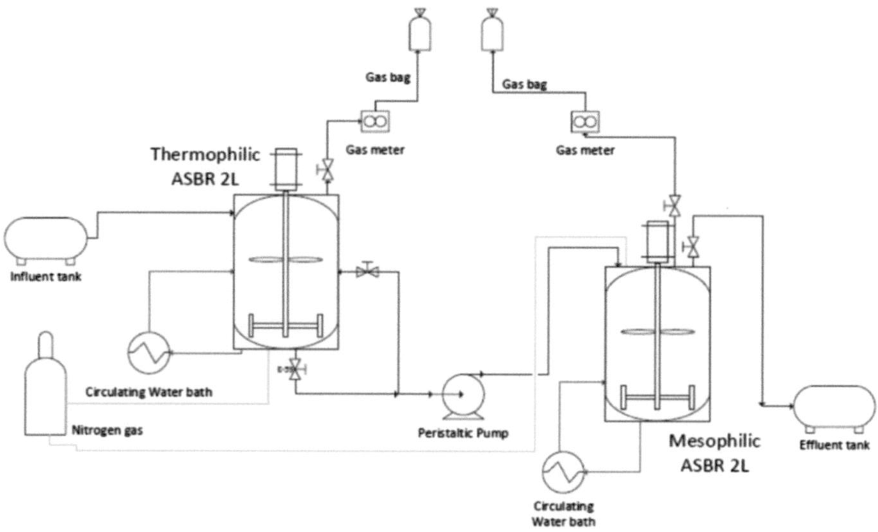

Fig. 9 Schematic of a two-stage thermophilic and mesophilic ASBR system (Maaroff et al., 2019)

cost and limited availability restrict their full application to reforming at a large scale. The potential of transition metals, such as nickel (Ni), iron (Fe) and cobalt (Co), which have relatively high activities and cost efficiency, has motivated studies in Thrust Area 3A to be applied to CH_4 reforming for converting CH_4 into H_2.

Transition metal (e.g. Ni, Fe and Co)-based catalyst was supported on magnesium aluminate ($MgAl_2O_4$) and titanium dioxide (TiO_2) and synthesised via sol–gel, impregnation and precipitation methods. A one-pot sol–gel method was used for the synthesis of monometallic Ni/TiO_2 catalysts. A series of Ni/TiO_2 catalysts under various loadings of Ni from 10 to 50 wt% was prepared by varying the amount of Ni nitrate during catalyst preparation. As a support for Ni-, Co- and Fe-based catalysts, $MgAl_2O_4$ was prepared by using the citrate fusion method. Nitrate salts of magnesium, aluminium and citric acid with a mole ratio of 1:2:3 were physically mixed in a mortar by using a pestle. The prepared $MgAl_2O_4$ was then loaded with the metals. Ni-, Co- and Fe-loaded catalysts were prepared via an impregnation method. Steam reforming, dry reforming and decomposition studies were performed in a fixed-bed reactor. Reaction was carried out at temperatures between 650 and 800 °C. H yield and selectivity were measured via gas chromatography.

Raw gas from the gas cylinders was directed to the reactor after passing through purification tanks. These tanks were used to remove moisture, sulphur, oxygen and suspended solid particles in gas. The system was designed using biogas as raw material. Therefore, the filter system was installed to remove sulphur, oxygen, CO_2 and water separately. Different types of materials were installed as adsorbents for this process. A gas filter was also installed prior to the gas entering the mass flow controller. This filter was used to remove solid particles.

6 Thrust Area 4: Biomass Pretreatment for Biohydrogen Production

EFB contains polysaccharide materials, such as glucan (25–55 wt%), hemicellulose (25–50 wt%) and lignin (15–30 wt%). Glucan and hemicellulose are structural carbohydrates that can be depolymerised into high-value C5 and C6 sugar intermediates (Abdul Rahman et al., 2018). In contrast with starch wherein carbohydrates are easily depolymerised into fermentable sugars, carbohydrate fractions in EFB are not readily available for downstream utilisation and conversion due to the complex and recalcitrant cell wall structure of EFB. Therefore, a pretreatment process is required to alter and open up the cell wall structure of EFB, particularly the lignin matrix, to hydrolyse hemicellulose, reduce the crystallinity of cellulose and make cellulose more accessible to enzymes during the enzymatic hydrolysis process to produce fermentable sugars.

Thrust Area 4 focused on developing a clean pretreatment technology for producing sugar intermediates from pressed, shredded EFB. The pretreatment process involved mechanical size reduction and physical–chemical pretreatment to deconstruct the EFB cell wall structure prior to enzymatic hydrolysis to produce C5 and C6 sugar intermediates for co-utilisation with POME in Thrust Area 3A during biohydrogen production.

Several categories of EFB were evaluated in this study to determine the influence of EFB morphology on pretreatment. EFB has three major types: fresh fibre (NFF), native dried fibre (NDF) and native ground fibre (NGF), with an average milled size of 2 mm. Anhydrous ammonia pretreatment (AAP) was conducted to pretreat the three types of EFB samples to overcome their recalcitrant cell wall structure (Marlowe, 2014), making cellulose more accessible to enzymes to increase enzymatic hydrolysis efficiency (Shamsudin et al., 2011). AAP is a dry-to-dry pretreatment process that consumes only a small amount of solvent (<3% anhydrous ammonia) and produces no liquid waste stream because ammonia is easily vaporised at room temperature. In pretreating EFB samples via AAP, the pretreatment conditions were based on the reported literature (Lau et al., 2010) of ammonia fibre expansion fibre pretreatment at 135 °C. The ratio of ammonia to water to EFB was 1:1:1 and pretreatment time was 45 min.

The morphology of the native and pretreated samples of different EFB types is presented in Fig. 10. The structural carbohydrate of the pretreated dried fibre (PDF) and pretreated ground fibre (PGF) slightly increased from the carbohydrate composition of the native EFB. AAP under the specified pretreatment conditions led to the incorporation of ammonia into the EFB cell wall structure, efficiently causing structure expansion, followed by structure disruption during the depressurisation of AAP (Shamsudin et al., 2011). Most ester linkages in the lignin–carbohydrate complex were broken during this process, significantly reducing lignin (~45%) in the pretreated samples. During AAP, anhydrous ammonia primarily deconstructed the ester bond and solubilised the hemicellulose and lignin of the EFB samples.

Fig. 10 Morphology of the native and pretreated samples of different EFB categories

The enzyme cocktail for the enzymatic hydrolysis of pretreated EFB was prepared using a combination of different enzyme species: (1) Celluclast 1.5 L, Viscozyme L and Novozyme 188; (2) Ecozyme Cell and Novozyme 188 and (3) Cellic CTec2 and Cellic HTec2. The Cellic CTec2 and Cellic HTec2 enzyme cocktail exhibited the greatest capability to hydrolyse the deconstructed EFB. Enzymatic hydrolysis was further optimised with the use of the Cellic CTec2 and Cellic HTec2 enzyme cocktail at a ratio of 1:1, loading with 32.5 filter paper units/g of glucan at a hydrolysis temperature of 50 °C, 150 rpm agitation speed and 48 h hydrolysis time. The yield of fermentable sugar per dry weight of EFB under this optimised condition was 83% glucan and 68% xylan. The enzymatic hydrolysis under 1% glucan loading (GL) of PFF, PDF and PGF indicated that the glucan conversion of these samples was 67.2%, 69.4% and 70.2%, respectively, as shown in Fig. 11.

On the basis of the result, the pretreatment conditions of AAP were further optimised using the PDF sample. In particular, the ratio of ammonia to water to EFB was currently 2:1:1 and pretreatment time was 30 min. The PDF sample pretreated under these optimised conditions showed that glucan conversion at 1% GL enzymatic hydrolysis improved glucan conversion from 69.4 to 83.4%. Hydrolysis at 3% GL (equivalent to 10% solid loading) indicated that the PDF sample significantly achieved a glucan conversion of 75% and the glucose concentration in the fermentable sugar yielded 25 g/L within 30 h of the hydrolysis period. Further evaluation of this fermentable sugar for co-utilisation with POME during biohydrogen production presented a complementary strategy for sustaining zero-waste technology oil palm milling.

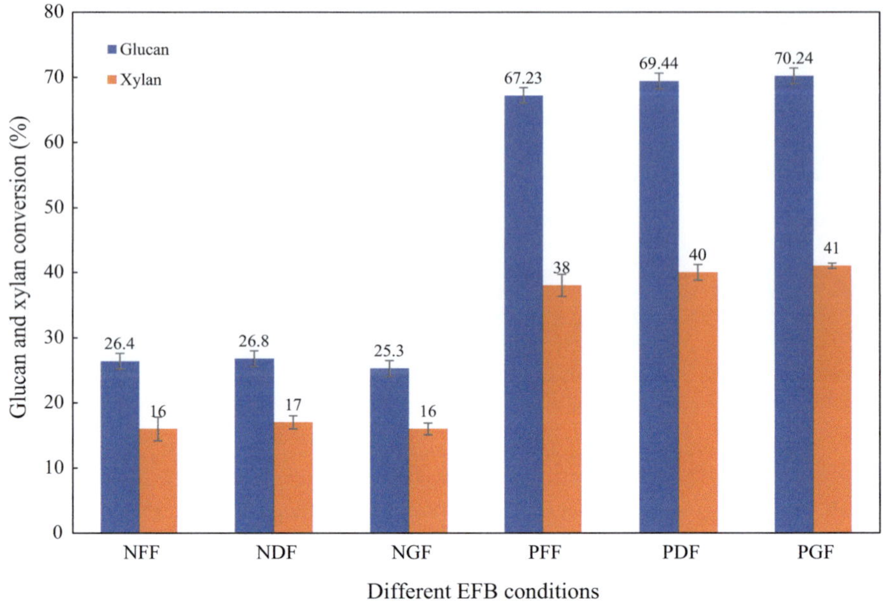

Fig. 11 Enzymatic hydrolysis performance of different EFB categories (under 1% GL)

7 Thrust Area 5: Algae CO_2 Sequestration and Effluent Treatment

The research in Thrust Area 5 focused on integrating the microalgal bioremediation of POME and the sequestration of CO_2. Local species of microalgae were used to treat the partially treated POME to avoid biosafety issues. The microalgae species were isolated from POME ponds to enhance their adaptation. POME that was partially treated in Thrust Area 3A through anaerobic digestion was further treated using microalgae in heterotrophic growth mode in a raceway pond. The microalgae utilised the residual nutrients in the partially treated POME for their growth and multiplication, further treating POME. The nutrient level in POME was further reduced by up to 80%.

The mature microalgae biomass used for bioremediating POME was harvested and the treated POME was used as a growth medium for CO_2 sequestration by other microalgae species. The CO_2 gas that was separated from biohydrogen during the purification step in Thrust Area 1 was used as feedstock for the microalgae CO_2 sequestration unit. Microalgae were used because of their ability to assimilate CO_2 through photosynthesis (Cuellar-Franca & Azapagic, 2015). Other native microalgae strains that were isolated from a POME treatment pond were used to capture CO_2 gas. The microalgae were cultivated in a series of photobioreactors in mixotrophic growth mode. The mature microalgae biomass in POME treatment and CO_2 sequestration were harvested and sent to Thrust Area 2 to be processed into organic fertiliser. The

N and P contents of the microalgae biomass will enhance fertiliser quality in terms of N and P compositions.

Figure 12 presents the isolated microalgae strains from the POME treatment ponds investigated in this research. By introducing microalgae for POME treatment, the positive symbiotic relationship between microalgae and other microorganisms, such as bacteria, can enhance contaminant reduction in POME. *Scenedesmus* sp. (UKM9) is the most efficient strain in POME treatment with 58%, 60%, 50% and 80% reduction in COD, total nitrogen, total phosphorus and total organic carbon (TOC), respectively. *Chlorella* spp. (UKM2) is the most promising microalgae strain for CO_2 sequestration. It can tolerate high CO_2 concentration (Kastanek et al., 2010) and successfully capture up to 25% CO_2 in a single pass. The optimised operational conditions for the best CO_2 fixation and total nitrogen reduction by this strain were achieved at 10% v/v CO_2 concentration, 1670 mL/min aeration rate and 24.8% v/v inoculum concentration with CO_2 fixation at 0.1208 g CO_2/L day and 80.9% total nitrogen reduction. Low CO_2 concentration is preferred to obtain a high dissolution rate of gas-to-liquid-phase CO_2 available for microalgae assimilation and to stabilise the pH of the suspension (Hariz et al., 2018). The microalgae assimilated nutrients from POME and CO_2 from flue gas. By the end of microalgae treatment, these components were converted into biomass (Hariz & Takriff, 2017). Mature algae biomass was harvested and used as a biofertiliser enhancer in Thrust Area 2.

8 Thrust Area 6: Water Recycling and Reuse

The integrated membrane technology for water recycling and reuse in Thrust Area 6 is the final stage in the UKM-YSD zero-waste research programme for POME tertiary treatment. The primary objective is to meet boiler feedwater standards and thus, allow water to be recycled and reused. The integrated membrane filtration system shown in Fig. 13 consists of a sand filter, an AC pretreatment unit and a crossflow membrane module used for further treating partially treated POME after passing through the microalgae reactor in Thrust Area 5 (Teow et al., 2016). The membrane modules used in the integrated membrane filtration system were reverse osmosis (RO), BW30 (https://www.lenntech.com/replacement/filmtec-bw30-4040.htm); low-pressure RO, XLE (https://www.lenntech.com/products/Filmtec-Membranes/154546/XLE-4040/index.html) and nanofiltration (NF) membrane, NF270 (https://www.lenntech.com/products/Filmtec-Membranes/149987/NF270-4040/index.html) purchased from DOW FILMTEC™.

Table 1 summarises the performance of the integrated membrane filtration system for different commercial membrane modules. Amongst the commercial membrane modules, BW30 exhibited the greatest performance in producing treated permeate that complied with the boiler feed standard for recycling and reusing. The BW30 membrane module for POME tertiary treatment presented the following removal efficiency: 98.54% ammonia nitrogen (NH_3-N), 99.93% COD, 99.98% total suspended solids (TSS), 99.51% total dissolved solids (TDS), 100% colour, 99.96% phosphorus,

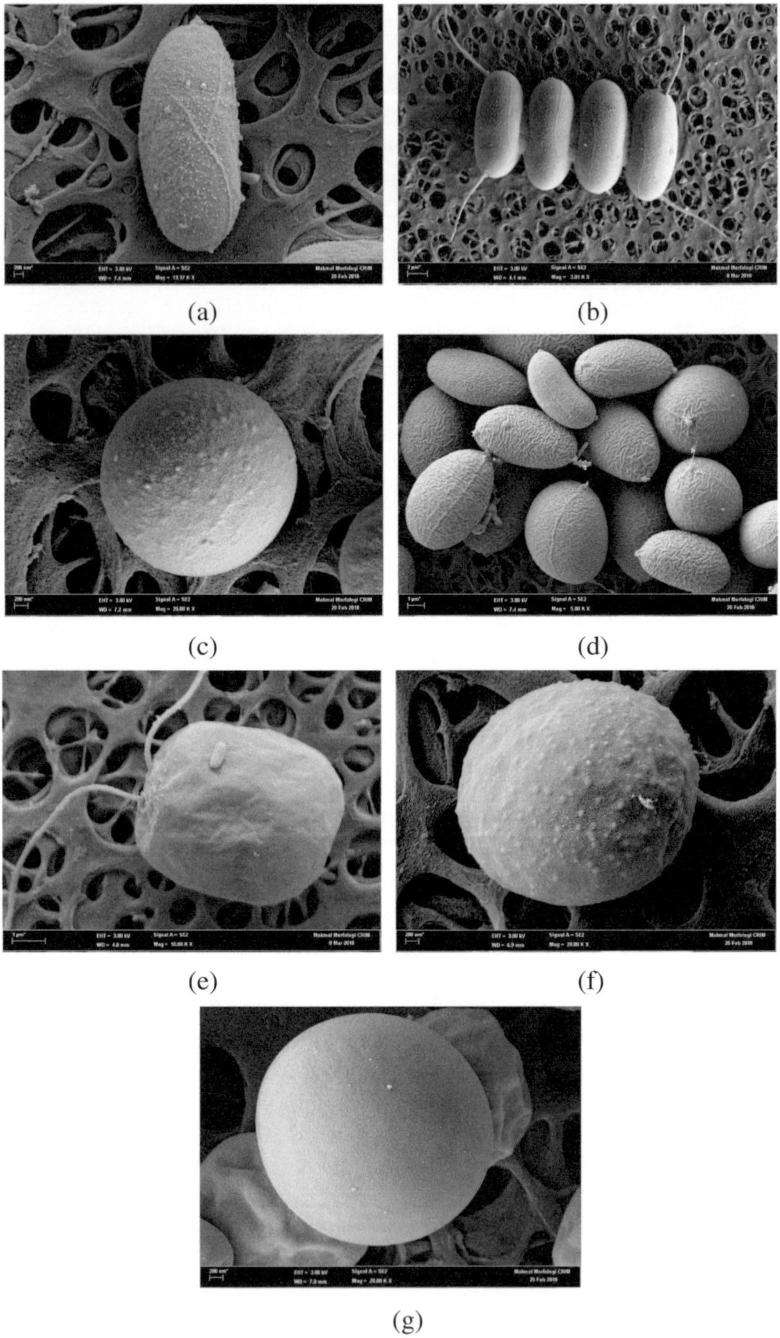

Fig. 12 Isolated microalgae strain from POME environment **a** *Characium* sp. UKM1 (13 k×), **b** *Scenedesmus* sp. UKM9 (3 k×), **c** *Chlorella sorokinia* UKM3 (20 k×), **d** *Coelastrella* sp. UKM4 (5 k×), **e** *Chlamydomonas* sp. UKM6 (10 k×), **f** *Chlorella pyrenoidosa* UKM7 (20 k×), **g** *Chlorella sorokinia* UKM8 (20 k×)

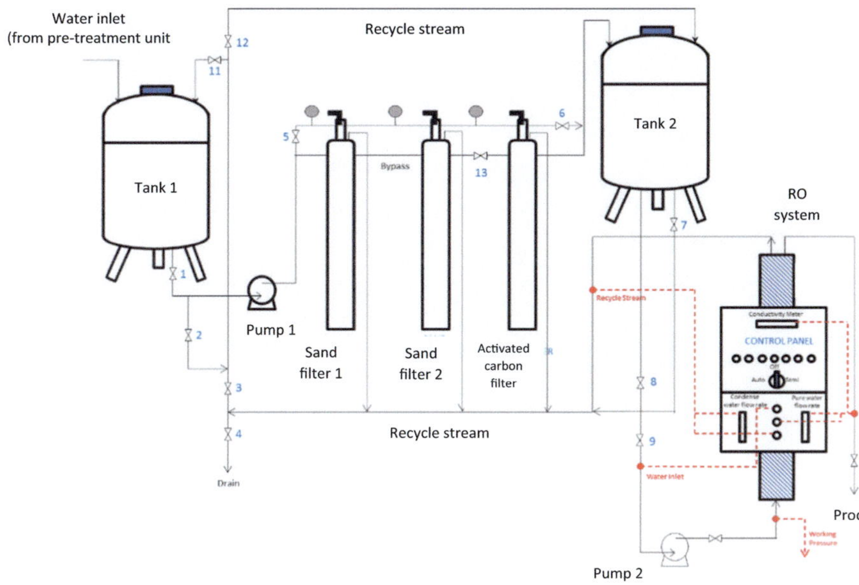

Fig. 13 Integrated membrane filtration system

99.81% turbidity, 99.50% conductivity, 55.62% hardness (Mg) and 68.49% hardness (Ca) (Ghani et al., 2017; Teow et al., 2018).

Figure 14 shows that odourless, crystal-clear permeate water is obtained from the integrated membrane filtration system by using BW30. POME treated with the integrated membrane filtration system can be fed into a boiler for internal plant

Table 1 Performance of the integrated membrane filtration system that uses different commercial membrane modules

Parameter	Feed solution	BW30	NF270	XLE	Boiler feedwater quality
NH_3-N (mg/L)	90.00	1.31	10.00	21.33	–
COD (mg/L)	152.00	0.11	68.33	15.00	<5
TSS (mg/L)	170.00	0.04	8.89	5.56	<10
TDS (mg/L)	5436.67	26.63	1660.00	293.00	<100
Colour (PtCo)	626.00	0.00	266.67	300.00	–
P (mg/L)	80.33	0.03	12.00	54.67	30–50
Turbidity (NTU)	46.90	0.09	1.74	2.74	0–3
Conductivity (μs)	10,590.00	53.40	3376.67	578.67	–
pH	7.68	8.60	8.49	9.08	7–10
Hardness, Mg (mg/L)	447.67	198.67	279.67	108.60	<500
Hardness, Ca (mg/L)	110.00	34.66	42.65	19.56	–

Fig. 14 Visualisation of water samples from Tank 1, Tank 2 and the permeate

usage. This outcome was supported by the study of Ahmad et al. (2003). The two-stage POME treatment system that combined pretreatment and membrane filtration processes produced high-quality treated effluent that can be used as boiler feedwater or drinking water. An oil palm mill requires a continuous supply of a large volume of steam from high-quality water for continuously supplying the production line and thus, maintaining high-efficiency operation. The alternative of reusing treated and purified POME from the integrated membrane filtration system will eventually help reduce total freshwater consumption and minimise the amount of wastewater discharge (Mavrov & Belieres, 2000).

9 Technology Readiness for Commercialisation

The developed zero-waste technology was fully demonstrated in a pilot-scale facility installed at Sime Darby's Tennamaram Oil Palm Mill, Selangor Darul Ehsan, Malaysia. Figures 15 and 16 show the schematic and photograph of the pilot-scale facility of the UKM-YSD zero-waste technology project, respectively. This research programme promises an alternative sustainable management practice for the oil palm industry.

Amongst the seven thrust areas of the UKM-YSD zero-waste technology project, Thrust Areas 2, 3B and 6 are ready for commercialisation. Thrust Area 2 is the bioconversion system for the co-composting of EFB and POME into biofertiliser. The optimisation provided by this system is to accelerate decomposition and reduce dependence on inorganic fertilisers. Given that the production of EFB is approximately 20% of that of FFB used in mill processes for oil extraction, this technology is

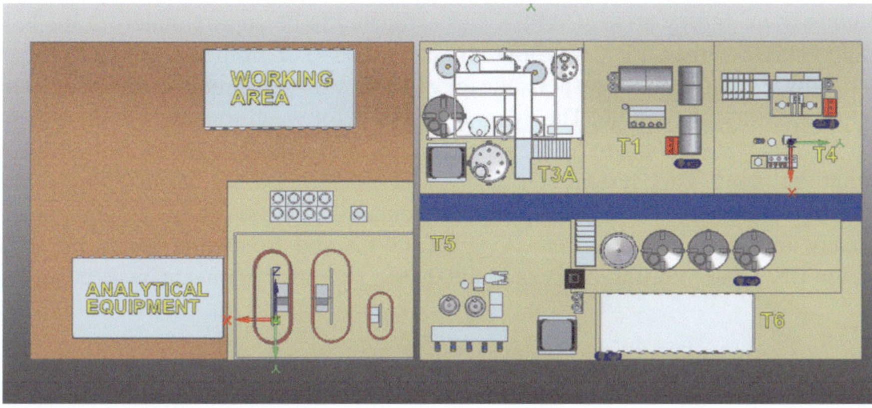

Fig. 15 Schematic of the pilot-scale facility

Fig. 16 Photograph of the pilot-scale facility

applicable or can be further expanded to a large-scale operation with capital expenditures (CAPEX), operating expenditures (OPEX) and production costs of 1.64 million USD, 0.45 million USD/year and 7.79 USD/metric tonne, respectively.

Thrust Area 3A is a two-stage anaerobic fermentation system that utilises POME and biomass for biohydrogen and biomethane production. The production of biohydrogen can be combined with existing biogas infrastructure to produce green and sustainable biofuel, which is cleaner and eliminates CO_2 emissions compared with commonly used hydrocarbon fuels. The system can be commercialised due to the high potential of alternative fuel or energy sources that can offer a climate-friendly solution to an energy crisis. CAPEX was estimated to be approximately 2.36–3.30 million USD, whilst OPEX was approximately 0.06–0.08 million USD/year.

Finally, Thrust Area 6 is responsible for water recycling and reuse by the integrated membrane filtration system. The integrated system for water treatment and purification reduces water footprint by recycling and reusing water to achieve zero effluent within the plant. This system can also reduce operating costs because the use of external water sources is reduced. From the economic analysis of this system, which is ready to be commercialised, the CAPEX, OPEX and production costs were 0.55 million USD, 0.07 million USD/year and 0.49 USD/tonne permeate, respectively.

10 Conclusion

The zero-waste research programme has successfully demonstrated that oil palm milling process by-products can be used to produce renewable energy and organic fertilisers by applying circular economy principles. In addition, they can be used to recover water resources. The environment is not only protected but new revenue streams can be created. The UKM-YSD zero-waste technology project serves as a pioneering initiative that can potentially change perceptions, particularly amongst international buyers and consumers, that oil palm production is environment-friendly. In addition, the research programme also demonstrates a successful university-industry partnership in addressing sustainability challenges. This zero-waste demonstration plant with the technology to produce biogas, biofertiliser and recycled water from oil palm processing by-products presents an excellent alternative for a sustainable management practice in the oil palm industry that is aligned with the principles of RSPO regarding environmental responsibility and the conservation of natural resources and biodiversity.

Acknowledgements The authors gratefully acknowledge the funding for this work from Yayasan Sime Darby (YSD), which provided seed funds through the establishment of the Universiti Kebangsaan Malaysia (UKM)-YSD Chair for Sustainability (PKT 6/2012) and the Sime Darby Plantation Research Sdn Bhd for providing additional grants and technical assistance. The authors also acknowledge all UKM postgraduates, researchers and Sime Darby Plantation Research Sdn Bhd who have contributed to the development of this integrated system.

References

Abdul Rahman, N. F., Harun, S., Sajab, M. S., Zubairi, S. I., Markom, M. M., Jahim, J. M., Mohd Nor, M. T., Abdullah, M. A., & Hashim, N. (2018). Boosting enzymatic hydrolysis of pressurized ammonium hydroxide pretreated empty fruit bunch using response surface methodology. *Journal of Engineering Science and Technology, 13*(8), 2421–2445.

Ahmad, A. L., Ismail, S., & Bhatia, S. (2003). Water recycling from palm oil mill effluent (POME) using membrane technology. *Desalination, 157*(1–3), 87–95.

Álvarez, M. A., Centeno, M. Á., & Odriozola, J. A. (2016). Ru-Ni Catalyst in the combined dry-steam reforming of methane: The importance in the metal order addition. *Topics in Catalysis, 59*(2–4), 303–313.

Arisht, S. N., Abdul, P. M., Jasni, J., Mohd Yasin, N. H., Lin, S. K., Wu, S. Y., Takriff, M. S., & Md. Jahim, J. (2020). Dose-response analysis of toxic effect from palm oil mill effluent (POME) by-products on biohydrogen producing bacteria—A preliminary study on microbial density and determination of EC50. *Ecotoxicology and Environmental Safety, 20*, 110991.

Chew, L. W., Masdar, M. S., & Kamarudin, S. K. (2015). Purification of bio-hydrogen using absorption techniques and the effect of CO_2 impurity on the PEMFC performance. *Journal of Chemistry, 5*(3A), 24–34.

Cuellar-Franca, R. M., & Azapagic, A. (2015). Carbon capture, storage and utilisation technologies: A critical analysis and comparison of their life cycle environmental impacts. *Journal of CO_2 Utilization, 9*, 82–102.

Das, D., & Veziroglu, T. N. (2008). Advances in biological hydrogen production processes. *International Journal of Hydrogen Energy, 33*, 6046–6057.

Ghani, M. S. H., Teow, Y. H., Ang, W. L., Mohammad, A. W., Ngteni, R., & Mohamed Yusof, K. M. (2017). Fouling assessment of tertiary palm oil mill effluent (POME) membrane treatment for water reclamation. *Water Reuse and Desalination, 8*(3), 412–423.

Ghani, M. S. H. (2019) *Pre-treatment, fouling behaviour, and cleaning property of membranes applied for tertiary palm oil mill effluent treatment* [Master Degree Dissertation]. Universiti Kebangsaan Malaysia.

Harahap, F., Leduc, S., Mesfun, S., Khatiwada, D., Kraxner, F., & Silveria, S. (2019). Opportunities to optimize the palm oil supply chain in Sumatra, Indonesia. *Energies, 12*(420), 1–24.

Hariz, H. B., & Takriff, M. S. (2017). Palm oil mill effluent treatment and CO_2 sequestration by using microalgae-sustainable strategies for environmental protection. *Environmental Science and Pollution Research, 24*, 20209–20240.

Hariz, H. B., Takriff, M. S., Ba-Abbad, M. M., Mohd Yasin, N. H., & Mohd Hakim, N. I. N. (2018). CO_2 fixation capability of Chlorella sp. And its use in treating agricultural wastewater. *Journal of Applied Phycology, 30*, 3017–3027.

Kastanek, F., Sabata, S., Solcova, O., Maleterova, Y., Kastanek, P., Branyikova, I., Kuthan, K., & Zachleder, V. (2010). In-field experimental verification of cultivation of microalgae Chlorella sp. using the flue gas from a cogeneration unit as a source of carbon dioxide. *Waste Management Resource, 28*, 961–966.

Kramannandita, R., Bantacut, T., Romli, M., & Makmoen, M. (2014). Utilizations of palm oil mills wastes as source of energy and water in the production process of crude palm oil. *Chemistry and Materials Research, 6*(8), 46–53.

Lau, M. J., Lau, M. W., Gunawan, C., & Dale, B. E. (2010). Ammonia fiber expansion (AFEX) pretreatment, enzymatic hydrolysis, and fermentation on empty palm fruit bunch fiber (EPFBF) for cellulosic ethanol production. *Applied Biochemistry and Biotechnology, 162*(7), 1847–1857.

Lay, C. H., Vo, T. P., Lin, P. Y., Abdul, P. M., Liu, C. M., & Lin, C. Y. (2019). Anaerobic hydrogen and methane production from low-strength beverage wastewater. *International Journal of Hydrogen Energy, 44*(28), 14351–14361.

Maaroff, R. M., Jahim, J. M., Azahar, A. M., Abdul, P. M., Masdar, M. S., Nordin, D., & Abd Nasir, M. A. (2019). Biohydrogen production from palm oil mill effluent (POME) by two stage anaerobic sequencing batch reactor (ASBR) system for better utilization of carbon sources in POME. *International Journal of Hydrogen Energy, 44*, 3395–3406.

Mahmod, S. S., Md. Jahim, J., Abdul, P. M., Luthfi, A. A. I., & Takriff, M. S. (2021) Techno-economic analysis of two-stage anaerobic system for biohydrogen an biomethane production from palm oil mill effluent. *Journal of Environmental Chemical Engineering, 9*(4), 105679.

Marlowe, A. S. (2014). *Enzyme optimization for lignocellulose hydrolysis using mechanistic modelling* [Master thesis]. Department of Chemical Engineering, Michigan Technological University, Michigan, United States of America.

Mavrov, V., & Belieres, E. (2000). Reduction of water consumption and wastewater quantities in the food industry by water recycling using membrane processes. *Desalination, 131*(1–3), 75–86.

Ministry of Agriculture. (2021). Third national agricultural policy (1998–2010)—Executive summary. Retrieved from https://www.pmo.gov.my/dokumenattached/Dasar/29THIRD_NATIONAL_AGRICULTURAL_POLICY_(1998_-_2010)_-_EXECUTIVE_SUMMARY.pdf

Mohamad, I. N., Rohani, R., Mastar, M. S., & Jahim, J. M. (2016). Permeation properties of polymeric membranes for biohydrogen purification. *International Journal of Hydrogen Energy, 41*(7), 4474–4488.

Mohamad, I. N. (2016). *Biohydroge upgrading from fermentation of palm oil mill effluent* [Master dissertation]. Universiti Kebangaan Malaysia.

MPOC. (2020). Nearly 1M work in oil palm industry. Retrieved from http://mpoc.org.my/nearly-1m-work-in-oil-palm-industry/

Ramirez-Moreno, M. J., Romero-Ibarra, I. C., Landeros, J. O., & Pfeiffer, H. (2013). Alkaline and alkaline-earth ceramic oxides for CO_2 capture, separation, and subsequent catalytic chemical conversion. In C. d. R. V. Morgado & V. Paulo (Eds.), *CO_2 sequestration and valorization* (pp. 403–442).

Roundtable on Sustainable Palm Oil. (2018). *Compilation of best management practices to reduce total emissions from palm oil production* (pp. 1–157).

Ru, F. Y., Zulkefli, N. N., Mt Yusuf, N. Y., & Masdar, M. S. (2018). Effect of operating parameter on H_2/CO_2 gas separation using electrochemical cell. *International Journal of Applied Engineering Research, 13*(1), 505–510.

Sajjad, A.-A., Teow, Y. H., & Mohammad, A. W. (2018). Sustainable approach of recycling palm oil mill effluent (POME) using integrated biofilm/membrane filtration system for internal plant usage. *Jurnal Teknologi, 80*(4), 165–172.

Shamsudin, S., Mohd Shah, U. K., Zainudin, H., Abd-Aziz, S., Mustafa Kamal, S. M., Shirai, Y., & Ali Hassan, M. (2011). Effect of steam pretreatment on oil palm empty fruit bunch for the production of sugars. *Biomass and Bioenergy, 36*, 280–288.

Sidek, M. Z., Cheah, Y. J., Zulkefli, N. N., Yusuf, N. Y. M., Isahak, W. N. R. W., Sitanggang, R., & Masdar, M. S. (2019). Effect of impregnated activated carbon on carbon dioxide adsorption performance for biohydrogen purification. *Materials Research Express, 6*, 1–15.

TDM Berhad. (2019). Recycling resources recycling empty fruit bunches (EFB). Retrieved from https://sustainability.tdmberhad.com.my/planet/recycling-resources-recycling-empty-fruit-bunches-efb/

Teow, Y. H., & Takriff, M. S. (2021). Zero waste technology for sustainable development in palm oil mills. *Journal of Oil Palm, Environment & Health, 12*, 55–68.

Teow, Y. H., Ghani, M. S. H., & Mohammad, A. W. (2018). Physical and chemical cleaning for nanofiltration/reverse osmosis (NF/RO) membranes in treatment of tertiary palm oil mill effluent (POME) for water reclamation. *Jurnal Kejuruteraan SI, 1*(4), 51–58.

Teow, Y. H., Mohammad, A. W., Wan Mohammad Hamdan, W. N. A., Ghani, M. S. H., Ngteni, R., & Mohamed Yusof, K. M. (2016). Pilot-scale integrated pretreatment/membrane filtration system for aerobic palm oil mill effluent (POME) treatment. In *The 2016 World Congress on Advances in Civil, Environmental, and Materials Research (ACEM16)*, pp. 1–15.

The Institution of Engineers, Malaysia (IEM). (2019). *Zero waste palm oil processing: An industry-university initiative* (p. 32). Jurutera.

Turchetti, L., Murmura, M. A., Monteleone, G., Lemonidou, A. A., Angeli, S. D., Palma, V., Ruocco, C., & Annesini, M. C. (2016). Kinetic assessment of Ni-based catalysts in low-temperature methane/biogas steam reforming. *International Journal of Hydrogen Energy, 41*(38), 16865–16877.

Wang, F., Li, W. Z., Lin, J. D., Chen, Z. Q., & Wang, Y. (2018). Crucial support effect on the durability of $Pt/MgAl_2O_4$ for partial oxidation of methane to syngas. *Applied Catalysis B: Environmental, 231*, 292–298.

Wu, S. Y., Hsiao, I. C., Liu, C. M., Mt Yusuf, N. Y., Wan Ishak, W. N. R., & Masdar, M. S. (2017). A novel bio-cellulose membrane and modified adsorption approach in CO_2/H_2 separation technique for PEM fuel cell applications. *International Journal of Hydrogen Energy, 42*(45), 27630–27640.

Zainal, B. S., Zinatizadeh, A. A., Chyuan, O. H., Mohd, N. S., & Ibrahim, S. (2018). Effects of process, operational and environmental variables on biohydrogen production using palm oil mill effluent. *International Journal of Hydrogen Energy, 43*, 10637–10644.

Advancement in Various Stages of Palm Oil Mill Effluent (POME) Treatment Process

Sabeeha N. B. A. Khadaroo, Phaik Eong Poh, Darwin Gouwanda, and Hui Min Tan

Abstract Palm oil milling has progressed over the years with modernization and technology development. However, treatment methods for palm oil mill effluent (POME) remain archaic with little improvement in technology; due to the inherent challenges associated with the characteristics of POME along with microbial population sensitive to operational changes in the existing treatment process. It is imperative for the resolutions taken in order to advance POME treatment to fulfill the 3'E's criteria—Efficient, Ease of operability, and Economic feasibility. This chapter presents an overview of three established methods that were tested at laboratory scale to modernize POME treatment—(i) implementing artificial intelligence for POME treatment; (ii) thermal pretreatment of POME coupled with dewatering device and (iii) post-treatment of anaerobically treated POME with coagulant assisted microbubble flotation. These enhancements have proven to improve the treated effluent quality substantially and can be integrated into existing plants.

Keywords Palm oil mil effluent · Anaerobic digestion · Automation · Dewatering · Microbubble flotation · Artificial intelligence

1 Introduction

Palm oil milling is a proven and reliable process that has been advancing over the years with the mechanization and development of modern technology (Menon, 2011). The milling process has improved considerably from being labor-intensive to the utilization of modern equipment. Meanwhile, the approach for treating wastewater generated from the milling process, i.e., palm oil mill effluent (POME) has remained unchanged for the past decades. The opted treatment process for POME is simple to operate and has a low maintenance cost, primarily to conform to environmental regulations. Additionally, a lack of skilled labor prevents alterations to the POME treatment process. A typical POME treatment system uses a series of ponds capable

S. N. B. A. Khadaroo · P. E. Poh (✉) · D. Gouwanda · H. M. Tan
School of Engineering, Monash University Malaysia, Jalan Lagoon Selatan, 47500 Bandar Sunway, Selangor, Malaysia
e-mail: poh.phaik.eong@monash.edu

D. C. Y. Foo et al. (eds.), *Sustainable Technologies for the Oil Palm Industry*,
https://doi.org/10.1007/978-981-19-4847-3_11

of treating POME to meet effluent discharge standards stipulated by authorities. Nevertheless, there are impediments associated with the implementation of ponding systems explicitly, which include (i) long treatment duration; (ii) low treatment efficiency; (iii) emission of odor and harmful greenhouse gases, and (iv) having a large footprint (Chin et al., 2013).

A vast range of new methods and technologies was proposed and developed to resolve the conundrums related to POME treatment. However, only a few of them have been adopted by the mills. In recent years, many new mills in Malaysia have implemented anaerobic digesters to replace the conventional anaerobic ponds in order to comply with the regulations for energy and environmental conservation. In short, ponding system is still the best alternative for POME treatment. Many new mills would choose to cover their ponds to capture greenhouse gas emissions from their open surfaces as it requires least work and investment. Question remains as how to convince the mill owners and other stakeholders to improve and even alter the existing POME treatment.

Considering the advantages and disadvantages of the ponding treatment system and the complications faced in the mill, solutions for advanced POME treatment should meet the 3'E's criteria:

i. **Efficient** and adaptive: This is essential due to the possibility of enforcing a stricter discharge standard in the future. The increasing demand for crude palm oil (CPO) also contributes to a higher generation of POME. Therefore, any new technologies should work efficiently and effectively in handling the augmented flow rate of POME.
ii. **Ease** of operation: Having a dearth of skilled workers entails that the POME treatment process should be relatively straightforward in its monitoring and operation. Any existing employees in the mill should be able to remedy process anomalies and problems to lessen productivity loss.
iii. **Economically** feasible: Many advances in wastewater treatment technology do not contemplate the economic feasibility in their implementation. Hence, they are always "good to have" instead of a "must have". It is advised that new advancements should take into account the implementation cost.

Two fundamental challenges in POME treatment are the high concentration of organic matter and erratic POME characteristics, which depend on many external factors. Anaerobic digestion is the primary treatment method used to eliminate most organic matter. However, the unpredictable characteristics of POME cause instability in the digestion process and the discrepancy in treated effluent quality. Moreover, the bacteria population for anaerobic digestion is problematic and susceptible to alterations in operating parameters. Therefore, treating POME anaerobically in a large pond for a prolonged time appeared practical to ensure the removal of contaminants from the wastewater. The large anaerobic pond allows equalization of POME and infers the minimum disruption to the consortium of bacteria present.

The ponds function at a moderately slow pace. POME is held in a single pond for 30 days. Hence, numerous ponds are needed to carry out the treatment process. With increased CPO production, more ponds are required to handle the higher generation of POME, if the treatment duration is to be remained. Implementation of high-rate anaerobic digesters is more feasible, considering their shorter POME treatment duration, higher process proficiency, and the capability for biogas capture (Poh et al., 2020). The remaining question is that "How to convince mill owners that high-rate anaerobic digesters are the better option for POME treatment, given the inherent and operating challenges?" The above question can be addressed from two facets: (i) automated control of high-rate anaerobic digester using least instrumentations, and (ii) pre-treating POME prior to anaerobic digestion. Both are discussed in the following sub-sections.

2 Implementation of Artificial Intelligence (AI) in POME Treatment

Due to the many variables to be monitored in the wastewater treatment process, proper quality control and monitoring system are necessary. Implementing AI in the wastewater treatment process can provide several advantages. With the right configurations and minimal measured variables, AI can provide reasonably accurate estimation of the process characteristics in real-time. Knowing the possible results of a treatment process allows the operators to put preventive measures in place to counteract any undesired effect (Zhao et al., 2020). To demonstrate the feasibility of AI in treating POME, Tan et al. (2018b) used adaptive neural fuzzy interference to monitor and automate the high-rate anaerobic digester. Shocks were introduced to study its effect on the process and to further validate this method against the abrupt changes in POME. In the following sub-section, an experiment along with necessary process instrument control was set up to study how an adaptive neural fuzzy inference can enhance the treatment process of POME.

2.1 Automating an Up-Flow Anaerobic Sludge Blanket-Hollow Centered Packed Bed (UASB-HCPB) Reactor in POME Treatment

A laboratory-scale hybrid reactor, i.e., up-flow anaerobic sludge blanket-hollow centered packed bed (UASB-HCPB) was used in this study. The capacity of the reactor was 5L, while the inoculated seed sludge was taken as 10% of the operating volume of the reactor. The water bath was connected to the reactor to ensure the temperature of the reactor was set at 55 °C (Poh & Chong, 2014). A Prominent DF4a pump was fitted at the bottom of the reactor to feed diluted POME at the bottom of the reactor, and to augment the contact time between POME and inoculated seed sludge. This setup also ensures thorough mixing within the reactor. A water displacement column was mounted to measure the volume of biogas produced (Tan et al., 2018a).

The microcontroller selected for this study was Arduino Mega 2560. The flow rate of the Prominent DF4a pump was controlled by Arduino. An additional dosing pump (Prominent Solenoid Metering Pump) was utilized to add sodium bicarbonate to the wastewater if the reactor's pH dropped below 7.0. Several sensors were installed and connected to Arduino to measure several main characteristics of the POME. pH probes were used to monitor the pH of feed, reactor, and effluent online. A temperature probe was installed to monitor the temperature of mixed liquor within the reactor. Lastly, a methane sensor was mounted to detect the presence of methane in the biogas produced (Tan et al., 2018a). Figure 1 illustrates the process and instrumentation diagram of the above-described experimental setup.

2.2 *Adaptive Neuro-Fuzzy Inference System (ANFIS) for POME Anaerobic Digester*

The input and output variables chosen to build the ANFIS model for the UABS-HCPB reactor were pH, organic loading rate (OLR), chemical oxygen demand (COD), and the total suspended solids (TSS). These variables were selected since they provide information on the condition of the reactor. For instance, pH describes the reactor's physicochemical characteristics, while COD offers an insight into the biological properties and the efficiency of the reactor. These variables can be easily monitored using sensors, while the laboratory procedures to test these parameters are relatively simple. The mentioned parameters are also crucial in order for the wastewater to comply with the discharge standards (Tan et al., 2018b).

The reactor pH, temperature, COD, OLR, and biogas production were constantly monitored and controlled to ensure system's stability throughout the study. Closely monitoring of these parameters also offers several benefits: (1) It allows smooth operation and the detection of anomalies that could cause reactor failure; (2) It can reduce number of operators and tasks involved; (3) It can reduce sampling and testing frequency; and (4) It allows the operators to take preventive measures when an irregularity is detected. Table 1 lists the importance and actions taken for the parameters mentioned above to ensure the proper functioning of the reactor (Poh et al., 2020).

The simulation and modeling in this study were undertaken in the ANFIS editor available in MATLAB R2015b Fuzzy Logic Toolbox. A total of 282 days of data were collected and used in this study. The data were partitioned into two parts. The first part consists of 214 days of experimental data, which were used as the training set. The remaining part (68 days) was used for validation. The data were normalized using Eq. 1.

$$Y_{norm} = \frac{Y - Y_{min}}{Y_{max} - Y_{min}} \tag{1}$$

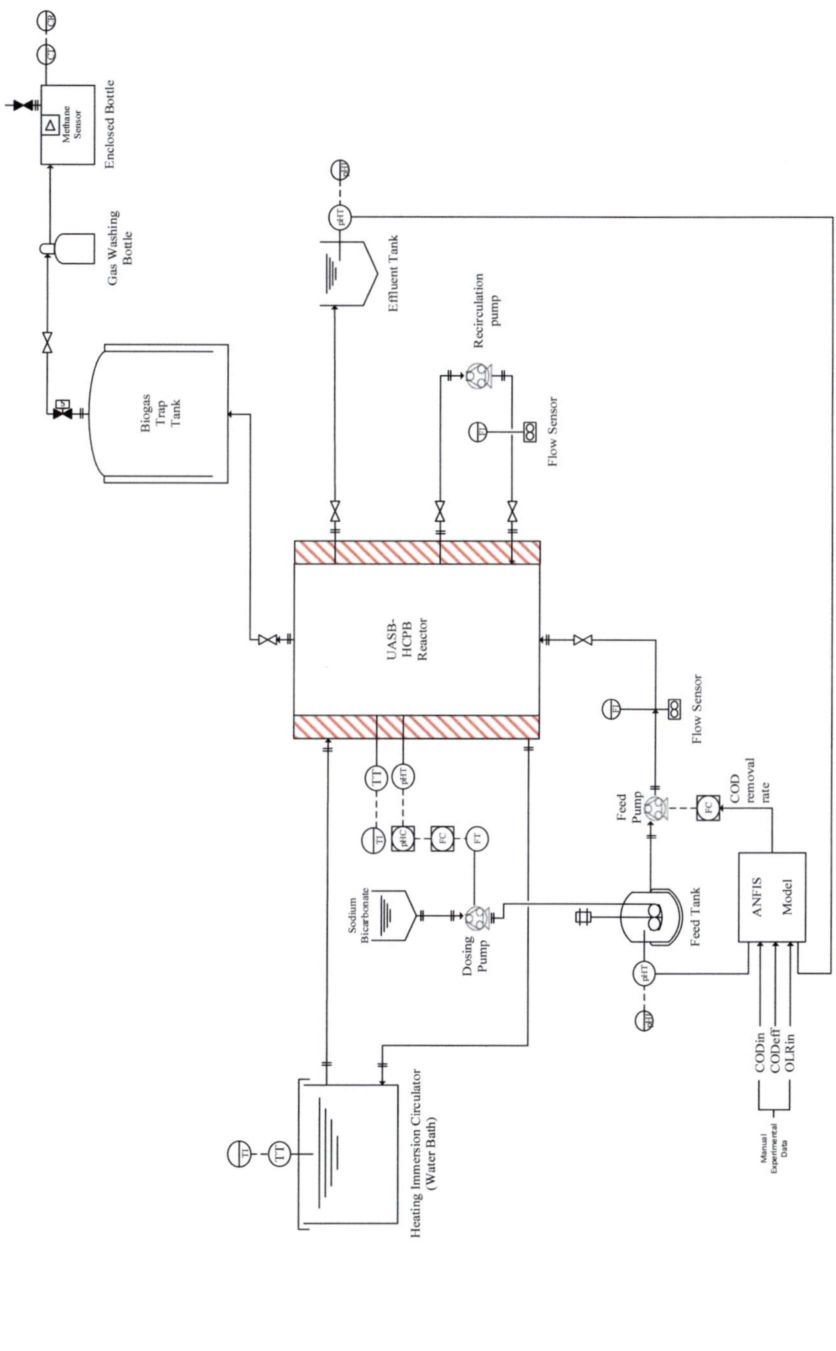

Fig. 1 Process and instrumentation diagram (P&ID) of the UASB-HCPB reactor after model implementation

Table 1 Parameters controlled and monitored in anaerobic digestion of POME

	Parameter	Importance	Countermeasures
Controlled parameters	pH	• Ensure digester balance • Provide ideal conditions for bacteria population	• Daily monitoring of feed, effluent, and reactor pH • Alkaline dosing
	OLR	• Biochemical reaction representation	• Daily measurement and monitoring • Included in ANFIS prediction
Monitored Parameters	COD	• Signifies efficacy of substrate utilization and metabolic activity • Dependent on influent flow rate and reactor working volume	• Removal efficiency is predicted using ANFIS, which allows corrective measures to be taken before reactor failure • Measurements (for feed and effluent) were conducted daily
	Biogas Quality	• Indicates performance of the reactor	• Real-time measurement daily using methane sensor to ensure biogas quality is maintained
	Temperature	• Provide ideal conditions for bacteria consortia • Ensure a high rate of reaction	• Daily monitoring at both water bath and within the reactor

where Y_{norm} is the normalized parameters, Y is the measured variables for the pH_{in}, COD_{in}, and OLR_{in}, Y_{max}, and Y_{min} are the maximum and minimum values of the variables.

The denormalization was performed using the same equation to compute the actual values of the predicted pH_{eff}, COD_{eff}, and TSS_{eff}.

Three fuzzy interference models designated as M1, M2, and M3 were built (Table 2). Each model was assigned distinct input and output parameters. Using four input parameters, M1 was constructed to predict the pH of the effluent (pH_{eff}). Meanwhile, using five and six input parameters, M2 and M3 were developed to predict the COD (COD_{eff}) and TSS of the effluent (TSS_{eff}), correspondingly. The effluent data was measured every 2 days since the reactor's hydraulic retention rate was set to two days. The effluent data measured 2 days prior (t − 2), known as historical data, were added to the models to enhance the prediction ability (Tan et al., 2018b).

Various measures were computed to assess the predictive ability of models M1, M2, and M3. These measure parameters were the average and standard deviations of the estimate errors, the determination coefficient denoted by R^2, the root mean square error denoted by RMSE, and the index of agreement denoted by IA. R^2, RMSE, and IA are parameters that indicate the discrepancy between the estimated and actual values. These parameters were evaluated using Eqs. 2, 3, and 4.

Table 2 ANFIS model setup

Model	Inputs	Output
M1	pH_{in}, COD_{in}, OLR_{in}, pH_{eff} (t − 2)	pH_{eff}
M2	pH_{in}, COD_{in}, OLR_{in}, pH_{eff} (t − 2), COD_{eff} (t − 2)	COD_{eff}
M3	pH_{in}, COD_{in}, OLR_{in}, pH_{eff} (t − 2), COD_{eff} (t − 2), TSS_{eff} (t − 2)	TSS_{eff}

$$R^2 = \frac{\left(\sum_{i=1}^{n}(A_i - A_m)(P_i - P_m)\right)^2}{\sum_{i=1}^{n}(A_i - A_m)^2 \sum_{i=1}^{n}(P_i - P_m)^2} \quad (2)$$

$$\text{RMSE} = \sqrt{\left(\frac{1}{n}\sum_{i=1}^{n}(P_i - A_i)^2\right)} \quad (3)$$

$$\text{IA} = 1 - \frac{\sum_{i=1}^{n}(P_i - A_i)^2}{\sum_{i=1}^{n}(|P_i - A_m| + |A_i - A_m|)^2} \quad (4)$$

where *A* represents actual measured values, *P* is the predicted values, *m* is the mean value, *i* is the initial value, and *n* is the number of data.

2.3 *Results and Discussion*

Table 3 presents the results obtained for the quantitative analysis to evaluate the performance of each model. Based on Table 3, M3 performed better as compared to M2 and M1 with the highest R^2, RMSE, and *IA* values. Meanwhile, Fig. 2 depicts the ability of M1 model in predicting the pH_{eff} values. Tan et al. (2018b) found that the average error and the standard deviation between the actual and predicted pH_{eff} were 2.06 and 1.68% congruently.

Albeit it can be observed that M1 was able to mimic the proclivity of the measured pH values, some incongruities were noted on days 10–14. The higher prediction of data was attributed to the drop in pH due to the decrease in alkalinity as the reactor was running at higher OLR of 36.32 kg COD/m^3day. However, this occurrence was not alarming since the alkalinity remained within a consistent range as the reactor

Table 3 ANFIS modeling results for anaerobic digestion of POME

Model	Input	Output	R^2	RMSE	IA
M1	pH_{in}, COD_{in}, OLR_{in}, pH_{eff} (t − 2)	pH_{eff}	0.63	0.0390	0.88
M2	pH_{in}, COD_{in}, OLR_{in}, pH_{eff} (t − 2), COD_{eff} (t − 2)	COD_{eff}	0.66	0.0618	0.89
M3	pH_{in}, COD_{in}, OLR_{in}, pH_{eff} (t − 2), COD_{eff} (t − 2), TSS_{eff} (t − 2)	TSS_{eff}	0.82	0.0377	0.95

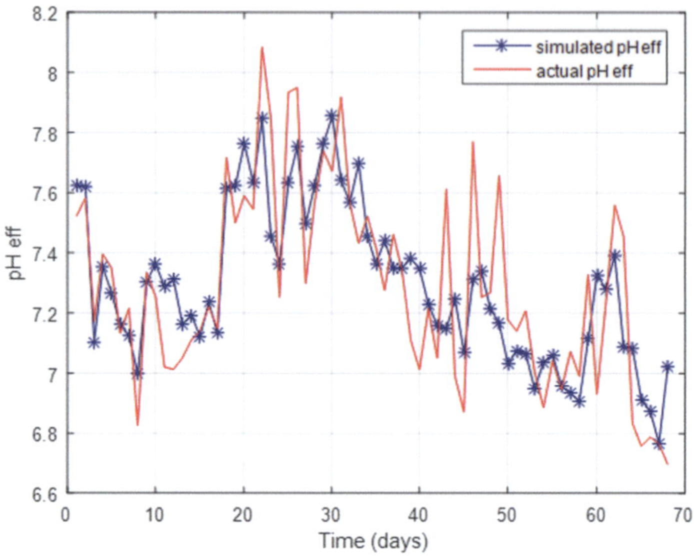

Fig. 2 Actual Effluent pH and simulated pH estimated by M1

achieved a steady state. Meanwhile, the disparities recorded on days 47–52 could be explained by the lag time taken for the bacteria consortium to adapt to the changes in pH. Since the reactor's optimum performance could not be attained instantly, the predicted pH_{eff} was lower than the actual values. The quantitative assessment in Table 3 shows that the R^2, RMSE, and IA values for the M1 model were computed to be 0.63, 0.0390, and 0.88, respectively.

On the other hand, the M2 model attained an average error and standard deviation of 8.32% and 7.65%, respectively. Albeit the average error and the standard deviation were greater than those in M1, due to the reactor's biological nature. The results achieved were considered satisfactory, and are in concurrence with studies conducted by (Hamawand & Baillie, 2015). Figure 3 depicts that M2 is relatively capable of providing compelling predictions, and can efficiently simulate the measured COD_{eff}. Moreover, it has the ability to forecast any unexpected spike in COD_{eff}, such as the one detected on day 18. In uncommon instances, it was observed that on day 8 and day 30, the model predicted almost consistent COD_{eff} for several consecutive days. A plausible explanation for this incident is ascribable to the variations in POME's physicochemical characteristics. (Poh et al., 2010) reported that low crop or high crop seasons considerably influence COD concentration, acidity, solid, as well as oil and grease contents of POME.

Additionally, the formation of a layer of scum was observed on day 8, which in turn caused an increase in the actual effluent values. Scum formation can decrease the reactor performance and make the modeling of the POME anaerobic digester difficult due to clogging in the pipes or solids washout from the bioreactor. The quantitative assessment for the M2 model was calculated to be 0.66 for the R^2, 0.0618 for the

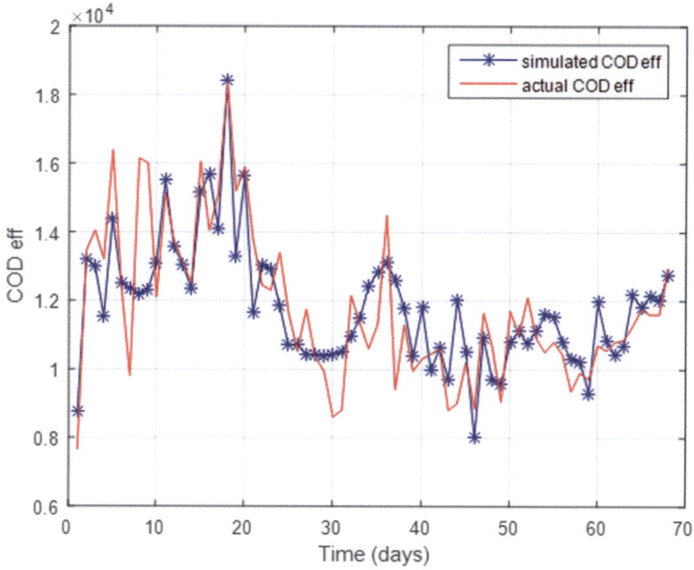

Fig. 3 Actual Effluent COD and simulated COD estimated by M2

RMSE, and 0.89 for the IA. Since the M1 and M2 models achieved R^2 and IA values at the lower side, the results indicate that M1 and M2 are capable of predicting the pH_{eff} and COD_{eff} correctly, nonetheless with a lack of precision.

Figure 4 demonstrates that M3 is able to predict the TSS_{eff} correctly with an average error and standard deviation of 6.93 and 6.28%, congruently. A more significant discrepancy was noted in M3 in comparison to M1. Meanwhile, it was noted that the deviation in M3 prediction was lower than in M2. This may occur due to the model's incapacity to predict the abrupt fall in TSS_{eff} on day 33, along with the sharp surge on day 37. These incidences are frequent in the anaerobic digester (Moreno, 2004; Murphy, 2007) and occur when POME concentration in the feed increases. A high concentration of POME triggers a higher OLR, which leads to larger particles flowing into the effluent tank.

Nonetheless, M3 can adequately predict TSS_{eff}, as shown in Fig. 4. M3 is also adept at simulating the abrupt rises and plunges, as illustrated on days 10–20. For quantitative analysis, M3 demonstrates superior prediction than M1 and M2 with R^2 of 0.82, and IA value of 0.95. The results achieved indicate that M3 has a superior prediction ability and higher accuracy than M1 and M2.

As theorized, the ANFIS model has the ability to predict the effluent's pH, COD and TSS accurately. The average errors computed ranged between 2.06 and 8.32%, while the standard deviation fluctuated between 1.38 and 7.65%. The trend predictions were deemed to be appropriate. The predicted tendencies were comparable to the measured data with minor variations. Moreover, the ANFIS model demonstrated

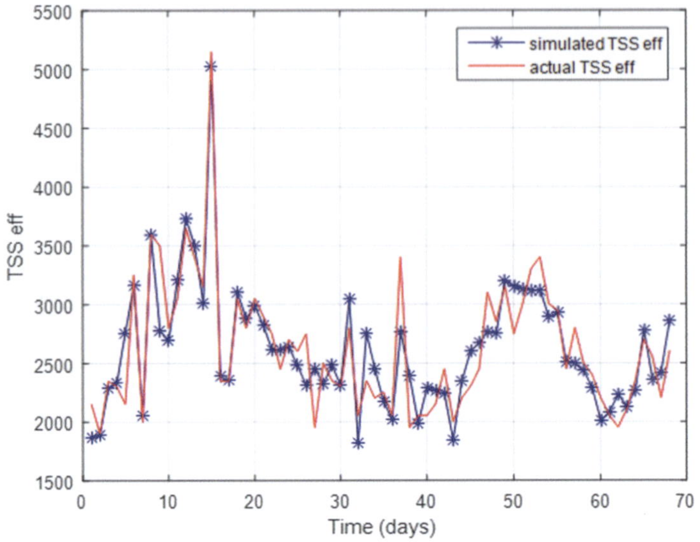

Fig. 4 Actual Effluent TSS and simulated TSS estimated by M3

the ability to manage unexpected fluctuations in the measured and predicted pH, COD, and TSS, as indicated by the results achieved.

3 Integration of Thermal Pretreatment and a Thickening Device

A lot of emphasis was placed on improving the efficiency of anaerobic digester for POME treatment. However, intrinsic properties of POME (i.e., high content of fibrous materials, oil and solids) and its varying characteristics throughout the year have rendered the digestion process in producing effluent and biogas of consistent quality. To address these issues, alteration of characteristics for the anaerobic digestion feed stream would be helpful.

The proposed alteration substitution of cooling pond with thermal pretreatment coupled with a dewatering device (Khadaroo et al., 2019b). Khadaroo et al. (2019b) reported that thermal pretreatment helps to induce the breakdown of complex molecules in POME. Thermal pretreatment enhances the anaerobic digestion performance and improves dewaterability of POME. Furthermore, it does not require re-neutralization of effluent after treatment, as compared to the commonly used chemical pretreatment techniques (Khadaroo et al., 2019b).

A dewatering device (e.g., a thickener) permits the removal of impurities and microorganisms from POME and aids in solid–liquid separation process. The thickener enhances the anaerobic digester's performance by integrating a configuration to control the load of the digesters since the physicochemical properties of POME are influenced by the efficiency of the oil extraction process along with high and low crop seasons (Poh & Chong, 2014). Different studies were conducted to observe the effect of dewatering and thermal pretreatment on the anaerobic digestion of POME (Khadaroo et al., 2020a). An energy analysis was also conducted to investigate the potential electricity generated from the proposed treatment process (Khadaroo et al., 2021).

3.1 Materials and Methods

3.1.1 Thermal Pretreatment and Dewatering

Raw POME was collected at the Sime Darby East Oil Mill, Malaysia. At the sampling location, the temperature of POME was measured to be 65 °C. The inoculum, anaerobic seed sludge was collected at the same mill. A total of 5 L of raw POME is placed in a beaker covered with aluminum foil and was thermally pretreated at 120 °C in an oven. The medium was stirred occasionally. The temperature was measured using an infrared thermometer at different heights along the beaker to ensure a uniform temperature.

After thermal pretreatment, the pretreated POME was placed in a settling column of 0.7 m height consisting of multiple sampling points to allow the removal of the different phases (oil, clarified liquid, and settled solids). The solid flocs in POME were allowed to settle for 24 h, at room condition. Samples of settled solids and clear liquid were extracted and hereafter denoted by "solid, S" and "liquid, L", respectively.

3.1.2 Anaerobic Digestion and Standard Methods for Parameters Testing

The chosen mode of anaerobic digestion was thermophilic batch anaerobic digestion. Once the different phases were extracted, they were recombined to make up the desired ratio. The sought ratio was poured in a 250 mL Schott bottle having two outlets. The operating volume was set at 100 mL. The inoculum volume was carefully selected and retained at 20% of the working volume during the course of the experiments. A reactor with a temperature of 55 °C under anaerobic conditions was used to cultivate and acclimatize the inoculum for 30 days to enable the bacteria consortium to be acclimatized to the conditions at which the anaerobic digestions experiments were conducted (Poh & Chong, 2014).

The inoculum volume was kept constant in all experimental runs for the evaluation of thermal pretreatment effects on anaerobic digestion performance of different solid: liquid ratios. A hot plate magnetic stirrer was used to heat the digesters and to ensure the medium for proper homogenization. The digesters were linked to a water displacement column using silicone tubes to facilitate the measurement of the

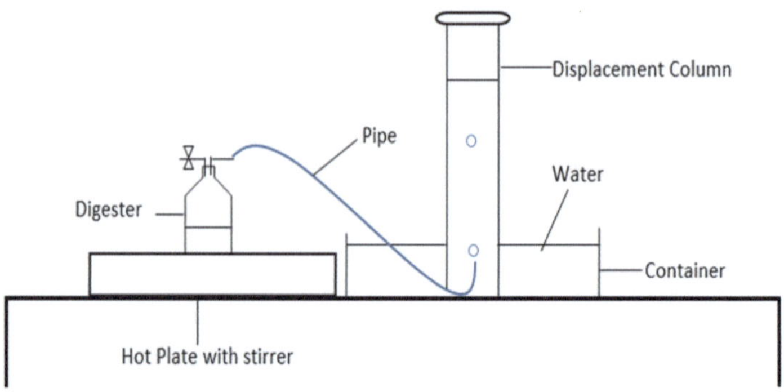

Fig. 5 Setup of thermophilic anaerobic digestion of POME

volume of the biogas generated as shown in Fig. 5. The pH of water was dosed to 2.0 using 1 M H_2SO_4 to avert the dissolution of carbon dioxide into the water, as per ASTM D5511 standard for the displacement column set up (Khadaroo et al., 2020b). This technique ascertains that the biogas mensuration using water displacement provides a more accurate result (Müller et al., 2004). The pH of the system was maintained between 6.8 and 8.0 by adjusting with 1 M $NaHCO_3$ to ensure optimum conditions. The pH was measured daily to assure that it lied within the mentioned range. The biogas composition in terms of methane, hydrogen sulfide, and carbon dioxide concentrations was evaluated using Binder COMBIMASS Gas Analyzer. The experiments were stopped when no more methane was measured in the digesters by Binder COMBIMASS Gas Analyzer (Khadaroo et al., 2020b).

3.2 *Results and Discussion*

Results presented in Fig. 6, Tables 4 and 5 show that thermally pretreated solid loadings in all tested conditions show increased biogas production, methane composition, as well as higher removal efficiencies of BOD, COD, TSS, and O&G. The best performing ratio was identified as 40 solid: 60 liquid (denoted as 40S:60L).

3.2.1 Biogas Generation and Methane Yield

The biogas production for the thermally pretreated 20S:80L POME reached 1470 mL which accounted for 960 mL more biogas than its untreated counterpart. The thermally treated 40S:60L POME produced 1886 mL of biogas, which is 456 and 415 mL more biogas in comparison to the untreated 40S:60L POME and the thermally pretreated 20S:80L POME, respectively.

The treated 50S:50L POME yielded 1509 mL biogas. The treated 50S:50L POME produced 187 mL more biogas than its untreated counterpart and 377 mL less biogas

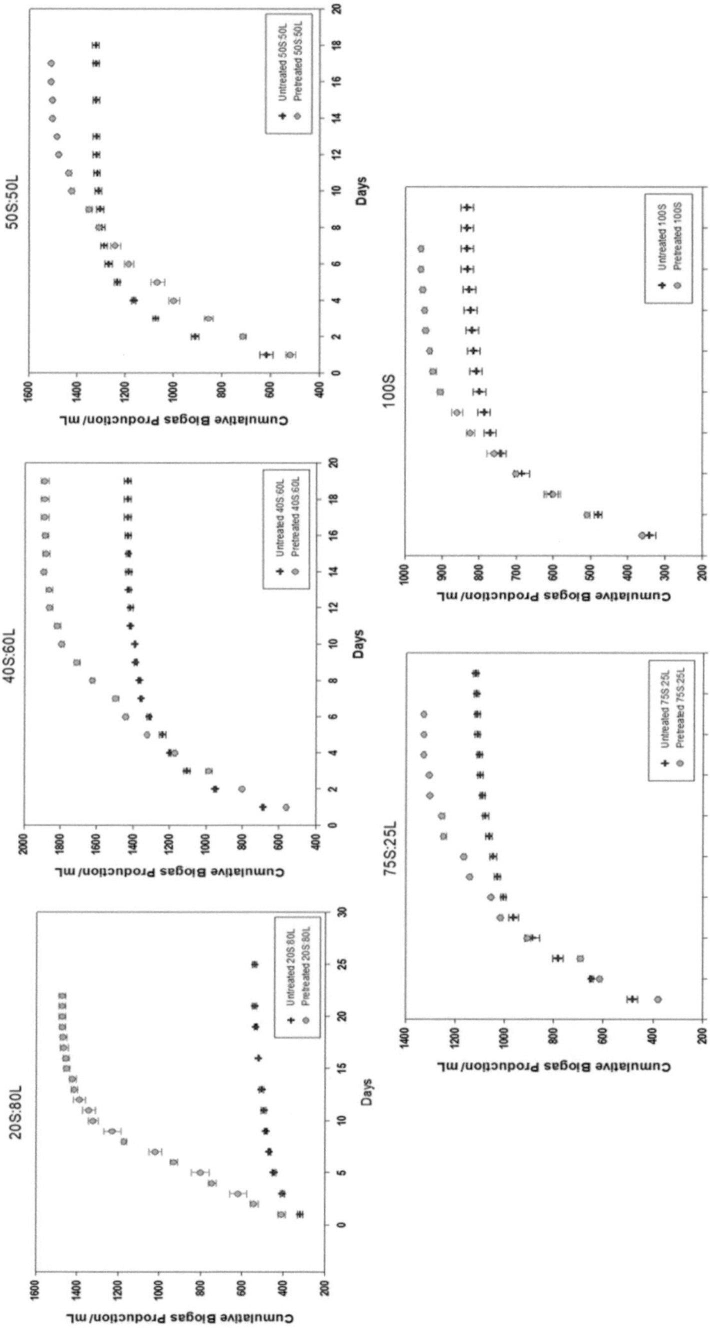

Fig. 6 Cumulative biogas production graphs of the thermally pretreated and untreated POME ratios (20S:80L, 40S:60L, 50S:50L, 75S:25L, 100S)

Table 4 Results obtained for the thermophilic anaerobic digestion of untreated POME at different solid loadings

Ratios	Dry solids content/%TS	Initial pH	Final pH	Cumulative biogas production/mL	Maximum methane composition /%	Minimum methane composition /%	H_2S composition/mg/L	Total anaerobes/100 mL	Duration of experiment /days	COD removal/%	BOD removal/%	TSS removal/%	O&G removal/%
20S:80L	3.29	7.23 ±0.05	7.52 ±0.03	539.44 ±10.29	73.83 ±2.42	50.36 ±2.07	204±4	1.2×10^6	≈25	62.53 ±1.14	58.07 ±1.76	55.44 ±3.46	26.20 ±0.46
40S:60L	4.02	7.21 ±0.02	7.74 ±0.09	1431.67 ±17.56	77.33 ±1.20	57.80 ±2.67	341±16	2.1×10^6	≈20	48.89 ±1.12	32.64 ±1.66	39.57 ±2.26	25.05 ±0.62
50S:50L	5.25	7.22 ±0.12	7.56 ±0.04	1322.78 ±13.62	64.26 ±2.71	54.13 ±0.87	560±21	7.5×10^5	≈18	33.26 ±0.71	23.97 ±1.30	29.40 ±1.55	19.24 ±0.45
75S:25L	6.4	7.27 ±0.10	7.44 ±0.04	1122.70 ±9.94	57.73 ±1.62	40.67 ±0.58	1823±23	1.5×10^5	≈17	23.11 ±0.41	24.10 ±0.52	20.76 ±0.57	26.57 ±0.18
100S	7.86	7.23 ±0.07	7.75 ±0.07	833.88 ±17.11	33.17 ±1.30	20.87 ±3.04	2968±52	9.3×10^4	≈17	7.67 ±1.05	21.78 ±1.84	19.06 ±2.85	31.16 ±0.62

Table 5 Results obtained for the thermophilic anaerobic digestion of thermally pretreated POME at different solid loadings

Ratios	Dry solids content/%TS	Initial pH	Final pH	Cumulative biogas production/mL	Maximum methane composition/%	Minimum methane composition/%	Total anaerobes/100 mL	Duration of experiment /days	COD removal/%	BOD removal/%	TSS removal/%	O&G removal/%
20S:80L	2.88	7.23 ±0.04	7.53 ±0.02	1471.10 ±15.23	79.23 ±1.34	71.30 ±2.71	1.5×10^6	≈ 22	84.50 ±1.01	84.41 ±0.15	83.03 ±0.91	82.88 ±0.34
40S:60L	3.98	7.27 ±0.02	7.57 ±0.07	1886.11 ±21.63	83.40 ±0.31	78.83 ±1.31	4.6×10^6	≈ 19	81.63 ±0.46	81.01 ±1.16	80.72 ±0.16	80.02 ±0.11
50S:50L	5.14	7.30 ±0.05	7.48 ±0.02	1509.43 ±4.43	76.97 ±0.73	71.40 ±0.79	1.1×10^6	≈ 17	65.38 ±0.04	62.72 ±0.36	67.20 ±0.75	64.81 ±0.40
75S:25L	6.32	7.20 ±0.02	7.71 ±0.06	1326.13 ±4.74	51.97 ±2.03	43.00 ±2.35	2.4×10^5	≈15	51.51 ±1.62	50.88 ±0.56	53.32 ±0.36	50.56 ±1.12
100S:0L	7.29	7.22 ±0.01	7.58 ±0.06	970.00 ±2.89	38.20 ±0.75	23.24 ±1.25	1.5×10^5	≈15	41.40 ±1.39	40.12 ±2.16	40.59 ±1.84	50.65 ±1.54

than the thermally treated 40S:60L solid loading. Similar observations could be found with the 75S:25L and 100S POME. Based on the results of the study, increasing the solid fraction of POME improved biogas production but up to 40% solids only. Meanwhile, the introduction of thermal pretreatment to POME is advantageous, as it improved the biogas yield as compared to the untreated POME.

The methane yield was computed to be 36.20 and 313.18 mL CH_4/g $COD_{removed}$ for the untreated and thermally treated 20S:80L solid loading. This accounted for a ninefold increase in the methane yield in the 20S:80L solid loading. The untreated and treated 40S:60L achieved a methane yield of 58.40 and 328.73mLCH_4/g$COD_{removed}$, resulting in a sixfold increase in the methane yield. The 50S:50L solid loading attained a methane yield of 40.66 and 89.88mLCH_4/g$COD_{removed}$ for the untreated and treated counterparts, respectively, accounting for a twofold increase in the methane yield. The 75S:25L recorded a methane yield of 27.84 and 54.06 mLCH_4/g$COD_{removed}$, for the non-pretreated and thermally treated assays. The methane yield calculated for the untreated and treated 100S solid loading were 16.69 and 31.52mLCH_4/g$COD_{removed}$, respectively. Both the treated 75S:25L and 100S had a twofold increase in the methane yield compared to their untreated counterparts. This further asserts the need for POME to undergo thermal pretreatment prior to anaerobic digestion.

3.2.2 Removal Efficiencies

The removal efficiencies of BOD, COD, TSS, and O&G were also investigated for all tested conditions. Post anaerobic digestion of the thermally treated 20S:80L POME, the COD, BOD, TSS, and O&G drastically decreased to 4696 ± 305, 1986 ± 150, 1866 ± 100, and 34 ± 1 mg/L resulting in a conspicuous percentage removal of 84.50 ± 1.01, 84.41 ± 0.15, 83.03 ± 0.91 and 82.88 ± 0.31% of COD, BOD, TSS, and O&G, respectively (Khadaroo et al., 2020b).

Before anaerobic digestion, the COD, BOD, TSS, and O&G of pretreated 40S:60L solid loading were measured as 40,800 ± 100, 20,090 ± 130, 16,000 ± 150, and 208 ± 12 mg/L. After anaerobic digestion, the COD, BOD, TSS, and O&G radically declined to 8155 ± 44, 4015 ± 67, 3162 ± 65, and 42 ± 1. These correspond to a remarkable removal percentage of 80.63 ± 0.46, 81.01 ± 1.16, 80.72 ± 0.16, and 80.02 ± 0.11% of COD, BOD, TSS, and O&G, respectively. The treated 40S:60L POME had a higher removal efficiency of 32.74, 48.37, 41.15, and 54.97% in terms of COD, BOD, TSS, and O&G congruently compared to the untreated 40S:60L assays (Khadaroo et al., 2020a). The above-mentioned results are summarized in Tables 4 and 5 where similar trends of biogas production were observed.

3.2.3 Energy Analysis for the Proposed Treatment System

Assuming that a medium-capacity plant treats 567.4 m^3 of POME per day (Sarwani et al., 2019). Electricity consumed in thermophilic anaerobic digestion and thermal pretreatment was evaluated to be 1.61 × 104 and 1.53 × 104 kWh/d, congruently. These values were determined by upscaling the results obtained from laboratory-scale experiments.

Some assumptions made for the energy analysis calculations. Firstly, no heat is lost to the environment. Secondly, there is only very insignificant (if any) transfer of energy to the walls of the tank, since the latter was further insulated. Finally, it was presumed that the specific heat capacity is similar to that of water which is 4.18 J/g °C (The density of POME was measured to be equal to that of water (Khadaroo et al., 2019a)).

In Table 6, it can be noted that thermophilic anaerobic digestion is able to bring forth nearly three times more electricity compared to the existing process. Meanwhile, the proposed treatment process consisting of the integration of thermal pretreatment and dewatering with an optimum 40S:60L solid loading achieved a total electricity generation of 75-fold greater compared to the existing treatment. Then again, the treated 50S:50L solid loading generated 14.6 times more electricity than that of the current treatment process. The high electricity generation is due to a higher volume of biogas produced with a superior methane purity compared to the 50S:50L solid loading and the other testes conditions. When extrapolating to an industrial scale, the outcome can be considerable concerning electricity generation. The findings above proved that the integration of thermal pretreatment and dewatering can substantially improve POME's treatment process while making the process more sustainable.

4 Post-Treatment of Anaerobically Treated POME with Assisted Microbubble Flotation

While the limitations of anaerobic digestion of POME could be solved via automation of anaerobic digester, adoption of thermal pretreatment, and dewatering processes, the effluent produced from anaerobic digestion still does not meet the regulatory standards. Therefore, it is essential to identify a treatment process that could replace the facultative ponds which have long treatment periods and occupy a large space in the palm oil mill. In this study, microbubble flotation was as a method to post-treat the anaerobic digested POME and the sub-sections below describe the methodology and the results of the study.

Table 6 Energy analysis case study on a medium-capacity plant

	Mesophilic AD without pretreatment (Sarwani et al., 2019)	Thermophilic AD without pretreatment (Khadaroo et al., 2021)	Thermophilic AD with pretreated 40S:60L (Khadaroo et al., 2021)	Thermophilic AD with pretreated 50S:50L (Khadaroo et al., 2021)
Biogas produced/m^3/d	28.36	2.61×10^3	5.45×10^3	4.53×10^3
Percentage CH_4 in biogas/%	55.0	61.9	77.8	72.5
CH_4 produced/m^3/d	15.6	1.62×10^3	4.24×10^3	3.29×10^3
Energy produced from CH_4/kJ/d	5.78×10^5	5.98×10^7	1.57×10^8	1.22×10^8
Electricity generated/kWh/d	1.60×10^2	1.66×10^4	4.35×10^4	3.38×10^4
Electricity consumed/kWh/d	–	1.61×10^4	3.15×10^4	3.15×10^4
Extra electricity generated/kWh/d	–	4.71×10^2	1.21×10^4	2.34×10^3
Extra electricity generated compared to conventional process/fold	–	2.9	75.1	14.6

4.1 Materials and Methods

4.1.1 Imaging and the Bubble Treatment of Anaerobically Treated POME

Figure 7 depicts the experimental setup for the microbubble flotation rig. The experimental setup consisted of a centrifugal pump (LangYou, LYQB-60, 370 W), microbubble generator (venturi), and two ball valves. In the flotation chamber, the water was continuously pumped from the lowermost section of the column using the venturi system. Air was drawn and inserted into the column, which created pressure difference between the inlet and throat of the venturi. The chosen flow rate of water was 0.331 and 0.44 L/s. A bubble generating device was integrated at the bottom of the column. Sampling was undertaken using a syringe which was coupled to a tube, in which water was drawn out from the midsection of the column. All experiments were undertaken in batches with a working volume of 1.5L.

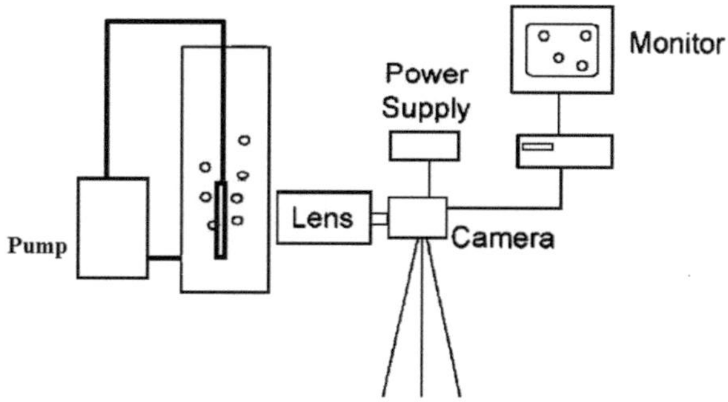

Fig. 7 Charge-coupled device (CCD) setup to capture the microbubble

4.1.2 Jar Test

Poly-aluminum chloride (PAC) coagulant of 30% concentration in white powder form was used in the experiments. After bubbling, the anaerobically treated POME was kept and utilized to undertake the jar test. The coagulation-flocculation examination was conducted using the traditional jar apparatus (VELP Scientifica Flocculator, JLT4). A 4-steel spindless paddle flocculator was utilized to stir the medium at a uniform speed. A 300 mL volume of 2.22 times diluted POME was placed in each beaker. Various concentrations of PAC coagulant ranging from a concentration of 0.5–6 g/L were used. The medium was mixed at a steady mixing rate of 150 rpm for a duration of five min. The setup was then altered to provide a slower mixing speed of 10 rpm with the steel paddle lifted halfway up the beaker to ameliorate the formation of flocs. The jar test was undertaken at a temperature of 25 °C. The sedimentation time was set to 30 min and supernatant of the treated POME was collected for further analysis.

4.2 Results and Discussion

4.2.1 COD (Chemical Oxygen Demand)

After multiple trials, it was observed that a minimum flow within the pump was required to prevent axial load on the pump shaft to be surpassed. Table 7 indicates the pH recorded prior to and post microbubble flotation for medium and high flow rates using the anaerobically treated POME. After microbubble flotation, the pH of the effluent was observed to increase ranging from 7.33 to 8.07. Table 7 shows that, after treatment, the pH of effluent increases with extended bubbling duration. However, in Fig. 8, it can be observed that at pH less than 8, the COD removal efficiency declined. An explanation for the latter is that the generation of bubbles between pH 7–8 tends to exhibit low negative zeta potential. Microbubbles with low charge cannot efficiently get rid of suspended solids in the wastewater. Furthermore, oxidation owing to the rupture of microbubbles tends to form hydroxyl radicals. The adherence of suspended matter to the low-charged microbubbles promotes to the removal of organic matter within the wastewater.

From Fig. 8, it can be seen that when the bubbling time was prolonged from 2.5 to 12.5 min, the COD removal efficiency increased from 9.8 to 53.7% and from 14 to 45.9% for bubbles of sizes 469 and 379.92 μm, congruently. In comparison to the bubbles generated at two different flow rates, $D_{32} = 379.92$ μm with smaller diameter showed a higher COD removal rate for bubbling time of 2.5–10 min. The latter occurs since the smaller bubbles accelerate the hydroxyl radical's formation due to elevated inner pressure which can break the conjugated carbon–carbon double bonds in melanoidins. The melanoidins are accountable for POME's brown color. In another study, it was stated that hydroxyl radicals produced in a small amount promote the breakdown of the organic matter present in the wastewater (Liu, et al., 2012).

Table 7 pH and temperature reading for different bubbling time

Parameters [a]	Anaerobically digested POME (ADPOME)	Effluent after microbubble flotation (19.8 L/min)	Effluent after microbubble flotation (26.4 L/min)	Effluent after microbubble flotation (19.8 L/min) + PAC coagulation	DOE discharge standard
pH	7.05	8.06	8.07	6.28	5–9
Temperature	18	26	27	25	45
COD	21,025	9725	11,375	1407	–
BOD	2220	510	1065	Not detected	100
TSS	17,995	7685	10,080	22	–
O&G	235	60	120	Not detected	50

[a] All units measured in mg/L except pH and temperature (°C)

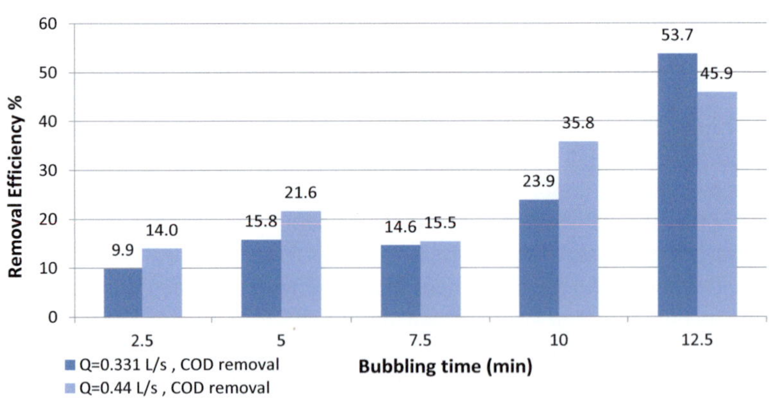

Fig. 8 Plot of COD removal at different bubbling times for $Q_L = 0.331$ L/s and $Q_L = 0.44$ L/s

Color change was observed at minutes 10 and 12.5 bubbling time; this occurred due to the hydroxyl radicals generated chemical reactions as the microbubbles are removed from the system. In general, smaller bubbles have a larger surface area. This is demonstrated in Eq. 5 for the relationship between surface area (*S*) and volume (*V*).

$$\frac{s}{v} = \frac{4\pi r^2}{4/3\pi r^3} = \frac{3}{r} \tag{5}$$

Small bubbles have larger surface area, which enhances bubble-particles collision efficiency and resulting in an augmented COD removal efficiency. Nonetheless, as bubbling time was extended, bubbles of size $D_{32} = 469\ \mu m$ were noted to achieve a superior removal rate with a highest COD removal of 53.7%. It was determined that floated microbubbles tend to produce a foam that cannot be broken down easily. It was observed when the water was recirculated at a higher flow rate, a larger amount of foam was brought forth. Due to foaming, the water level in the column decreased substantially, making shorter distances accessible for the bubbles. This phenomenon indicated that at higher flow rates, the generation of smaller bubbles with larger surface areas lessened the displacement room for the bubble to move owing to the generation of a large amount of foam. Subsequently, the possibility of contact between the bubble and the contaminants in the medium is therefore drastically reduced. Hence, COD removal rate decreases when bigger bubbles are used.

4.2.2 BOD (Biochemical Oxygen Demand)

The BOD removal rate was observed to progressively increase with bubbling time, ranging from 51.8 to 77% for $D_{32} = 469\ \mu m$ and 29.7 to 52% for $D_{32} = 379.92\ \mu m$. The trend indicates that a prolonged bubbling time will enhance the BOD removal efficiency. Nonetheless, the latter was not observed for higher flow rates generated microbubbles. Figure 9 shows the BOD removal efficiency of this method.

It was deduced upon those smaller microbubbles having larger total surface areas, would impart higher flotation efficiency of particles in a fluid body owing to a more elevated flux in mass transfer. However, the opposite was observed in the BOD removal trend. In turn, it was noted that smaller microbubbles (0.44 L/s) attained lesser removal efficiency. It can therefore be hypothesized that the influence of surface area on the removal efficiency took precedence over the impact of rising velocity. An explanation for this is that more particles are attracted to smaller bubble sizes, however, low rising velocity allows less bubbles to reach the surface in the set period of time. The latter causes the removal rate to decrease. This phenomenon is especially conspicuous due to the presence of large complex organic masses, resulting in a further curbed rising velocity and BOD removal efficiency.

The BOD parameter primarily relates to the concentration of organic particles in the POME colloid while the COD pertains to the conglomerate concentrations

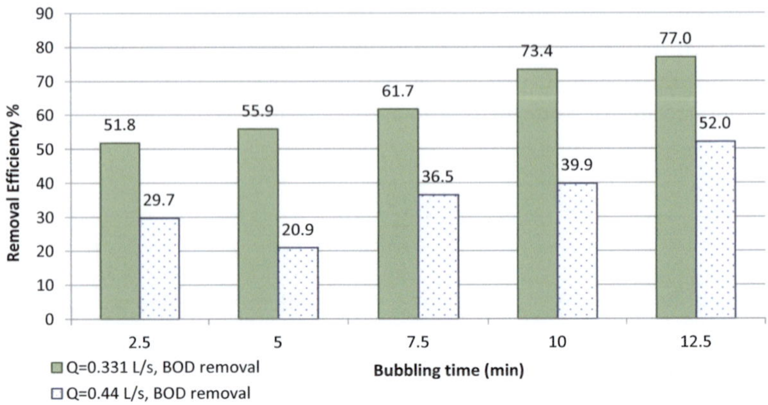

Fig. 9 Plot of BOD removal at different bubbling times for $Q_L = 0.331$ L/s and $Q_L = 0.44$ L/s

of both inorganic and organic molecules. From the results, it was found that the bubbling treatment yielded higher BOD removal efficiency compared to the COD removal rate, indicating that micro bubbling flotation treatment is more efficient in degrading organic particles than inorganic particles (Nguyen & Evans, 2004).

4.2.3 Oil and Grease (O&G)

Figure 10 shows the results obtained for the removal efficiency of O&G insoluble organic compounds in POME. It can be noted that as increasing bubbling time enhanced O&G removal efficiency. The removal efficiencies of microbubbles with bubble diameter of $D_{32} = 469$ μm and $D_{32} = 379.92$ μm were found to be 10.6% to 74.5% and 3% to 48.9%, respectively. From the study, it was noted that as increased temperature and pH enhanced the removal of oil. The latter can be explained by the influence of several physicochemical characteristics of oil and grease content in POME (Ahmad et al., 2003). The rise in temperature occurs due to cavitation and friction of the pump. While the increase in pH is related to the dissociation of OH^- ions in water, spurring the OH^- ions to react with residual oil bringing forth saponification. The latter produced soap that is soluble in water, resulting in a larger O&G removal efficiency at prolonged bubbling times (Ahmad et al., 2003).

Ahmad et al. (2003) reported that at pH levels greater than 7, oil compounds display a higher chemical attraction (affinity) for negatively charged surfaces. The oil droplets are attracted to negative ions which give rise to negatively charged particles (Ghernaout & Ghernaout, 2012). The latter is a potential explanation for the elevated O&G removal efficiency depicted in Fig. 9 with extended bubbling time. It is therefore sensible to conclude that adsorption on the surface of the bubbles is more conducive as bubbling time is extended.

When comparing the 10 and 12.5 min bubbling times in Fig. 10, treated POME with a bubble of size 469 μm resulted in a higher O&G removal efficiency. The

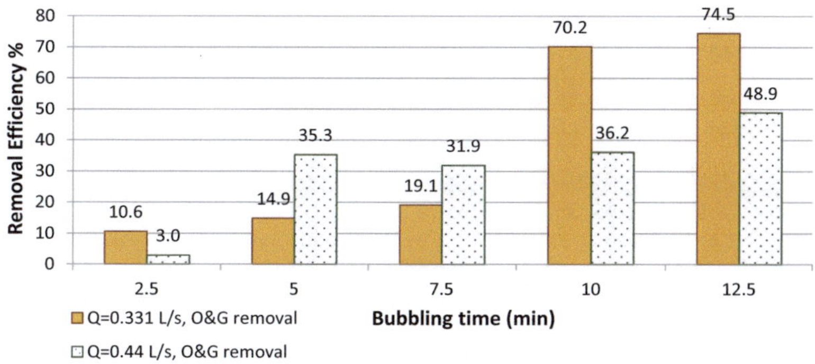

Fig. 10 Plot of O&G removal at different bubbling times for $Q_L = 0.331$ L/s and $Q_L = 0.44$ L/s

smaller bubbles with larger surface areas (at a higher flow rate) had a lower O&G removal rate, as compared to larger microbubbles when the bubbling time was increased. The smaller-sized microbubbles attained a lower O&G removal rate since low rising velocity was prevailing as a fundamental parameter in removal efficiency in comparison to the total surface area. The latter can be explained by the long carbon chains found in oil and grease molecules; thus, the molecular masses tend to be heavier. Therefore, bigger microbubbles with greater rising velocity can float and adhere to the heavy oil particles, efficiently removing the latter.

5 Conclusion

The advancement in technologies available for the treatment of POME has rapidly evolved over the years owing to the intrinsic challenges surrounding the treatment process of POME, the fluctuating characteristics of POME, the lack of skilled labor to manage the treatment process as well as more stringent regulations when it comes to the discharge of the latter. Solutions employed to mitigate the aforementioned predicaments must be efficient, technically, economically, and environmentally feasible. In other words, the technology has to be practically and efficiently utilized in the mills. Besides, it has to be economically viable and environmental-friendly. This can be achieved through the reduction in greenhouse gases emission and treated effluent that meets the environmental standards. The introduction of an "artificial brain", thermal pretreatment coupled with a dewatering device, and coagulant-aided microbubble flotation for POME treatment process enhance the treatment efficacy and increase biogas generation. In the future, pilot-scale studies which incorporate all three aspects mentioned above will be carried out in collaboration with palm oil mill practitioners to ensure effective implementation of these technologies, in order to enhance the POME treatment processes.

References

Ahmad, A., Ismail, S., Ibrahim, N., & Bhatia, S. (2003). Removal of suspended solids and residual oil from palm oil mill effluent. *Journal of Chemical Technology and Biotechnology, 78*, 971–978.

Chin, M. J., Poh, P. E., Tey, B. T., Chan, E. S., & Chin, K. L. (2013). Biogas from palm oil mill effluent (POME): Opportunities and challenges from Malaysia's perspective. *Renewable and Sustainable Energy Reviews, 26*, 717–726.

Ghernaout, D., & Ghernaout, B. (2012). Sweep flocculation as a second form of charge neutralization—A review. *Desalination and Water Treatment, 3*(44), 15–28.

Hamawand, I., & Baillie, C. (2015). Anaerobic digestion and biogas potential: Simulation of lab and industrial-scale processes. *Energies, 8*, 454–474.

Khadaroo, S. N. B. A. (2021). *Integration of thermal pretreatment and dewatering in the anaerobic co-digestion of palm oil mill effluent (POME) for enhanced treatment performance in terms of biogas production and treated effluent quality*. Monash University.

Khadaroo, S. N. B. A., Poh, P. E., Gouwanda, D., & Grassia, P. (2019b). Applicability of various pretreatment techniques to enhance the anaerobic digestion of palm oil mill effluent (POME): A review. *Journal of Environmental Chemical Engineering, 7*, 103310. https://doi.org/10.1016/j.jece.2019.103310

Khadaroo, S. N. B. A., Grassia, P., Gouwanda, D., & Poh, P. E. (2020a). The impact of thermal pretreatment on various solid-liquid ratios of palm oil mill effluent (POME) for enhanced thermophilic anaerobic digestion performance. *Journal of Cleaner Production, 261*, 121159. https://doi.org/10.1016/j.jclepro.2020.121159

Khadaroo, S. N. B. A., Grassia, P., Gouwanda, D., & Poh, P. E. (2019a). Is the dewatering of palm oil mill effluent (POME) feasible? Effect of temperature on POME's rheological properties and compressive behavior. *Chemical Engineering Science, 202*, 519–528. https://doi.org/10.1016/j.ces.2019.03.051

Khadaroo, S. N. B. A., Grassia, P., Gouwanda, D., & Poh, P. E. (2020b). The influence of different solid-liquid ratios on the thermophilic anaerobic digestion performance of palm oil mill effluent (POME). *Journal of Environmental Management, 257*, 109996. https://doi.org/10.1016/j.jenvman.2019.109996

Liu, S., et al. (2012). The effect of different types of micro-bubbles on the performance of the coagulation flotation process for coke waste-water. *Journal of Chemical Technology & Biotechnology, 87*, 206–215.

Menon, N. R. (2011). Development of palm oil mill technology. *Palm Oil Engineering Bulletin, 100*, 39–53.

Moreno PA (2004). Evaluation of factors responsible for high effluent suspended solids events in the Kuwahee wastewater treatment plant. Masters Theses. University of Tennessee, Knoxville "Trace: Tennessee Research and Creative Exchange".

Murphy, S., 2007. General information on solids. *City of Boulder/ USGS Water Quality Monitoring.* http://bcn.boulder.co.us/basin/data/BACT/info/TSS.html

Müller, W. R., Frommert, I., & Jörg, R. (2004). Standardized methods for anaerobic biodegradability testing. *Re/Views in Environmental Science & Bio/Technology*, 3(2), 141–158. https://doi.org/10.1007/s11157-004-4350-6

Nguyen, A., & Evans, G. (2004). Attachment interaction between air bubbles and particles in froth flotation. *Experimental Thermal and Fluid Science, 28*(5), 381–385.

Poh, P. E., & Chong, M. F. (2014). Upflow anaerobic sludge blanket-hollow centered packed bed (UASB-HCPB) reactor for thermophilic palm oil mill effluent (POME) treatment. *Biomass and Bioenergy, 67*, 231–242. https://doi.org/10.1016/j.biombioe.2014.05.007

Poh, P. E., Yong, W. J., & Chong, M. F. (2010). Palm oil mill effluent (POME) characteristic in high crop season and the applicability of high-rate anaerobic bioreactors for the treatment of POME. *Industrial & Engineering Chemistry Research, 49*(22), 11732-11740. https://doi.org/10.1021/ie101486w

Poh, P. E, Khadaroo, Sabeeha. N. B. A, Tan, H. M., & Gouwanda, D. (2020). An official publication of the Malaysian Palm Oil Council (MPOC) advancing on palm oil mill effluent (POME) treatment—What is in store for you? *Journal of Oil Palm, Environment & Health, 11*, 6–12. https://doi.org/10.5366/jope.2020.02

Sarwani, M. K. I., Fawzi, M., Osman, S. A., & Nasrin, A. B. (2019). Bio-methane from palm oil mill effluent (POME): Transportation fuel potential in Malaysia. *Journal of Advanced Research in Fluid Mechanics and Thermal Sciences, 1*, 1–11.

Tan, H. M., Gouwanda, D., & Poh, P. E. (2018a). Adaptive neural-fuzzy inference system versus anaerobic digestion model No.1 for performance prediction of thermophilic anaerobic digestion of palm oil mill effluent. *Process Safety and Environment Protection, 117*, 92–99. https://doi.org/10.1016/j.psep.2018.04.013

Tan, H. M., Poh, P. E., & Gouwanda, D. (2018b). Resolving stability issue of thermophilic high-rate anaerobic palm oil mill effluent treatment via adaptive neuro-fuzzy inference system predictive model. *Journal of Cleaner Production, 198*, 797–805. https://doi.org/10.1016/j.jclepro.2018.07.027

Zhao, L., Dai, T., Qiao, Z., Sun, P., Hao, J., & Yang, Y. (2020). Application of artificial intelligence to wastewater treatment: a bibliometric analysis and systematic review of technology, economy, management, and wastewater reuse. *Process Safety and Environmental Protection, 133*, 169–182. https://doi.org/10.1016/j.psep.2019.11.014

Performance and Stability of Pre-commercialized Integrated Anaerobic–Aerobic Bioreactor (IAAB) for the Treatment of Palm Oil Mill Effluent (POME)

Yi Jing Chan, Roy Jun Wei Chong, Mei Fong Chong, Denny Kok Sum Ng, and Lian Keong Lim

Abstract This chapter aims to evaluate the performance of a pre-commercialized Integrated Anaerobic–Aerobic Bioreactor (IAAB) (3000 m^3) under variable organic loadings and environmental conditions with respect to effluent quality and methane yield. During the steady state operation of IAAB, the system achieved 99% of removal efficiency for Chemical Oxygen Demand (COD), and Biochemical Oxygen Demand (BOD) and methane yield up to 0.26 L CH_4/g COD at organic loading rate (OLR) of 2.0–20.0 g COD/L day. Achievement of BOD <100 mg/L throughout 200 operational days with 45% of compliance was reported. The system could significantly reduce 70% of footprint and 78% of hydraulic retention times compared with current conventional treatment systems (e.g., cover lagoon, anaerobic bioreactor, etc.). During the operation, there are number of issues such as scum formation and foaming in the system. By manipulating the sludge recirculation rate within the range of 70–140 m^3/h, the scum and foaming issues are resolved in anaerobic compartment. Meanwhile, the foaming issues in aerobic compartment were successfully resolved by dosing the advanced biological formulation produced by Novozymes, namely Bioremove 5100 and Bioremove 3200 at optimum ratio of 50:50. Further work on optimization for the recirculation flow rate in the anaerobic compartment with consideration of fluid dynamics and microbiology is required to achieve 100% compliance of BOD <100 mg/L.

Y. J. Chan (✉) · R. J. W. Chong
Department of Chemical and Environmental Engineering, University of Nottingham Malaysia, Broga Road, 43500 Semenyih, Selangor, Malaysia
e-mail: Yi-Jing.Chan@nottingham.edu.my

M. F. Chong
628, Jalan Pulau Tioman U10/94, 40170 Taman Greenhill, Shah Alam, Selangor, Malaysia

D. K. S. Ng
School of Engineering and Physical Sciences, Heriot-Watt University Malaysia, Putrajaya, Malaysia

L. K. Lim
Havys Oil Mill Sdn Bhd, C/O Paramount Estate, KM31 Jalan Bahau-Keratang, Mukim Bera, Pahang Darul Makmur, Malaysia

D. C. Y. Foo et al. (eds.), *Sustainable Technologies for the Oil Palm Industry*,
https://doi.org/10.1007/978-981-19-4847-3_12

Keywords Palm oil mill effluent (POME) · Anaerobic · Aerobic · Integrated anaerobic–aerobic bioreactor (IAAB) · Biogas · Methane yield

Nomenclature

Abbreviation

BOD	Biochemical Oxygen Demand
COD	Chemical Oxygen Demand
DO	Dissolved Oxygen
MLSS	Mixed Liquor Suspended Solid
MLVSS	Mixed Liquor Volatile Suspended Solid
RAS	Returned activated sludge
OLR	Organic Loading Rate
TSS	Total Suspended Solids
UASB	Upflow Anaerobic Sludge Blanket
UASFF	Upflow Anaerobic Sludge Blanket Fixed Film
POME	Palm Oil Mill Effluent

1 Introduction

Malaysia has been experiencing a rise in economic growth as it is one of the largest contributors in palm oil products worldwide. The massive growth of oil palm tree is blessed by Malaysia's tropical climate that facilitates these successful agricultural schemes. Lately, Malaysia contributed 33% of world exports and 28% of world palm oil production (MPOC, 2020). However, the production of crude palm oil (CPO) has also generated significant amount of wastewater, namely palm oil mill effluent (POME). POME is recognized as an agro-output that exhibits high acidity with average values of 50,000 mg/L chemical oxygen demand (COD) and 25,000 mg/L biochemical oxygen demand (BOD) (Yacob et al., 2005). Fresh fruit bunches (FFBs) harvested from oil palm plantation are sent to the palm oil mill for extraction of CPO. Then, CPO is further purified in refinery to be palm oil products (e.g., styrene, soap noodle, etc.). POME are generated during the steam condensate during the sterilization process, decanter, etc. Such waste is required for proper treatment prior discharging into the environment. It was reported that most palm oil mills in Malaysia operate at a capacity of 45 t/h of FFB which produces 29.25 t/h POME (Akhbari et al., 2020). Undoubtedly, this has triggered the need for better agricultural, industrial, and sustainability practices to be implemented in palm oil mills to combat the depletion in environmental quality posed by this industry.

Generally, POME treatment involves both anaerobic and aerobic degradation, as applying former method alone is insufficient to meet the current effluent discharge legislation. Chan et al. (2009) has reviewed that the organic matter in anaerobic effluent are not completely stabilized and an additional post treatment step is required to treat the ammonium ion and hydrogen sulfide found in effluent (Chan et al., 2009). Anaerobic digestion is proven to render low energy cost, reactor volume, high organic load treatment while producing biogas for energy supply; however, further polishing of POME effluent is required (Tchobanoglous et al., 2004). Hence, aerobic degradation is implemented to further treat the anaerobic effluent to meet the stringent discharge standard. Moreover, the anaerobic–aerobic process is capable of providing higher overall treatment efficiency, while effectively reducing energy consumption, as well as sludge disposal (Cervantes et al., 2006; Frostell, 1983). The anaerobic–aerobic system using high rate bioreactors have been extensively studied to identify better options in treating POME. For instance, the *upflow anaerobic sludge blanket* (UASB) reactor has demonstrated various positive features such as its capability in treating high organic loadings with short hydraulic retention time (HRT) while consuming less energy (Tchobanoglous et al., 2004).

Since then, various novel designs which incorporate the features of UASB have been proposed to enhance bioreactor performance in treating POME. Najafpour et al. suggested to couple upflow fixed film (UFF) with a UASB reactor to shorten the start-up period at low HRT. The hybrid reactor, i.e. upflow anaerobic sludge-fixed film (UASFF) bioreactor was proven to be able to remove 97% COD at HRT of 3 days with organic loading rate (OLR) of 11.58 kg COD/m^3 day Najafpour et al. (2006). Poh and Chong (2014) also demonstrated that an upflow anaerobic sludge blanket-hollow centered packed bed (UASB-HCPB) reactor could remove 90% COD and BOD, and 80% suspended solid, while producing 60% methane (Poh & Chong, 2014). Nonetheless, these high-rate bioreactors are still in their infancy in term of up-scaling due to high cost and unstable performance.

In recent years, the idea of utilizing a compact high-rate bioreactors has garnered attentions in overcoming space limitations, odor problems and biosolids production (Chan et al., 2009). The high rate integrated anaerobic–aerobic bioreactors (IAAB) have been developed to replace the conventional treatment methods mentioned above. IAAB can be defined as a breakthrough in innovation as it exploits both benefits of both the anaerobic and aerobic degradation processes while providing better biodegradation. The capability of IAAB in treating POME within a short hydraulic retention time at reduced space utility has been demonstrated at previous study. The overall COD, BOD, and total suspended solids (TSS) removal efficiencies greater than 99% were attained for OLR up to 18.5 g COD/L day with methane yield of 0.32 L CH_4/g $COD_{removed}$ (Chan et al., 2012). Prior to industrial application and commercial adoption, basic knowledge and technological advancement ought to be conducted via pilot or demonstration plants. In addition, the initiative of pilot plants is subjected to create a balance between establishing technologies and constructing a first commercial market. Previous pilot studies on IAAB have shown to exhibit good stability, high performance in terms of POME substrate removal efficiency >99% and methane yield of 0.24 L CH_4/g COD removed with OLR of 10.5 g COD/L day as compared to conventional system (Chan et al., 2012). Similarly, evaluation of biokinetic coefficient on IAAB suggested that OLR range of 10.5–22.5 g COD/L day results in POME substrate removal > 99% while producing up till 64% of CH_4 gas (Chan et al., 2017). The study has shown encouraging results and hence, the current study focuses on utilizing the similar IAAB design at a pre-commercialized scale to further evaluate the efficiency of the unit in treating POME. The aforementioned studies are deemed successful and promising, yet only applicable in pilot scale applications. Hence, the present study was undertaken to investigate the performance of the pre-commercialized scale IAAB at different OLRs under mesophilic condition. The determined maximum sustainable OLR will indicate whether the proposed novel design is practical for industrial utilization. In addition, the study on the effect of mixed liquor suspended solids (MLSS) and F/M ratio are of great importance as it will affect the overall performance of IAAB if not taken into appropriate measurements. Furthermore, poor mixing behavior has shown to impose large positive impact in the digestion efficiency when considering up-scaling of IAAB, since most commercial digester inhibit continuous mixing process (Kobayashi et al., 2013). Thereafter, operational issues encountered by pre-commercialized IAAB such as foaming in aerobic compartment and scum formation in anaerobic compartment will be emphasized. The current study provides the recommended operating conditions of OLR, MLSS, and F/M ratio which give the highest methane yield, total removal efficiency of COD and BOD while mitigating operational issues faced by pre-commercialized IAAB, thus to maintain high performance and stability for the long run. This work is significant as these development activities are not only addressing pure technical challenges, but also reducing the organizational, market, and institutional risks and uncertainties that key stake-holders might face in this advancing new technology i.e. IAAB.

2 Materials and Methods

2.1 Wastewater Preparation

In this project, the pre-commercialized IAAB is built at a palm oil mill located in Pahang, Malaysia. Therefore, the POME is taken directly from the palm oil mill. Over the period, the characteristics of the POME are evaluated and presented in Table 1.

2.2 Reactor Configuration and Operating Procedures

The pre-commercialized IAAB is built to treat the raw POME based on IAAB system. The simplified process flowsheet is shown in Fig. 1, while the photo for the unit is

Table 1 Characteristics of POME

Parameter	Units	Average	Range
pH	–	4.5 ± 0.10	4.18–4.7
BOD	mg/L	35,100 ± 10,391	4100–86,700
COD	mg/L	74,016 ± 37,738	8500–176,400
TSS	mg/L	31,611 ± 15,531	31,200–34,300

shown in Fig. 2. This IAAB unit has a design capacity of 10 m^3/h (maximum inlet flowrate) which is about 20% of the full-scale plant based on a 60 t/h of FFB mill capacity. As shown in Fig. 1, the IAAB system consisted of a transfer sump, IAAB unit, treated effluent tank, and sludge holding tank.

Raw POME is fed to the transfer sump where a grip trap with bar screens are installed to filter out coarse solids and debris. The raw POME is stored in transfer sump to ensure a constant supply of POME into the anaerobic compartment, in which the degradation of complex organic matter without oxygen occurs. The feeding system is designed in an upflow manner and the POME is fed into a liquid distribution system through an inverter controlled feed pump. This anaerobic compartment with an effective volume of 1125 m^3 is the major equipment in the biogas plant. The anaerobic compartment is inoculated with the anaerobic sludge obtained from the ponding system in the same palm oil mill. The anaerobic bacteria activity, MLSS, pH profile and upflow velocity were maintained for an efficient performance. Recirculation system is introduced in the anaerobic compartment to ensure homogenous and uniform distribution of the POME. For the purpose of sampling and the removal of scum, POME can be released from the side of the anaerobic compartment to the

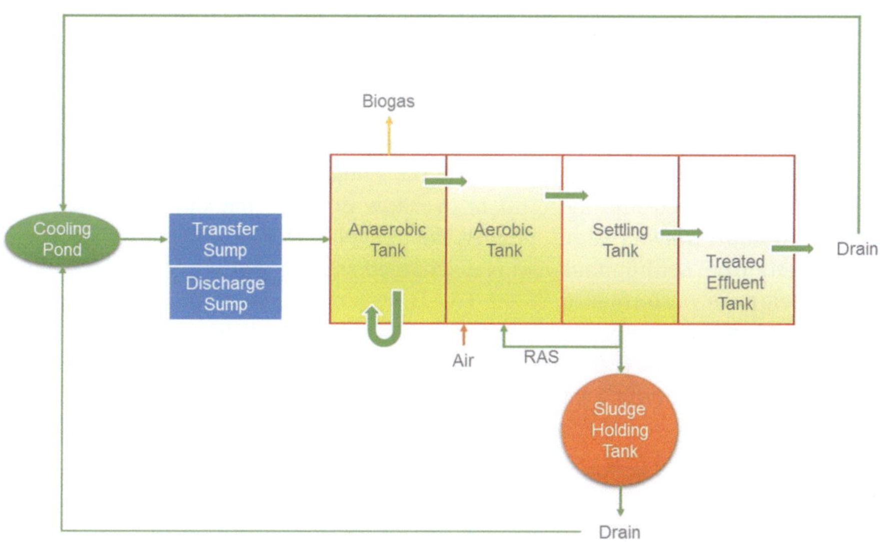

Fig. 1 The simplified IAAB system for the treatment for POME (Chan et al., 2020)

Fig. 2 Pre-commercialized scale of IAAB system (*Courtesy of* Havys Oil Mill Sdn. Bhd.) (Chan et al., 2020)

drain sump. Biogas was collected from the geo-membrane installed at the top of the anaerobic compartment (Fig. 2) and be sent to a moisture separator (MS) for moisture removal. A flame separator (FS) is also used to prevent fires or explosions before releasing the biogas from the blower at 40 °C and 1 atm. The pressure of the biogas in the geo-membrane was controlled by using blowers at 0.4 mbar. The safety of the biogas geo-membrane was ensured with manual relief valve. After being treated anaerobically, POME overflows into the aerobic compartment (977 m^3) where the waste is further digested in the presence of oxygen. Blower is used to supply air from the surroundings with a dissolved oxygen (DO) concentration of 3 ppm to the aerobic compartment. This is mainly to allow complete sludge mixture in the reactor and to supply adequate dissolved oxygen for biological processes.

From aerobic compartment, the POME overflows into the settling tank where the sedimentation process occurs, allowing the sludge to settle by gravity. The sludge is then transferred to the returned activated sludge (RAS) pump system, where a portion of the solids is returned to the aerobic compartment to maintain the desired biomass concentration in the aerobic compartment. The remaining portion of the sludge is transferred to a sludge holding tank where the sludge is kept in storage. The sludge can either be pumped into the effluent overflow drain by the sludge pumps or be transferred to the transfer sump. Finally, the treated POME flows into the treatment effluent tank from settling tank and is discharged as clear water.

The treatment performance of the IAAB system was monitored and analyzed. The stability of the reactor was assessed in terms of pH, MLSS concentration, and food-to-microorganism (F/M) ratio. Samples were collected at different sampling points along the anaerobic and aerobic compartments. The achievement of more than 65% of COD removal efficiency for three consecutive days was taken as stable performance of the IAAB.

2.3 Analytical Methods

For anaerobic process, several monitoring parameters were evaluated during the entire operation, including COD, BOD, and TSS concentrations of the effluent, as well as pH, temperature, $MLSS_{an}$, of the anaerobic compartment together with methane yield and methane composition. Whereas for aerobic process, the COD, BOD, TSS concentrations, and pH of the treated effluent, as well as temperature, DO, and $MLSS_a$ of the aerobic compartment were analyzed. Analytical determinations of BOD, COD, and TSS were carried out in accordance with the Standard Methods for the Examination of Water and Wastewater (Association et al., 1912). BOD_3 was analyzed on samples incubated for 3 days at 30 °C according to the EQA 1974. COD was analyzed by using the colorimetric method with a HACH spectrophotometer (DR2000, Loveland, CO). TSS and MLSS was determined by filtering a sample through a glass fiber filter (Whatman grade GF/A, 1.6 μm, UK) and the residue retained on the filter is dried in an oven (Memmert, Germany) at 105 °C whereas VSS and MLVSS were determined by ashing the dry sample in a 550 °C muffled furnace (Carbolite, UK) for 15 min. The composition of biogas was measured using a biogas analyzer (GFM 416 series, UK).

3 Results and Discussion

3.1 Steady State Performance of IAAB at Different OLRs

The performance of the anaerobic–aerobic system for POME treatment was monitored to evaluate its performance from the aspect of COD and BOD removal efficiencies at anaerobic and aerobic compartments respectively, as well as methane composition and yield at various loading rates. The stability of the IAAB was evaluated from the aspect of pH, MLSS concentration, and F/M ratio. The optimum operating conditions for IAAB were determined within the range implemented in this study as presented in Table 2. It is expected that the most appropriate OLR and MLSS concentrations will neither be too low or too high, as both conditions will deteriorate the performance of IAAB. Generally, the IAAB achieved high treatment efficiency with overall COD, BOD, and TSS removals of up to 99%, at average OLR of 6.3 ± 4.7 kg COD/m^3 day and total HRT of 10.0 ± 5.9 days. The outcomes of the current study are presented and discussed in the following sections.

The composition of biogas produced by the IAAB is presented in Table 3. The average methane composition of biogas ranges between 60.2–64.4%, compared to those obtained by Malakahmad et al. (2014), where the methane content was found to be 54–75%. Poh and Chong (2014) have also reported that UASB-HCPB reactor have successfully produced biogas with methane content of 42.5–76.1% and the remaining being carbon dioxide when treating POME under thermophilic condition (Poh & Chong, 2014). The high methane content is desirable as it represents the heating value

Table 2 Operating conditions for the anaerobic and aerobic compartment of the IAAB

Operating conditions	Anaerobic	Aerobic
OLR (kg COD/m^3 day)	0–20	0–9.5
HRT (days)	4.59–27.7	4.1–22.7
MLSS (mg/L)	9000–49,600	9000–40,500
DO (mg/L)	–	≥2
pH	6.5–7.4	7.5–8.5

Table 3 The composition of the biogas generated by the IAAB

Component	Average	Range
CH_4 (%)	63.2	60.2–64.4
CO_2 (%)	31.3	30.2–35.1
O_2 (%)	0.35	0.2–4.0
H_2S (ppm)	852	814–1902

of the gas. It is noticeable that <1 vol% of oxygen was picked up by the sensor. The oxygen content may be due to the exposure of the instrument to the atmosphere when the biogas was analyzed. Another possible cause is due to the leakage of oxygen into the anaerobic compartment from the aerobic section of the bioreactor. Nonetheless, the trace of oxygen content is negligible. The hydrogen sulphide generated is expected to be low when methane production is high. This phenomena is explainable by the competition between the sulphate-reducing bacteria with methane-producing bacteria for available hydrogen during anaerobic degradation (Eriksen et al., 2012). However, the traces of hydrogen sulfide in biogas is still unfavorable as it brings about corrosiveness and odor potential of the biogas which may consequent to hazardous situation when being ignited. Hence, it is essential to remove the acidic gas from the biogas before utilization.

From Fig. 3, it can be observed that the IAAB exhibited a stable methane yield which falls in the range of 0.190–0.257 L CH_4/g COD upon operating on the 10th day onwards. The increase in methane yield also corresponded with the increasing MLSS concentration (Sect. 3.3) as operational day increases until the 30th day as shown in Fig. 3.

This shows that the operational condition of the IAAB is conducive for the bacterial activity of the methanogens. The methane yield obtained in the present study is similar to the methane yield of 0.22–0.24 L CH_4/g COD obtained in mesophilic system at lab scale as reported by previous study (Chan et al., 2013). The alignment of the results achieved by the IAAB at pre-commercialized scale with lab scale shows the potential of the proposed technology in providing higher treatment efficiency once further work on optimization is conducted. These results are also used in the development of a simulation model (Chong et al., 2021). In the next section, the effect of OLR and MLSS concentrations would be evaluated to obtain the recommended operating conditions for IAAB.

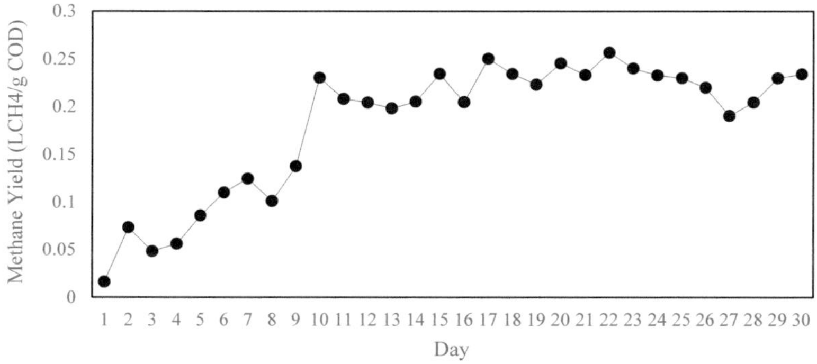

Fig. 3 Methane yield obtained by the IAAB throughout 30 operational days

3.2 *Effect of OLR*

Figure 4 shows the performance of the anaerobic compartment in terms of COD removal efficiencies at various OLRs. OLR was increased from 5 kg COD/m^3 day up to 20 kg COD/m^3 day, testing the maximum loading condition for the design of a potential full scale IAAB system. It can be seen that the anaerobic COD removal efficiency is stable and consistent over 71.70–77.30% as OLR range increases until OLR of 10–12 kg COD/m^3 day. Thereafter, the anaerobic COD removal efficiency experienced a drastic fall to 65.89% at OLR range of 18–20 kg COD/m^3 day. The trend of the results are expected as similar results were portrayed by the previous work conducted, where the anaerobic COD removal was reportedly portraying a drastic decreasing trend in efficiency at approximately OLR of 19.6 COD/L day onwards (Chan et al., 2017). Conversely, the COD removal efficiency in the aerobic compartment was relatively higher and more stable, with an average value of 91 $\pm$ 6.1% at OLR ranging from 0.48 to 9.53 kg COD/m^3 day. Similarly, the overall COD removal efficiency of the entire IAAB remained stable, averaged at 95.6 $\pm$ 6.6% regardless of the various OLR applied. The effect of OLR on the overall COD removal efficiency was not as straight forward. The overall COD removal efficiency provided by the IAAB system was contributed by the anaerobic and aerobic compartment of the bioreactor respectively. As reported previously, the increment of OLR resulted in lower contribution from anaerobic compartment to the overall COD removal efficiency (Chan et al., 2012, 2017). On the contrary, COD removal efficiency in the aerobic compartment was compensated for the reduction in efficiency posed by the anaerobic zone at high OLR, as the aerobic microorganism activity was promoted. In short, the COD removal efficiencies in the anaerobic and aerobic zones were inversely related.

Rectification on the importance of aerobic system in bioreactor can be proven via present study to ensure treated effluent meets the discharge standard especially when operating at a high loading rate. Nonetheless, an excessively high loading rate will consequent to failure of the final treated effluent in meeting the discharge limit. This

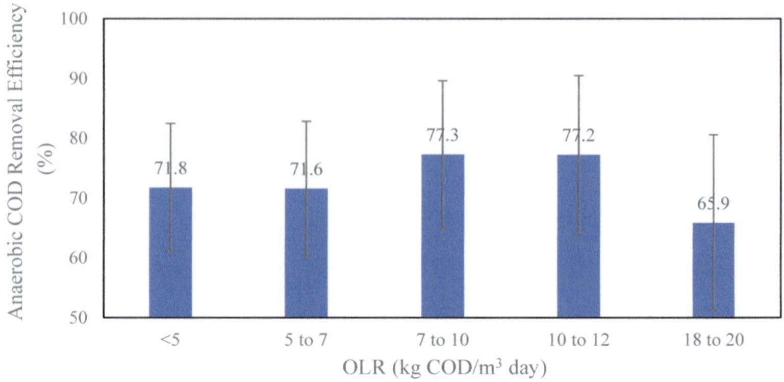

Fig. 4 The anaerobic COD removal efficiency of IAAB at different range of OLR (Chan et al., 2020)

is explainable by the dominant effect of OLR on the performance of the anaerobic zone as this compartment is heavily responsible in removing most of the COD in the wastewater. Additionally, inhibition of the aerobic biomass activity was reported when OLR was increased to 25.0 g COD/L day which consequent to the fall in the COD removal efficiency (Chan et al., 2012). Concurrent to this, some operational issues such as foaming and scum formation were also encountered in anaerobic compartment and this will be discussed in Sect. 3.4.

It is also noticeable that the anaerobic COD removal efficiency is not as high as reported previously. This is expected as the performance of present study have not been optimized to cater to the scaled-up volume of the bioreactor. Furthermore, it is suspected that the recirculation flowrate drawn from the IAAB (<70 m^3/h) by the external centrifugal pump is not sufficiently high enough in providing adequate mixing within the system. Generally, mixing in the anaerobic zone is essential as it promotes the distribution of substrates and microorganisms thoroughly within the digester in achieving homogeneity. On top of that, mixing can overcome the rheological behavior of the nature of slurry which poses complications in attaining turbulence while also forming dead zones inside the digester (Singh et al., 2020). The recommended recirculation flowrate is as suggested in Sect. 3.5. According to Singh et al. (2020), approximately 44% failures of the biogas plants occur due to mixing flaws. Therefore, further work can be done in determining the optimum recirculation flow rate in the anaerobic zone through a multidisciplinary approach which involves the expertise in fluid dynamics and microbiology.

Figure 5 depicts the capability of IAAB in treating POME that exhibits a large range of COD. It can be observed that the anaerobic COD removal efficiency increases as the strength of the wastewater increases. This shows that this technology is effective in treating POME which usually exhibits COD at the range of 85,000–100,000 mg/L (Yap et al., 2020). Undoubtedly, the IAAB technology holds potential for treating a large quantity of high strength wastewater as it is known as a valuable feedstock for anaerobic–aerobic treatment to harness its high COD content for

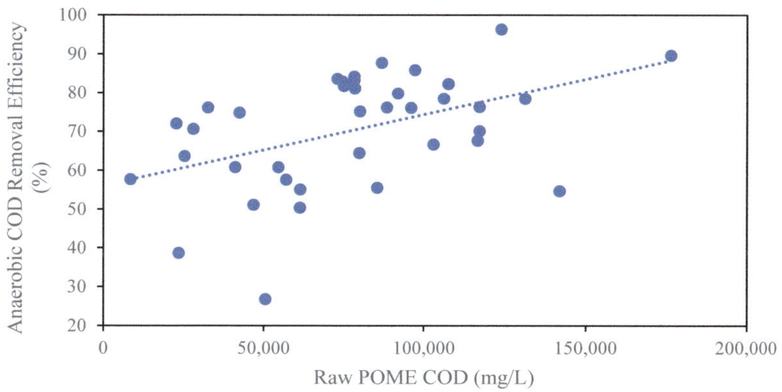

Fig. 5 Anaerobic COD removal efficiency for different COD influent (Chan et al., 2020)

energy generation. Moreover, the present innovation greatly reduces sludge production when it is treated at the anaerobic section of the system (Chan et al., 2009). Subsequently, the aerobic section will counter the fluctuations in the quality of the anaerobically treated effluent. The integration and optimization of both biological treatments are indeed a reliable innovation in overcoming the wastewater produced in this industry.

3.3 *Effect of MLSS Concentration and F/M*

It is essential to monitor the MLSS concentration and F/M ratio in understanding the ongoing biochemical activity and their variations during the operation, as it greatly signifies the performance of the IAAB. Figure 6 depicts the MLSS concentration in the aerobic and anaerobic compartment of the IAAB throughout the operational days. For the aerobic compartment, it can be observed that the highest MLSS concentration of 38,600 mg/L falls on the 19th day of the operation. The decrease in MLSS concentration is suspected to be brought upon by the increment in OLR. As reported from previous work, MLSS concentration reduces drastically when the anaerobic OLR was increased to 19.5–21.0 kg COD/m^3 day due to wash out of anaerobic sludge with poor settleability (Chan et al., 2017). The initial increase in MLSS concentration shows increasing microbial activity along the days which coincided with the increasing aerobic COD removal. The increase in the population of the microorganism shows that the IAAB is capable in maintaining high biomass concentration to provide efficient treatment on POME. On the other hand, the MLSS concentration in the anaerobic zone is consistent over the range of 15,000–19,000 mg/L which corresponded to the plateau achieved by the methane yield shown in Fig. 3 on 15th days onwards. This means that MLSS concentration at this range is sufficient in providing efficient treatment on the POME while yielding stable methane generation.

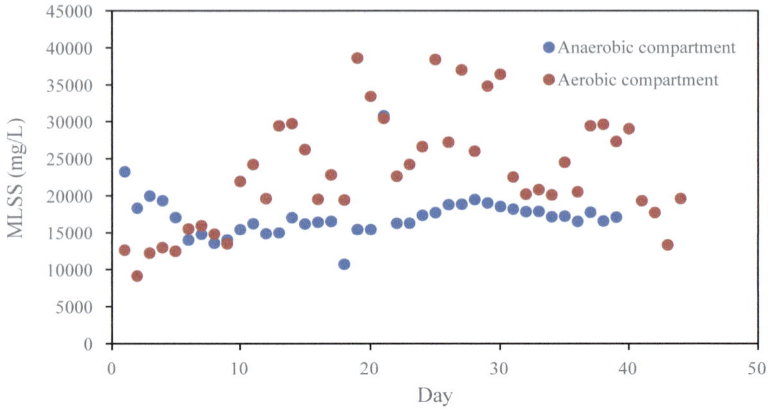

Fig. 6 MLSS concentration along the operational days of the anaerobic and aerobic compartment of IAAB

Figure 7 shows the effect of F/M ratio on the COD removal efficiency. It can be observed that the COD removal efficiency decreases along the increment of F/M ratio in the aerobic compartment. Thus, this occurrence has consequent to the reduction in the overall COD removal as the anaerobic effluent is no longer being further biodegraded at an optimum condition. Since F/M ratio is dependent on the OLR and MLSS concentration, the result is expected as the system experienced a shock loading where the sudden increase in the organic matter was unable to be treated effectively by the microorganisms. Similar results were also reported from previous work where a higher bacterial population is required to treat OLR higher than 11.7 kg COD/ m^3 day (Chan et al., 2013).

The decrease in the overall COD removal efficiency coincides with the final BOD effluent of the system as presented in Fig. 8. It can be observed that majority of the

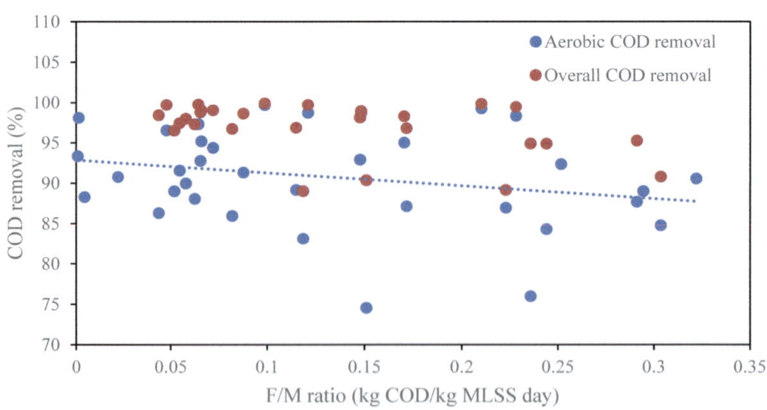

Fig. 7 The effect of F/M ratio on the aerobic and overall COD removal efficiency

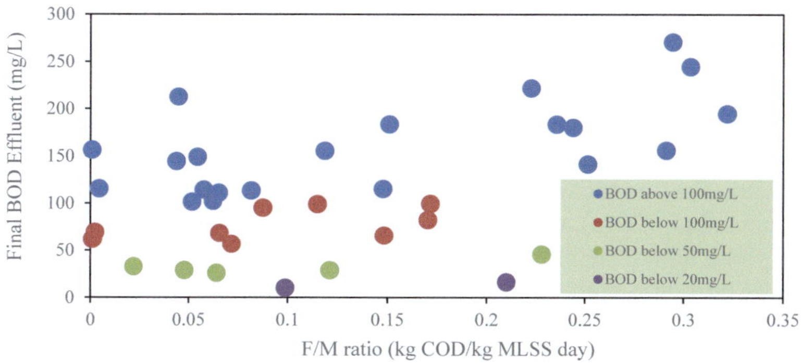

Fig. 8 The effect of F/M ratio on the final BOD effluent

sample exhibited BOD above 100 mg/L which indicates poor BOD removal when F/M ratio increases (Fig. 8). The present study outcome was also reflected by the results published by Lateef et al. (2013), where F/M ratio of 0.21–0.98 kg BOD/kg MLVSS day was accounted for the BOD removal efficiency of 98.3–84.5% (Lateef et al., 2013). Diez et al. (2002) has also reported a decrease in BOD removal at high F/M ratio above 0.2 g BOD/MLVSS day. It is also worth noticing that some samples exhibited BOD above 100 mg/L when F/M ratio is too low (<0.05 kg COD/kg MLSS day). This is explainable by the shortage of food for adequate bacterial growth in the system hence, poor biodegradation of the organic matter occurred. Nonetheless, the inverse relationship between F/M ratio and the organic removal efficiency proves that increasing F/M ratio does not favor the performance of the IAAB which further rectifies the effect of the OLR increment on treatment efficiency as discussed in the previous section.

3.4 *Operational Issues Faced by IAAB*

During the start-up and steady state operation of 200 days, there are few operational problems faced by anaerobic and aerobic compartment. For instance, foaming issues faced by both anaerobic and aerobic compartment, whereas scum formation occurred in anaerobic compartment. This is possibly due to the inconsistent or changes in influent conditions which created a shock loading state to the plant. In addition, the sudden change in POME influent can result in lower effluent quality as insufficient contact time is observed between the microbial community and the POME substrate. The detailed operational problems, their root causes and rectifications methods will be discussed in the following sub-sections.

3.4.1 Anaerobic Compartment

The anaerobic process and top-notch performance are affected by the presence of microorganism to the available substrates and nutrients, appropriate pH, temperature, HRT, Solid Retention Time (SRT), and the distribution of POME substrate (Kress et al., 2018). The former which all influenced by the mixing behavior. According to Kariyama et al. (2018) and Singh et al. (2019), various issues such as failure in methane yield, flawed stabilization of raw slurry, reduction in digester volume, high operational cost, non-homogenous distribution of temperature and substrate, short circuiting, occurrence of sediment at bottom of digester, dead zone, and especially scum formation are mainly due to inadequate mixing.

The phenomena of scum formation is of common problems affecting the full-scale anaerobic digester (Kariyama et al., 2018). In this study, scum formation was observed on days 56–59 (Fig. 9a) due to the high OLR (18 kg COD/m^3 day) applied and therefore, high COD effluent (38,000–73,800 mg/L) and low COD removal (25–55%) were observed. Some studies have shown that substrate with high concentration of fatty acids or high grease content in the influent are among the cause for the formation of scum which will diminish the efficiency of anaerobic digester (Halalsheh et al., 2005; Pagilla et al., 1997). This is true as the current treatment for POME contains high amount of oil and grease, ranging from 130 to 18,000 mg/L.

As suggested by Kshirsagar and Pawar (2018), the deformation of scum can be solved by increasing the surface velocity and designing concrete flaps at the baseline of the anaerobic digester. Furthermore, the geometry and configurations of anaerobic digester plays a vital role where an egg-shaped digester is much preferable as compared to cylindrical digester in terms of maintaining homogeneity, uniform mixing and reducing dead zones (Singh et al., 2020). A similar approach by Kobayashi et al. (2013) observed that there is no scum formation when the OLR is keeping low (5–10 kg COD/m^3 day) and stressed that potential of scum formation is mostly related to the increase in OLR and HRT shortening. Besides, too low HRT would result in an incomplete degradation of POME substrate (especially oil

(a) (b)

Fig. 9 **a** Scum formation, **b** foaming

and grease) or bacteria wash-out at an early stage. Therefore, to rectify this scum problem effectively, the OLR was immediately reduced to 5 kg COD/m^3 day. At the same time, the recirculation rate was increased to 140 m^3/h to breakdown the scum layer while providing better mixing to the digester as discussed earlier. Thereafter, the anaerobically treated effluent COD was gradually reduced to 13,800 mg/L on day 60.

Besides scum formation, foaming is also a serious operational problem usually occurs in full-scale biogas plants at wastewater treatment plants (WWTPs) (Nguyen et al., 2019). Foaming causes inefficient gas recovery, blockages of gas mixing, fouling of gas collection pipes and covering the digester wall with foam solid (Ganidi et al., 2009; Pagilla et al., 1997). Besides, it will limit the methane production and incur high manufacturing cost when taking into account of cleaning, repairing and maintenance (Ganidi et al., 2009; Moeller & Görsch, 2015; Westlund et al., 1998). Although numerous studies have addressed for this reoccurring phenomenon, the exact principle and concept of foaming in anaerobic digestion is still not fully understood (Barjenbruch et al., 2000).

In this study, foaming problem was also reported on day 66 (Fig. 9b) and low methane production was observed (0.11 L CH_4/g COD). When foam was beginning to build up significantly, the recirculation rate was immediately reduced to 70 m^3/h. This is because high recirculation rate of 140 m^3/h (which was implemented on day 56) has led to over mixing condition. This has significantly reduced the wastewater surface tension, and thus leading to this foaming problem. Apart from this, intermittent mixing of 140 m^3/h for every time interval of 30 min was also incorporated in the anaerobic compartment. Both approaches were deemed feasible, as foaming condition is reduced gradually and the performance of the anaerobic compartment was back to normal after 10–15 days of implementation of these approaches. Therefore, it can be concluded that adequate mixing (recirculation rate) is of utmost importance to maintain optimal anaerobic digester function while minimizing the operational problems.

3.4.2 Aerobic Compartment

In aerobic compartment, microorganism activity and adequate control of MLSS plays a significant role in the activated sludge process as unfavorable conditions would affect the effluent quality (de los Reyes, 2010). The most common problems during the aerobic treatment includes foaming, sludge bulking, and sludge rising (Khodabakhshi et al., 2015). As dictated by Fryer et al. (2011), large volume of foam can lead to undesirable operational conditions such as blockage in pipes, exposure of pathogens, reduced plant performance, reduction in oxygen transfer, walkway and plant monitoring equipment obstruction. The aforementioned issues are faced by 40% of overall wastewater treatment plants, 78% of existing activated sludge systems in the country and most major problem encountered in South Africa (Khodabakhshi et al., 2015). Generally, foaming and bulking problems are associated to various of filamentous bacteria, production of extracellular polymers, sludge treatment facilities which contains oil and grease, and synthesis of bio-surfactant in the presence of hydrophobic substrate (Pal et al., 2014). In addition, the three main components for stable foam formation required presence of air bubbles, surfactants, and hydrophobic cells (Petrovski et al., 2011).

There are many types of foaming formation either chemically or biologically, and both must undergo dispersion of gas in a liquid. First, chemical foams are derived mainly from excess surfactant (white foams), which applies to most pharmaceutical, cosmetics, textiles, food, paper and biotechnology industries (Collivignarelli et al., 2020). Nonetheless, this study follows the biological approach (brown foams) which is related to the growth of filamentous bacteria named "foam former" or bacterial-synthesized hydrophobic high-molecular weight substances in MLSS (Fig. 10a) which was detected on day 68 (30 March 2018). The foam appearance and characteristics in the current study are observed to be brownish-like which is associated to high F/M ratio of 0.22–0.32 kg COD/kg MLSS, high OLR of 4.23–9.53 kg COD/m^3 day, and high grease and oil content of 10,000–18,000 mg/L. This is due to the foaming problem occurred in anaerobic compartment (day 66) where excessive sludge/foam from the anaerobic compartment was carried over to the aerobic compartment. Overall, insufficient MLSS results in lesser microorganism population for the degradation process in aerobic compartment which also contributes to higher F/M ratio. Thus, inefficient degradation mechanism leads to higher oil and grease concentration which end up with increasing foaming issues.

Various ways for the treatment of foaming issues can be distinguished into short-term and long-term control methods. Short-term control involves chlorination, polymer and coagulation addition, while long-term control involves alteration in aeration, biomass concentration, influent waste septicity (H_2S and organic acids), and nutrient addition (Richard, 2003). According to Pal et al. (2014), short-term methods are of most favorable for rapid and effective regulation of activated sludge

(a) (b)

Fig. 10 Day 1: 30 March 2018 (**a**) 5 April 2018 (**b**) Foam condition after 5 days with Bioremove 5100

foaming, however the usage of chemical solutions could be expensive for long-term usage (Tsang et al., 2008) and potential of forming undesirable by-products such as trihalomethanes (THMs) (Caravelli et al., 2003). The former can be solved by alternating toward biological approach, Bioremove from Novozymes that can improve digester performance, resulting in cost-efficient and more sustainable to enhanced COD removal and stable plant operation (reduces foaming issues) (Novozymes, 2021b). Bioremove is defined as blend of microorganisms in powder state for the biological degradation of substrate, which can be utilized in most wastewater treatment plants (Chempoint, 2021). The solution of Bioremove is suitable to treat highly complex waste streams, choosing from a large spectrum of microorganisms that could cater a wide range of organic substrate to ensure the compliance of the plant. The simple approach of Bioremove greatly improved the plant stability, and approximately 16% COD reduction in effluent concentrations without the need to invest in capital investment (e.g. reactor volume) (Novozymes, 2021a). On the other hand, many operators have tried to resolve forming issues by reducing the aeration for long-term focus, yet results in higher filament growth rate (Richard, 2003). Therefore, in this experimental study, rectification has been decided upon aerobic compartment, in which Bioremove by Novozymes will be utilized. The application of Bioremove 5100 dosing has proven to reduce the foaming conditions after 5 days as seen in Fig. 10a, b.

While application of Bioremove seems to counteract the foaming problems, other parameters such as COD and BOD removal efficiencies will be affected as shown in Fig. 11. Therefore, comparison between Bioremove 3200, Bioremove 5100 and combination of both at 50:50 ratio will determine the best overall results for COD and BOD removals along with foaming reduction (Fig. 11). Bioremove 5100 is able to control the foaming but unable to effectively reduce COD and BOD. Besides, effluent produced by using Bioremove 5100 alone has a darker color with a thin layer of oil. On the other hand, Bioremove 3200 is able to reduce COD and BOD but produce excessive foam. Effluent produced by using Bioremove 3200 alone has a light brown color, however no oil layer is observed. Finally, combination of Bioremove 5100 and Bioremove 3200 at ratio of 50:50 is the optimum choice where lower effluent COD and BOD ranging from 1420 to 3910 mg/L and 109–310 mg/L, respectively were observed.

In conclusion, anaerobic compartment plays a vital role in the overall performance of IAAB as it removes most of the organic fraction from POME, and most importantly, any operational problems occurred in the anaerobic compartment will impact the performance of the subsequent aerobic compartment. This is because the scum or foam formed in the anaerobic compartment will eventually pass through the aerobic compartment, which will result in overall performance deterioration (approximately 40–70%). Therefore, it is important to maintain the IAAB at the recommended operating conditions so that the aforementioned operational problems could be minimized while achieving high treatment efficiency. The recommended operating conditions will be discussed in the next section.

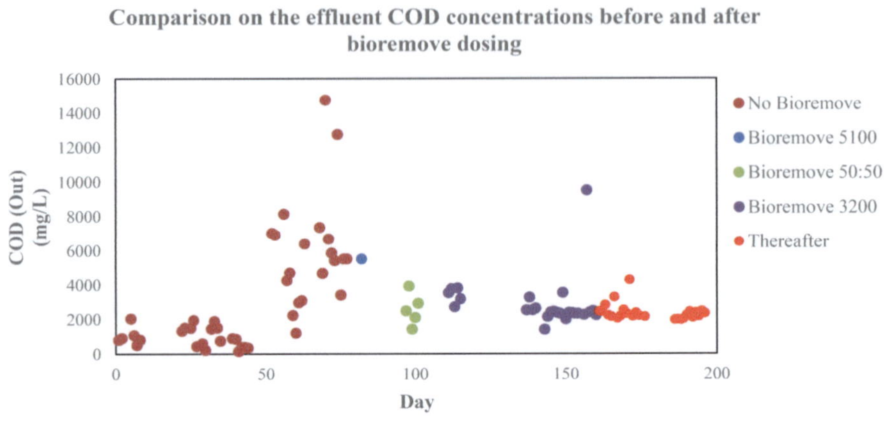

(a)

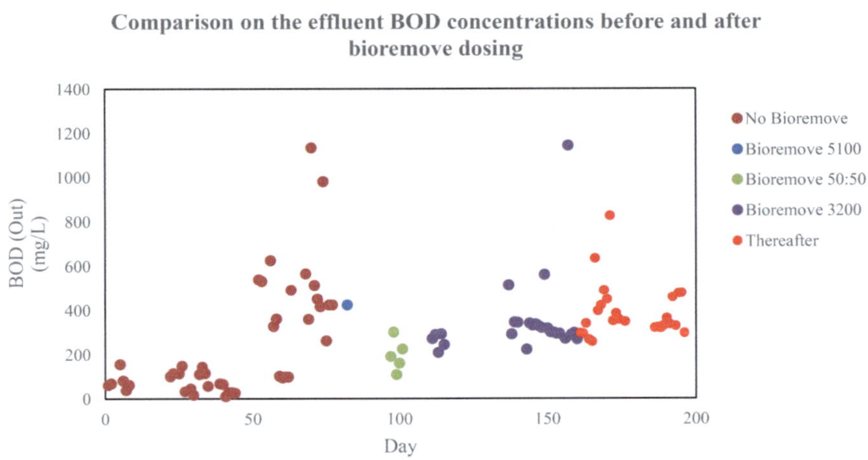

(b)

Fig. 11 Effluent quality with/without Bioremove dosing: **a** COD concentrations **b** BOD concentrations

3.5 Recommended Operating Conditions for IAAB

The operating conditions suggested for the anaerobic and aerobic compartments in IAAB are as presented in Tables 4 and 5 respectively. As discussed from the previous section, increment in OLR will result to lower anaerobic performance but better aerobic performance. Despite that, excessively high OLR will result to failure in the overall system in treating the effluent in meeting the stringent discharge standards as well as inhibiting methanogenesis which ultimately leads to lower methane production. This is explainable by the dominant effect of OLR on the performance of the anaerobic zone. In addition, the anaerobic compartment is heavily responsible in removing most of the organic matter found in the wastewater. Referring to the present work's outcome, OLR of 7–12 kg COD/m^3 day is the most appropriate range in attaining stable and consistent overall COD removal efficiency. The lower range of OLR is not selected as higher loading capacity is necessary in reducing the bioreactor volume as well as the capital costs. It was observed that consistent methane yield was attained over the MLSS range of 18,000–19,000 mg/L (Fig. 6). In previous work, low MLSS concentration was reported to be incapable in sustaining high bioactivity in the bioreactor. On the other hand, excessively high MLSS concentration will consequent to low COD removal, long sludge settling time and high concentration of suspended solid in the effluent (Chan et al., 2013). Hence, the MLSS concentration is proposed to be fixed at 19,000 mg/L. Mixing features are also incorporated to the anaerobic zone to promote intimate contact between the microorganisms with the substrate within the system. This can be done by providing either intermittent of 140 m^3/h for every time interval of 30 min or continuous recirculation rate of 70 m^3/h respectively (Sect. 3.5). The COD removal efficiency set for the anaerobic zone will be around 70–85% to ensure that the anaerobic effluent has adequate nutrients available for the aerobes to carry out its biodegradation under the presence of oxygen. Hence, the COD removal is set to be sufficient but not excessively high to ensure effective aerobic function in the next subsequent stage. The methane yield is expected to be 0.22 L CH_4/g COD with biogas production of 63% methane content. Based on this, the IAAB is expected to be generating power of 390 kW by utilizing the methane production of 6 m^3/h (Table 4).

The F/M ratio in the aerobic compartment is suggested to operate at below 0.17 kg COD/kg MLSS day as the treatment efficiency decreases above the value mentioned. The present work's outcome has also proven that the IAAB is able to achieve a final BOD effluent below 100 mg/L with 45% of compliance. As mentioned in Sect. 3.3, higher COD removal efficiency was attained when operating at lower F/M ratio. Moving on, the MLSS concentration is recommended to be within the range of 19,000–26,000 mg/L in ensuring that very high aeration is not required in the aerobic compartment as such implementation will increase the operating cost. DO

Table 4 Recommended operating conditions for anaerobic compartment

Anaerobic	Recommended range
OLR (kg COD/m^3 day)	7–12
MLSS concentration (mg/L)	19,000
Recirculation rate	140 m^3/h intermittent mixing for 30 min every hour 70 m^3/h continuous mixing
COD removal (%)	70–85
Methane yield (LCH_4/g COD)	0.22
Methane purity (%)	63%
Power generation	6 m^3/h, COD 80,000 mg/L, 390 kW @40% gas engine efficiency

Table 5 Recommended operating conditions for aerobic compartment

Aerobic	Recommended range
F/M (kg COD/kg MLSS day)	≤0.17
MLSS (mg/L)	19,000–26,000
DO (mg/L)	≥2
COD removal (%)	85–99.6
Energy consumption (kW/day)	35

is supplied at 2 mg/L and above to sustain the bacterial population of aerobes to provide decent biodegradation of the wastewater (>80% BOD removals). With the suggestions above, final COD removal of 85.0–99.6% has been achieved by this pre-commercialized IAAB. The energy consumption of the aerobic zone is estimated to be 35 kW per day. The operating conditions presented in Table 5 are proposed for the operation of IAAB in attaining high treatment efficiency (overall COD removal efficiency >99%) as well as high methane generation (>0.22 L CH_4/g COD).

4 Conclusion

The Integrated Anaerobic–Aerobic Bioreactor (IAAB) has been successfully scaled up into pre-commercialized scale while achieving high overall COD, BOD, and TSS removals of more than 99% along with high methane yield (>0.22 L CH_4/g COD) containing 63% of methane purity. Undoubtedly, the present study has proven that various factors including OLR and MLSS concentration posed significant effects on the performance of the IAAB. With consideration of the effects brought upon by different factors discussed in this chapter, the anaerobic zone of the IAAB is suggested to operate at OLR of 7–12 kg COD/m^3 day under MLSS concentration of 19,000 mg/L, and with intermittent and continuous mixing through recirculation flowrate of 140 m^3/h at interval of 30 min and 30 m^3/h respectively. The aerobic zone is proposed to operate below F/M ratio of 0.17 kg COD/kg MLSS day under MLSS concentration of 19,000–26,000 mg/L with DO supply of more than 2 mg/L. In commercial setting, anaerobic compartment is prone to problems such as scum formation and foaming issue due to insufficient mixing and over mixing, respectively and these can be solved by adjusting the recirculation rate. Besides, the occurrence of foaming in aerobic compartment has been successfully reduced by dosing Bioremove 5100 and Bioremove 3200 at ratio of 50:50. Results show that it is important to maintain the anaerobic compartment at the recommended operating conditions so that the aforementioned operational problems could be minimized. Any operational problems occurred in the anaerobic compartment will significantly impact the performance of the subsequent aerobic compartment. Therefore, future improvement and recommendations in terms of performance of IAAB would be focused on process optimization of the anaerobic compartment, particularly in the aspect of microbiology, fluid dynamics, and mass transfer in slurry viscous conditions. Satisfactory results from this study will enable a closer step toward the industrial revolution for the treatment of POME. The successful achievement of IAAB toward commercialization will assist palm oil industry in tackling climate change and maintaining water quality brought upon by the conventional settings which greatly depletes the ozone layer of the earth. Certainly, the innovation and high performance of IAAB will be brought upon safe water and sanitation for human beings and accessibility to green and sustainable energy.

Acknowledgements The authors gratefully acknowledge the financial support from The University of Nottingham Malaysia and Havys Oil Mill.

References

Akhbari, A., Kutty, P. K., Chuen, O. C., & Ibrahim, S. (2020). A study of palm oil mill processing and environmental assessment of palm oil mill effluent treatment. *Environmental Engineering Research, 25*(2), 212–221.

Association, A. P. H., Association, A. W. W., Federation, W. P. C., & Federation, W. E. (1912). *Standard methods for the examination of water and wastewater*. American Public Health Association.

Barjenbruch, M., Hoffmann, H., Kopplow, O., & Tränckner, J. (2000). Minimizing of foaming in digesters by pre-treatment of the surplus-sludge. *Water Science and Technology, 42*(9), 235–241.

Caravelli, A., Contreras, E. M., Giannuzzi, L., & Zaritzky, N. (2003). Modeling of chlorine effect on floc forming and filamentous micro-organisms of activated sludges. *Water Research, 37*(9), 2097–2105.

Cervantes, F., Pavlostathis, S., & van Haandel, A. (2006). *Advanced biological treatment processes for industrial wastewaters—Principles & applications*, IWA Publishing.

Chan, Y. J., Chong, M. F., & Law, C. (2013). Optimization of palm oil mill effluent treatment in an integrated anaerobic-aerobic bioreactor. *Sustainable Environment Research, 23*, 153–170.

Chan, Y. J., Chong, M. F., & Law, C. L. (2012). Start-up, steady state performance and kinetic evaluation of a thermophilic integrated anaerobic–aerobic bioreactor (IAAB). *Bioresource Technology, 125*, 145–157.

Chan, Y. J., Chong, M. F., & Law, C. L. (2017). Performance and kinetic evaluation of an integrated anaerobic-aerobic bioreactor in the treatment of palm oil mill effluent. *Environmental Technology, 38*(8), 1005–1021.

Chan, Y. J., Chong, M. F., Law, C. L., & Hassell, D. G. (2009). A review on anaerobic–aerobic treatment of industrial and municipal wastewater. *Chemical Engineering Journal, 155*(1–2), 1–18.

Chan, Y. J., Hue, F. S., Chong, M. F., Ng, D. K. S., & Lim, L. K. (2020). Pre-commercialized integrated anaerobic-aerobic bioreactor (IAAB) for palm oil mill effluent (POME) treatment & biogas generation. *Journal of Oil Palm, Environment & Health, 11*, 57–66.

Chong, J. W. R., Chan, Y. J., Chong, S., Ho, Y. C., Mohamad, M., Tan, W. N., Cheng, C. K., & Lim, J. W. (2021). Simulation and optimisation of integrated anaerobic-aerobic bioreactor (IAAB) for the treatment of palm oil mill effluent. *Processes, 9*, 1124.

Chempoint. (2021). Novozymes bioaugmentation.

Collivignarelli, M. C., Baldi, M., Abbà, A., Caccamo, F. M., Carnevale Miino, M., Rada, E. C., & Torretta, V. (2020). Foams in wastewater treatment plants: from causes to control methods. *Applied Sciences, 10*(8).

de los Reyes, F. L. (2010). In K. N. Timmis (Ed.), *Handbook of Hydrocarbon and lipid microbiology* (pp. 2401–2411), Springer.

Diez, M. C., Castillo, G., Aguilar, L., Vidal, G., & Mora, M. L. (2002). Operational factors and nutrient effects on activated sludge treatment of Pinus radiata kraft mill wastewater. *Bioresource Technology, 83*(2), 131–138.

Eriksen, J., Andersen, A. J., Poulsen, H. V., Adamsen, A. P., & Petersen, S. O. (2012). Sulfur turnover and emissions during storage of cattle slurry: Effects of acidification and sulfur addition. *Journal of Environmental Quality, 41*(5), 1633–1641.

Frostell, B. (1983). Anaerobic-aerobic biological treatment of starch industry waste waters. *Starch - Stärke, 35*(6), 185–189.

Fryer, M., O'Flaherty, E., & Gray, N. F. (2011). Evaluating the measurement of activated sludge foam potential. *Water, 3*(1), 424–444.

Ganidi, N., Tyrrel, S., & Cartmell, E. (2009). Anaerobic digestion foaming causes—A review. *Bioresource Technology, 100*(23), 5546–5554.

Halalsheh, M., Koppes, J., den Elzen, J., Zeeman, G., Fayyad, M., & Lettinga, G. (2005). Effect of SRT and temperature on biological conversions and the related scum-forming potential. *Water Research, 39*(12), 2475–2482.

Kariyama, I. D., Zhai, X., & Wu, B. (2018). Influence of mixing on anaerobic digestion efficiency in stirred tank digesters: A review. *Water Research, 143*, 503–517.

Khodabakhshi, N., Asadollahfardi, G., & Shahriarinia, E. (2015). Removal of foaming from industrial wastewater treatment plants. *Water Practice and Technology, 10*(3), 415–423.

Kobayashi, T., Wu, Y.-P., Xu, K.-Q., & Li, Y.-Y. (2013). Effect of mixing driven by siphon flow: Parallel experiments using the anaerobic reactors with different mixing modes. *Energ., 6*(8), 4207–4222.

Kress, P., Nägele, H.-J., Oechsner, H., & Ruile, S. (2018). Effect of agitation time on nutrient distribution in full-scale CSTR biogas digesters. *Bioresource Technology, 247*, 1–6.

Kshirsagar, V. S., Pawar, P. M. (2018). Mixing performance improvement by passive modifications in an anaerobic digester design. *Materials Today: Proceedings, 5*(9, Part 3), 20600–20607.

Lateef, A., Nawaz Chaudhry, M., & Ilyas, S. (2013). Biological treatment of dairy wastewater using activated sludge. *ScienceAsia, 39*, 179.

Malakahmad, A., Lahin, F. A., & Yee, W. (2014). Biodegradation of high-strength palm oil mill effluent (POME) through anaerobes partitioning in an integrated baffled reactor inoculated with anaerobic pond sludge. *Water, Air, & Soil Pollution, 225*(3), 1883.

Moeller, L., & Görsch, K. (2015). Foam formation in full-scale biogas plants processing biogenic waste. *Energy, Sustainability and Society, 5*(1), 1.

MPOC. (2020). Malaysian Palm Oil Industry.

Nguyen, L. N., Johir, M. A. H., Commault, A., Bustamante, H., Aurisch, R., Lowrie, R., & Nghiem, L. D. (2019). Impacts of mixing on foaming, methane production, stratification and microbial community in full-scale anaerobic co-digestion process. *Bioresource Technology, 281*, 226–233.

Novozymes. (2021a). BioRemove 1100 case study.

Novozymes. (2021b). Water & Waste.

Pagilla, K. R., Craney, K. C., & Kido, W. H. (1997). Causes and effects of foaming in anaerobic sludge digesters. *Water Science and Technology, 36*(6), 463–470.

Pal, P., Khairnar, K., & Paunikar, W. N. (2014). Causes and remedies for filamentous foaming in activated sludge treatment plant. *Global Nest Journal, 16*, 762–772.

Petrovski, S., Dyson, Z. A., Quill, E. S., McIlroy, S. J., Tillett, D., & Seviour, R. J. (2011). An examination of the mechanisms for stable foam formation in activated sludge systems. *Water Research, 45*(5), 2146–2154.

Poh, P. E., & Chong, M. F. (2014). Upflow anaerobic sludge blanket-hollow centered packed bed (UASB-HCPB) reactor for thermophilic palm oil mill effluent (POME) treatment. *Biomass and Bioenergy, 67*, 231–242.

Richard, M. (2003). Activated sludge microbiology problems and their control.

Singh, B., Szamosi, Z., & Simenfalvi, Z. (2020). Impact of mixing intensity and duration on biogas production in an anaerobic digester: A review. *Critical Reviews in Biotechnology, 40*(4), 508–521.

Singh, B., Szamosi, Z., & Siménfalvi, Z. (2019). State of the art on mixing in an anaerobic digester: A review. *Renewable Energy, 141*, 922–936.

Tchobanoglous, G., Burton, F. L., & Stensel, H. D. (2004). Wastewater engineering—Treatment and reuse (4th ed.). Metcalf & Eddy.

Tsang, Y. F., Sin, S. N., & Chua, H. (2008). Nocardia foaming control in activated sludge process treating domestic wastewater. *Bioresource Technology, 99*(9), 3381–3388.

Westlund, Å. D., Hagland, E., & Rothman, M. (1998). Foaming in anaerobic digesters caused by Microthrix parvicella. *Water Science and Technology, 37*(4), 51–55.

Yacob, S., Hassan, M. A., Shirai, Y., Wakisaka, M., & Subash, S. (2005). Baseline study of methane emission from open digesting tanks of palm oil mill effluent treatment. *Chemosphere, 59*(11), 1575–1581.

Yap, C. C., Chan, Y. J., Loh, S. K., Supramaniam, C. V., Soh, A. C., Chong, M. F., Chew, C. L., & Lim, L. K. (2020). Comparison of different industrial scale palm oil mill effluent anaerobic systems in degradation of organic contaminants and kinetic performance. *Journal of Cleaner Production, 262*, 121361.

3-MCPDE in Palm Oil Processing: Formation Factors, Transference to Food and Mitigation Approaches

Chien Lye Chew, Amirul Al Hafiz Abdul Hamid, Hemavathi Silvamany, and Soon Huat Tiong

Abstract Palm oil is the world most produced and consumed vegetable oil. Food safety officials have recently focused their attention on vegetable oil, including palm oil, due to the presence of 3-monochloro-1, 2-propanediol and related esters (3-MCPDE). The European Food Safety Authority (EFSA) proposed to limit the 3-MCPDE in palm oil to 2.5 ppm by year 2021. This heat-induced process contaminant is formed during the refining process through its precursors, such as chlorinated compound and diacylglycerol in vegetable oil. The consumption of foods containing 3-MCPDE that exceed the tolerable daily intake limit may cause health complications. Therefore, many research works have been conducted to mitigate 3-MCPDE formation during palm oil processing. This chapter discusses the 3-MCPDE's formation factors, occurrences during food processing, mitigation initiatives and industrial practices in addressing the 3-MCPDE issue. It is hoped that the chapter could provide an insight into 3-MCPDE formation factors and their mitigation strategies.

The original version of the chapter was revised: There is a minor error on page 336 where the text in the bracket has been missed inadvertently ("and 3-MCPDE content by 50% (). The studies…") and this has been corrected to "and 3-MCPDE content by 50% (Chew et al., 2021). The studies…". The correction to the book is available at https://doi.org/10.1007/978-981-19-4847-3_15

C. L. Chew (✉) · A. A. H. Abdul Hamid · H. Silvamany
Sime Darby Plantation Research, R&D Centre—Carey Island, Lot 2664 Jalan Pulau Carey, 42960 Pulau Carey, Selangor, Malaysia
e-mail: chienlyechew@gmail.com

A. A. H. Abdul Hamid
e-mail: amirul.alhafiz.hamid@simedarbyplantation.com

H. Silvamany
e-mail: hemavathi.s@simedarbyplantation.com

C. L. Chew
Monash-Industry Palm Oil Education and Research Platform (MIPO), Monash University Malaysia, Jalan Lagoon Selatan, 47500 Bandar Sunway, Selangor, Malaysia

S. H. Tiong
Sime Darby Plantation Technology Centre, UPM-MTDC Technology Centre III, 1st Floor, Block B, Lebuh Silikon, University Putra Malaysia, 43400 Serdang, Selangor, Malaysia
e-mail: tiong.s.h33@gmail.com

D. C. Y. Foo et al. (eds.), *Sustainable Technologies for the Oil Palm Industry*,
https://doi.org/10.1007/978-981-19-4847-3_13

Keywords Chlorinated compounds · Diacylglycerol · Free fatty acid · 3-monochloropropanediol esters · Oil processing · Deodorization

Abbreviations

3-MCPDE	3-Monochloro-1, 2-propanediol ester
CPO	Crude palm oil
DAG	Diacylglycerol
DCO	Diluted crude oil
EFB	Empty fruit bunch
EFSA	European food safety authority
FFA	Free fatty acid
FFB	Fresh fruit bunch
MAG	Monoacylglycerol
PORAM	Palm oil refiners association of Malaysia
SC	Sterilizer condensate
SFB	Sterilized fruit bunches
TAG	Triacylglycerol
TC	Total chlorine (combination of organic and inorganic chlorine)
TDI	Tolerable daily intake

1 Introduction

Palm oil is one of the world's most consumed vegetable oils, predominantly for food applications. Palm oil demand is expected to increase due to world population and economic growths, particularly in developing countries. In year 2020, the global edible oil production reached 208 million tons, with approximately 73 million tons of palm oil (USDA, 2020). The versatility of palm oil physicochemical properties enables the oil to be separated into two distinct fractions known as *palm olein* and *stearin*, depending on the food applications. Palm olein is primarily used for frying and domestic cooking, while palm stearin is used mainly to produce shortening and margarine. Therefore, the food safety and quality of the palm oil produced are essential to the industry.

The crude palm oil (CPO) comprises of triacylglycerol (TAG), diacylglycerol (DAG), monoacylglycerol (MAG), free fatty acids (FFA), and other minor components (e.g., micronutrients, oxidation products and metals) (Table 1) (Chew & Saparin, 2021). These components determine the overall quality of palm oil. Since CPO is primarily used for food, its quality criteria must be met. Typically, the millers and refiners use the specification set by Palm Oil Refiners Association of Malaysia

(PORAM) as the quality standard for palm oil trading (Table 2). The PORAM specifications cover three CPO's major quality parameters, which are (i) FFA (to measure oil hydrolytic stability), (ii) moisture and impurities (to measure solid impurities and moisture content) and (iii) degree of bleachability index (to measure the easiness of CPO to be refined) (Chong, 2012; PORAM, 2000).

Lately, the food safety authorities intensified their attention on the quality of vegetable oil, especially the process contaminant known as 3-monochloro-1, 2-propanediol esters (3-MCPDE). Studies showed that refined palm oil contained a considerable amount of 3-MCPDE (Destaillats et al., 2012; Rahn & Yaylayan, 2011). The Joint Food and Agriculture Organization and World Health Organization Expert Committee on Food Additives (JECFA) recommended a Provisional Maximum Tolerable Daily Intake of 4 μg/kg body weight for 3-MCPDE, either on its own or in combination with glycidyl esters, (JECFA, 2017). Subsequently, the European Food SafetyFood safety Authority (EFSA) has recommended that the limit of 3-MCPDE in palm oil to be kept at 2.5 ppm or below, starting in year 2021 (Union, 2020). Moreover, in year 2017, the EFSA published a comprehensive study on the benchmark dose approach in risk assessment of 3-MCPDE as a known human

Table 1 General composition of CPO

Composition	Value
Triacylglycerol (TAG) (%)	91.5 ± 3.5
Diacylglycerol (DAG) (%)	5.0 ± 2.0
Monoacylglycerol (MAG) (%)	0.5 ± 0.2
Free fatty acid (FFA) (%)	3.0 ± 2.0
Moisture (%)	0.15 ± 0.05
Solid Impurities (%)	0.02 ± 0.01
Tocopherols and tocotrienols (ppm)	800 ± 300
Carotenoids (ppm)	600 ± 200
Phytosterols (ppm)	450 ± 150
Squalene (ppm)	350 ± 150
Phospholipids (ppm)	71.5 ± 68.5
Metals (Copper and Iron) (ppm)	5.5 ± 4.5
Lipid oxidation products (ppm)	Traces

Table 2 Quality requirement for crude palm oil in PORAM specification (MPOB, 2019; PORAM, 2000)

Characteristics	Value
Free fatty acid (as palmitic) (% max)	5.0
Moisture and impurities (% max)	0.25
Degree of bleachability index (min)	2.3
[a]Total chlorine (ppm max)	2.0

[a]Additional proposed quality requirement for CPO

carcinogen and established its Tolerable Daily Intake (TDI) at 2 μg/kg of body weight (EFSA, 2018).

Several studies have shown that CPO with a high amount of FFA, DAG, chlorinated compounds (i.e., organochlorine and inorganic chlorine) during the refining process promotes the formation of 3-MCPDE (Craft et al., 2012; Šmidrkal et al., 2016). Presently, the overall chlorinated compounds in CPO are measured using a total chlorine analyzer, according to ASTM D4929-04 Method B (ASTM, 2010) and termed as 'total chlorine' (TC). Recently, Malaysian authorities proposed to limit the TC content in CPO by 2 ppm (Table 2). Meanwhile, the limit for 3-MCPDE in refined palm oil was proposed to be below 2.5 ppm and 1.25 ppm by years 2021 and 2022, respectively (MPOB, 2019). In this chapter, the factors contributing to the 3-MCPDE formation was first investigated. Subsequently, the influence of the 3-MCPDE during food processing was reviewed. This is then followed by recent initiatives to mitigate this process contaminant. This work provides industry insights into steps to reduce the content of 3-MCPDE (in ppm) to meet the allowable limits.

2 Factors Influencing 3-MCPDE Formation

The 3-MCPDE is formed in edible oils, including palm oil, during their refining process at the deodorizer. This is mainly due to the high-temperature process and the presence of chloride ions. A study shows that 3-MCPDE was formed and detected in refined oil but not in CPO (Ramli et al., 2011). Subsequently, Destaillats et al. (2012) found that thermal treatment (235 °C) of triolein oils in the presence of chlorinated compounds (i.e., organochlorine pesticides and $FeCl_2$) increased the formation of 3-MCPDE. The formation of 3-MCPDE had been proposed to proceed through both S_N2 nucleophilic substitution and free radical reaction initiated from acylglycerol such as TAG and TAG to for acyloxonium intermediate under high temperature condition, as shown in Fig. 1 (Rahn & Yaylayan, 2011; Zhang et al., 2013). Studies showed that high DAG and MAG in model oil increased 3-MCPDE formation after thermal treatment imitating the deodorization conditions (Freudenstein et al., 2013; Shimizu et al., 2012). The formation of 3-MCPDE was also observed in model oils with pure TAG, without partial acylglycerols (i.e., DAG and MAG). Therefore, TAG appears to be also the precursor for 3-MCPDE formation (Destaillats et al., 2012; Ermacora & Hrncirik, 2014). However, partial acylglycerols (i.e., DAG and MAG) were more reactive than TAG to form 3-MCPDE, with DAG being the most reactive precursor due to higher concentration compared to MAG (Freudenstein et al., 2013; Shimizu et al., 2012).

Since CPO contains a high percentage of acylglycerols (>95%), the 3-MCPDE formation in palm oil is constrained by the concentration of chlorinated compounds (Che Man et al., 1999; Ermacora & Hrncirik, 2014). Studies by Tiong et al. (2018); Lakshmanan and Yung (2021) showed that TC in CPO directly relates 3-MCPDE formed in refined palm oil with the r^2 of 0.73 and 0.91, respectively. Meanwhile, Nagy et al. (2011) found over 200 monochlorinated organic compounds and chlorinated

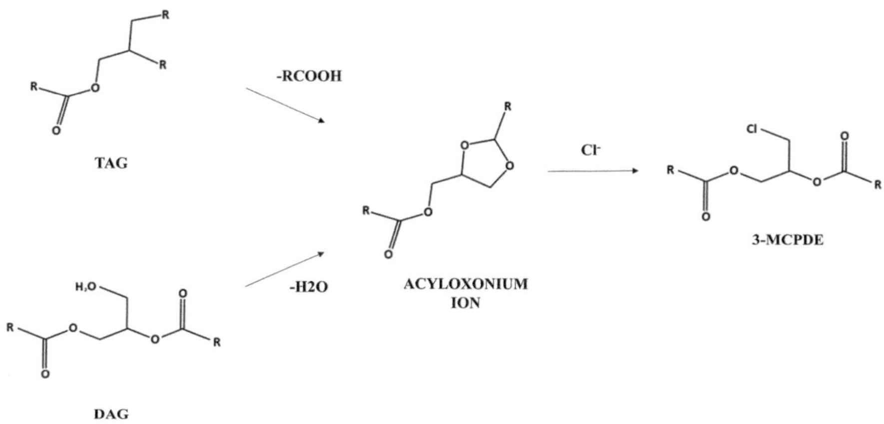

Fig. 1 Proposed formation mechanism of 3-MCPDE from DAG and TAG through acyloxonium ion

inorganic compounds (i.e., $FeCl_2$, $FeCl_3$, $MgCl_2$, and $CaCl_2$) in CPO. Organochlorine compounds, natural occurring lipid substances, could liberate chlorides during deodorization (Chew & Saparin, 2021). These free chlorides then react with acylglycerols to form 3-MCPDE (Nagy et al., 2011; Tiong et al., 2018). The organochlorine compounds as the precursor for 3-MCPDE were further verified by a model showing that oil enriched with organochlorine compounds under thermal treatment resulted in a higher 3-MCPDE formation (Tiong et al., 2018).

Organochlorine compounds are believed to be biosynthesized in all plant tissue not only in oil palm fruits (Tiong et al., 2018). Typically, the ripe palm fruits are harvested and delivered to oil mills for the oil extraction process. In the mill, the harvested palm fruits known as fresh fruit bunches (FFB) are subjected to a series of processing units (Chew et al., 2021b). First, the FFB is subjected to a heat treatment (known as sterilization process) with saturated steam (140 °C) to deactivate the lipase enzyme and loosen the palm fruits from its bunch (see Fig. 2). Then, the sterilized FFB undergoes a mechanical stripping process to separate the oil containing palm fruits from its bunch, producing empty fruit bunch (EFB) as by-products. In the pressing stage, the detached palm fruits are subjected to mechanical pressure to extract their pressed liquor, a mixture of oil, water and fine fibrous material from the palm mesocarp. Therefore, some organochlorine compounds could be extracted with the press liquor during pressing (Nagy et al., 2011; Tiong et al., 2018). The press liquor (containing $<50\%$ w/w oil) is then added with hot water (30% w/w) to facilitate the clarification process using a settling tank. This clarification process helps to enhance the separation of the oil from impurities (i.e., water and solid) to obtain the oil purity at 95% (w/w) and above. Subsequently, the recovered oil from the clarifier is delivered to a high-speed centrifuge and followed by vacuum drying to produce a final CPO with moisture and solid impurities of 0.25% (w/w) and below. Notably, the sterilization process causes some oil from the palm fruits to leech out, resulting in a small fraction of oil in the sterilizer condensate (SC) and EFB.

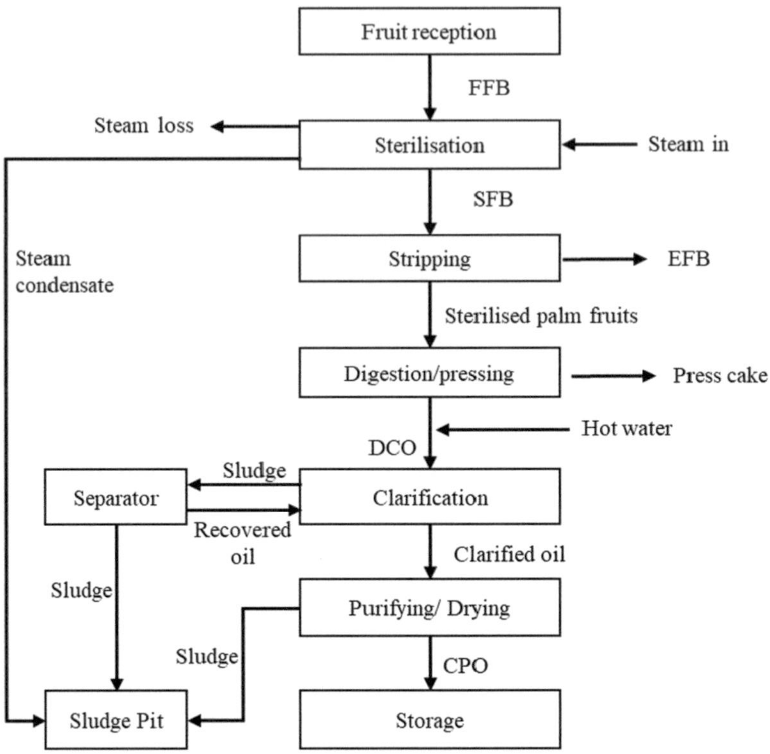

Fig. 2 Block diagram for palm oil mill processing

Therefore, some mills tend to recover SC and EFB oils to the CPO for maximizing their production.

A study was conducted to investigate the effect of each processing unit on the concentration of organochlorine compounds (Chew & Saparin, 2021). Samples for each process stream (i.e., EFB, SC, press liquor, clarifier, and CPO) were collected. Before analysis, the oil from each sample was mechanically extracted and purified to a similar purity with CPO. Interestingly, the organochlorine compounds are partitioned and concentrated in some oils' fractions during palm oil mill processing, particularly in EFB and SC oils (Fig. 3). This finding suggested that oil recovered from waste streams, mainly SC and EFB back into main CPO stream could increase the overall content of organochlorine in the CPO. Consequently, the 3-MCPDE in refined palm oil could be affected. Since the SC is predominant water (>98% w/w), it is used in some mills to replace hot water as dilution water and mixed into press liquor as an indirect method to recover the oil from SC (see Fig. 4). In recent years, many mills also tend to recover oils from EFB, where EFB press is used to extract EFB liquor containing water and some amount of oil (<20% w/w). As the volume of the EFB liquor is relatively low compared to SC (<15% v/v), it is then mixed with SC and used as dilution water in some mills. A separate study by Chew et al. (2018) investigated the

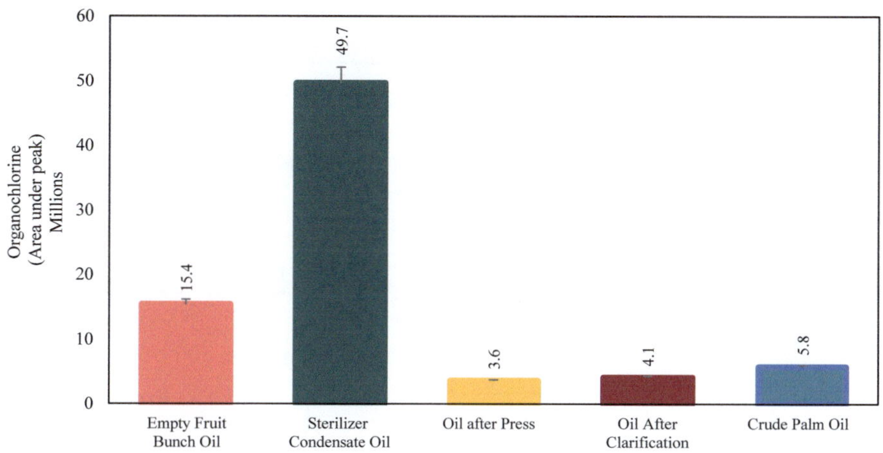

Fig. 3 Organochlorine content in various oil streams at palm oil mill

effect of dilution water on the amount of TC content in the oil. This study showed that the types of dilution medium used could affect the TC and 3-MCPDE in the CPO and its refined oil, where dilution medium using SC and EFB liquor mixture resulted in the highest TC level in the CPO (Fig. 5). This finding reaffirms the earlier postulation that mixing a chlorine-rich stream into a CPO processing line could increase its TC content and subsequently surge the 3-MCPDE content in its refined oil. Therefore, reducing 3-MCPDE's precursors, i.e., TC, is essential to mitigate this process contaminant.

Aside from the acylglycerols and chlorinated compounds as precursors, the composition of minor constituents in CPO and refining process conditions could also influence the 3-MCPDE formation. The acidity of CPO could encourage and trigger acylglycerols for chloride nucleophile attack to form 3-MCPDE (Rahn & Yaylayan, 2011). In a study by Freudenstein et al. (2013), increasing the alkalinity with disodium carbonate and sodium bicarbonate reduced the formation of 3-MCPDE from 7 ppm to less than 3 ppm. Contrarily, an addition of 10% lauric acid reduced the oil acidity and resulted in a higher 3-MCPDE formation. Although FFA is not a direct precursor for the formation of 3-MCPDE, it does influence the acidity of the oils that affect the 3-MCPDE formation (Zulkurnain et al., 2012).

During the refining process, chlorinated compounds could enter the CPO depending on the material and chemical used. Typically, the chloride in bleaching clay and water ranged from 50 to 150 ppm. The acid activation process using hydrochloric acid (HCl) in bleaching clay could introduce chloride into the oil. Therefore, bleaching clay manufacturers are now offering bleaching clay that is activated with non-chloride acid. Although another acid, such as sulfuric acid (H_2SO_4), contains no chloride ions, their higher acidity could also promote 3-MCPDE formation (Ramli et al., 2011). Therefore, natural bleaching earth that does not contain chloride and neutral pH is preferred to reduce 3-MCPDE formation. Furthermore,

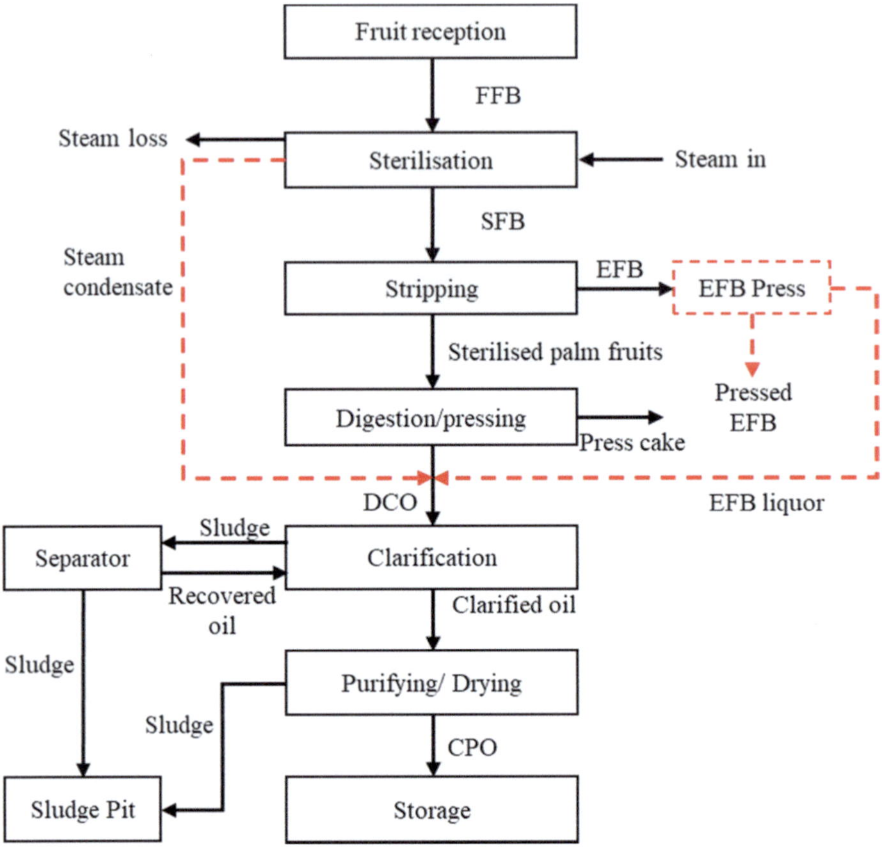

Fig. 4 Block diagram for palm oil mill processing with sterilizer condensate and EFB liquor as dilution medium

chloride is also detected in the acid used in the degumming process. Nevertheless, the amount of acid (<0.1% w/w to CPO) used for degumming is relatively low and negligible. Interestingly, CPO exposed to water, such as water degumming, shows a reduction of 3-MCPDE (Matthäus et al., 2011; Pudel et al., 2011; Ramli et al., 2011). As such, the oil palm industry must consider the precursors and factors that influence 3-MCPDE formation to effectively mitigate this issue and ensure the food prepared with oil does not exceed the allowable limit.

3 Transference of 3-MCPDE to Food

3-MCPDE is found in various household products, including foods that contain edible oils and fats. Additionally, domestic cooking methods such as frying, baking,

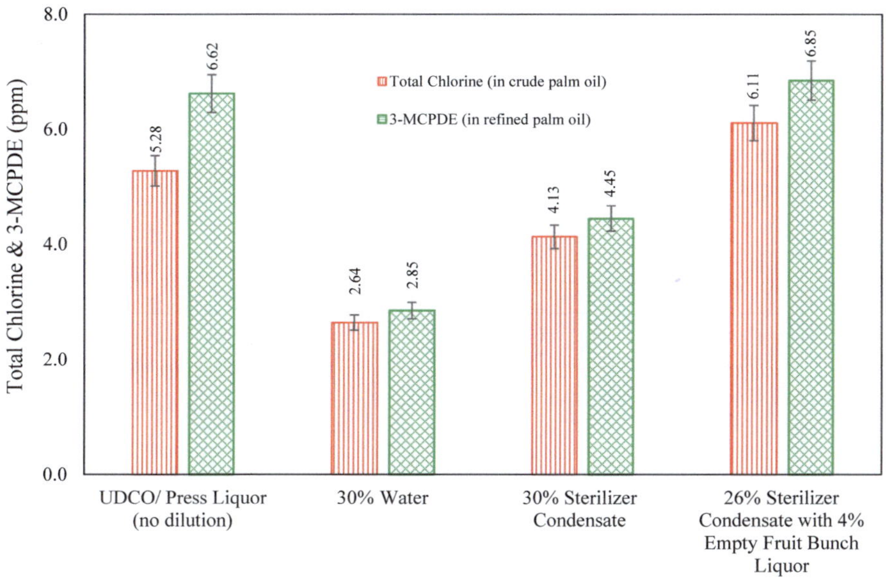

Fig. 5 Total chlorine in CPO and 3-MCPDE content in its refined oil for various types of dilution water—in the oil mill (Chew et al., 2018)

roasting, microwaving, and grilling have also increased 3-MCPDE levels in foods (Crews et al., 2001). Recent studies showed that the oil absorbed into food increases with the increase of the frying cycle (Lumanlan et al., 2019; Touffet et al., 2020). These studies further showed that the increase in oil absorption was due to prolonged heat treatment that degraded the oil. Under thermal treatment, the oil decomposed to produce total polar compounds that increase oil viscosity. In contrast, the 3-MCPDE content in refined oil during the deep-frying showed a decreasing trend over frying cycles (Chew et al., 2021c; Wong et al., 2019) (Fig. 6a). Therefore, a study by Chew et al. (2021c) showed that 3-MCPDEs absorbed into the food are averaged at 0.18 ± 0.04 ppm over 30 frying cycles (Fig. 6b). In another study, foods prepared with palm oil containing 1.64 ppm of 3-MCPDE resulted in 0.12–0.25 ppm of this contaminant in the fried foods (Arisseto et al., 2017). This study further reinforced the suggestion of Chew et al. (2021c) that 3-MCPDE in food was carried over from frying oil. However, the decomposition and formation of 3-MCPDE in the foods are influenced by the oil used (type, quality), cooking condition (frying temperatures, duration, cycles) and composition of food (salt and moisture content) (Dingel & Matissek, 2015; Wong et al., 2017; Zhou et al., 2014). Therefore, the risk of human exposure to 3-MCPDE in food products prepared with cooking oil containing 3-MCPDE is less of concern due to degradation of 3-MCPDE at high temperature cooking condition (i.e. during frying) compared to direct consumption of cooking oil.

In short, factors influencing oil absorption into food affect the 3-MCPDE concentration. Although the repeatedly use of oil shows a reduction in 3-MCPDE content,

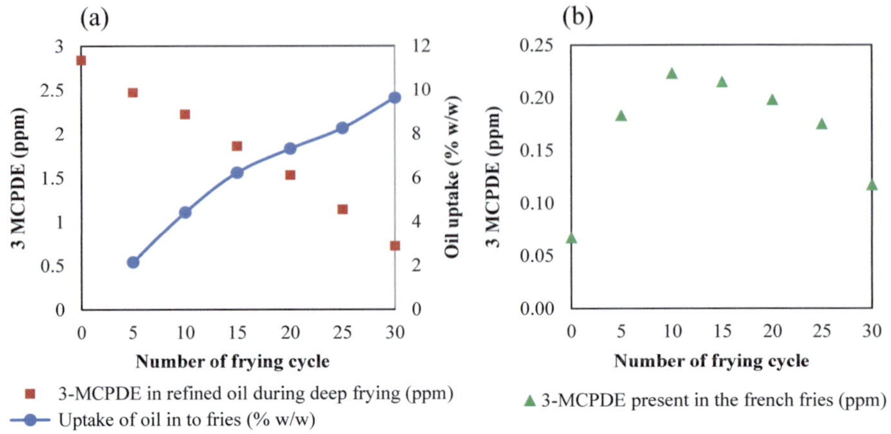

Fig. 6 3-MCPDE content in **a** refined oil and **b** in French fries during deep-frying (Chew et al., 2021c)

the prolonged exposure of frying oil to heat could deteriorate the oil quality. As a result, the nutritional value of the food is reduced. By taking Chew et al. (2021c) 's study with the average 0.18 ppm of 3-MCPDE in the fries, the maximum TDI for an average human body mass of 62 kg is equivalent to the consumption of 0.69 kg fries daily (EFSA, 2018; Walpole et al., 2012). However, it is essential to note that the 3-MCPDE is also detected in various heated foods. Therefore, minimizing the 3-MCPDE in foods is vital by reducing this contaminant in edible oil and ensure the overall consumed food is within the recommended limit of 3-MCPDE.

In recent studies, adding the antioxidants in the frying medium were suggested to reduce the 3-MCPDE formation during cooking (i.e., frying and baking) (Goh et al., 2020; Wong et al., 2019). Although the 3-MCPDE were reduced over the frying cycle (Chew et al., 2021c), the addition of antioxidants could further reduce the 3-MCPDE in the oil and food (Wong et al., 2019). This finding is in agreement with Huang et al. (2021), where frying oil added with tert-butylhydroquinone (TBHQ) resulted in a lower 3-MCPDE. It is well established that antioxidants could slow lipid oxidation and improve oil stability by scavenging the free radicals that oxidize the oil, especially the polyunsaturated fatty acid. Similarly, the antioxidants could minimize 3-MCPDE's precursors such as chloride and cyclic acyloxonium radicals (Zhang et al., 2016). Therefore, the addition of antioxidants in frying oil could slow down the formation of 3-MCPDE during frying and improve the overall oil quality.

4 Mitigation Strategies

4.1 Precursors Reduction in CPO

One possible mitigation is to identify and to reduce the 3-MCPDE's precursor, i.e., chlorine sources in CPO. The chlorinated compounds of the oil palm fruits at the plantation are affected by various external factors such as soil, irrigation, and pest control (Destaillats et al., 2012). Some factors that may cause chlorinated compounds contamination in FFB and CPO are harvesting and transportation. During harvesting and evacuation, some solid impurities are attached to FFB and processed together to produce CPO. These solid impurities consist of predominantly fibrous materials and soil that contain chlorinated compounds. Consequently, some of the chlorinated compounds are then carried over to CPO during processing. These compounds could be leached out into press liquor during the pressing stage (see Fig. 2). Therefore, the reduction of chlorine source at the feedstock (FFB) is expected to minimize the process contaminant in CPO. The FFB washing with water before oil extraction is perhaps an essential step in reducing the accumulation of chlorine sources during milling processing. A study showed that the FFB cleaning system produced CPO with a TC content as low as 0.5 ppm (Syed Hilmi et al., 2018). Results showed that TC in CPO produced from washed FFB was significantly lower compared to conventionally produced CPO. As a result, the 3-MCPDE of the refined oil was reduced considerably.

Besides, the milling process could also influence the TC in CPO. In the effort to maximize the oil recovery during the milling process, some mills tend to recover oil from SC and EFB by using these streams as dilution water (Fig. 5). These poor-quality oils were then blended with CPO during the palm oil processing and enriched the TC (Chew et al., 2021b). Therefore, the oils recovered from waste streams is suggested to be segregated from the main CPO processing line. The oil mills could use a sludge separator to recover SC and EFB oils in a separate processing line for non-food applications. Through these methods, the TC and 3-MCPDE contents could be reduced by 30% in CPO and refined oil, respectively (Rahmat et al., 2019). In addition, this method improved the CPO's overall quality with higher hydrolytic and oxidative stabilities.

The formation of 3-MCPDE requires chloride ions that can be easily detached from chlorine sources. Therefore, a method that reduces TC concentration in CPO is crucial to prevent the formation of 3-MCPDE during the refining process. According to Destaillats et al. (2012), most of the chlorinated compounds in plant material are naturally polar. Therefore, liquid-liquid extraction using a polar solvent solution through water washing of CPO could effectively strip out the chlorinated compounds in CPO. Studies showed that reducing the 3-MCPDE in refined oil between 20 and 84% could be achieved by washing the CPO using water (Chew et al., 2021d; Matthäus & Pudel, 2013; Matthäus et al., 2011; Zulkurnain et al., 2012). In response to this positive finding, verification trials are performed in both laboratory and demonstration-plant with the capacity of 1 L and 20 t/hr, respectively.

At the laboratory scale, the reduction of TC and 3-MCPDE was only up to 50% and 51%, respectively. Meanwhile, the demonstration plant with 20 t/hr capacity showed a higher reduction of TC and 3-MCPDE with an average of 80% and 78%, respectively. The demonstration plant exhibited a better TC reduction, owing to the high-speed mixing system that increases the surface contact between the oil and water. However, CPO washing uses clean water and generates more wastewater. In the light of sustainability, the author has recently developed an alternative CPO washing system using aerobic liquor, treated water from a palm oil mill (Chew et al., 2021d). This sustainable and straightforward method was reported to reduce the TC and 3-MCPDE content by 50% (Chew et al., 2021). The studies on CPO washing are summarized in Table 3.

Another possible mitigation strategy for TC reduction is dechlorination of CPO using sodium metabisulfite (Spaparin et al., 2018). The sodium metabisulfite was first mixed with heated CPO. Then the spent sodium metabisulfite is filtered out to avoid the presence of the dechlorination agent in the oil. As a result, the dechlorinated CPO contained a TC content below 2 ppm and subsequently reduced the 3-MCPDE in its refined oil. Overall, reduction of TC in CPO is effective using dechlorination agents. However, the high cost of the sodium metabisulfite and oil loss of this method remains a challenge for commercial applications. Therefore, the economic viability of the mitigation methods is an essential factor to be considered for commercialization.

Apart from chlorine source reduction, the authors suggest improving the overall CPO quality as one of the mitigation actions. In short, the focus should be given to minimize the hydrolysis of oil during processing. Consequently, other factors contributing to the 3-MCPDE formation, such as FFA (pH), DAG, and MAG, could be minimized. Ebongue et al. (2006) and Tagoe et al. (2012) suggested that the FFB should be immediately treated with heat after harvesting to deactivate its lipolytic hydrolysis that forms the FFA. Besides, the segregation of poor-quality fruits is essential to improve the quality of CPO produced. For instance, detached palm fruits collected from estates are highly oxidized and hydrolyzed and could deteriorate the overall CPO quality (). Based on the authors' experience, the FFA of oil extracted from detached palm fruits is between 4 and 18%, depending on the quality of the detached fruits. In addition, the oil recovered from the waste stream should be segregated from the CPO production line. These segregation processes (i.e., detached fruits and oil from the waste stream) enable the production of CPO with low contaminants and FFA. It is worth highlighting that Sime Darby Plantation has commercially implemented this multilevel segregation technique to produce high-quality CPO (Chew, 2021). Through this method, CPO with exceptionally low FFA known as Premium Quality oil and Superior Quality oil with the FFA below 1.2% and 1.5% are produced. These oils contain lower TC as compared to conventionally produced CPO (Table 4). Furthermore, the 3-MCPDE content in its refined oils are low, i.e., an average of 1.25 ppm and 1.97 ppm, respectively. These unique CPOs are used for speciality fat products (i.e., infant formula, frying oil, and red palm olein). This finding showed that high-quality CPO with low FFA could reduce the 3-MCPDE formation during the refining process. Nevertheless, the 3-MCPDE in these oils could be further reduced with refining process modification.

Table 3 Effect of CPO washing on TC and 3-MCPDE on its refined oil. Results represent the means ± standard deviation of the mean value ($n = 3$)

Methods	Capacity (size)	TC (ppm)	3-MCPDE (ppm), (%)	References
Washing with water	[a]NA	[a]NA	20	Matthäus et al. (2011);
	[a]NA	[a]NA	38	Matthäus and Pudel (2013);
	[a]NA	[a]NA	84	Zulkurnain et al. (2012)
Washing with distilled water	Laboratory (1 L)	50.04 ± 0.88%	51.26 ± 1.56	–
	Demo-plant (20 t/hr)	80.13 ± 0.63%	78.40 ± 1.75	
Washing with aerobic liquor	Laboratory (1 L)	50.22 ± 0.36%	50.61 ± 1.42	Chew et al., 2021d; Chew, Kong & Chan, 2021e

[a]NA indicates that the data is not available in the published work

Table 4 The level of TC and 3-MCPDE level in CPO and its refined oil, respectively

Type of CPO	CPO ($n = 60$)		RPO ($n = 60$)
	FFA (%)	TC (ppm)	3-MCPDE (ppm)
Normal	<5.0	4.06 ± 1.96	3.03 ± 0.87
Premium quality	<1.5	2.11 ± 0.61	1.97 ± 0.92
Superior quality	<1.2	1.77 ± 0.25	1.25 ± 0.6

4.2 Refining Process Modification

Besides precursor mitigation, modification of process conditions in the refinery could also address the 3-MCPDE formation issue. Water degumming has been proposed as an alternative to acid degumming in oil refining by some researchers (Ramli et al., 2011; Zulkurnain et al., 2012, 2013). Like the CPO washing technique, the chlorinated compounds that are polar can be extracted from the oil during this process. Compared to the conventional acid degumming technique of CPO, water degumming is less efficient in removing phosphatides, iron, and pigment, due to low concentration of these compounds in CPO compared to unrefined soft oils. These compounds cause color fixation issues during deodorization (Ramli et al., 2011). Surprisingly, using more water (up to 5%) during water degumming could remove impurities and achieves a red color specification of less than 3.0 (Zulkurnain et al., 2013).

In the bleaching step, the use of natural clay instead of acid-activated bleaching clay is recommended. Since natural clay is neutral in pH, it does not promote the formation of 3-MCPDE. However, natural clays typically have a lower surface area, which may affect the refined oil quality. The efficiency of bleaching clays in absorbing impurities is dependent on its absorptive capability, which in turn rely on the surface area. A study showed that a combination of water degumming and magnesium silicate as a bleaching clay resulted in the highest reduction of 3-MCPDE (Zulkurnain et al., 2012).

In the refining process, 3-MCPDE is formed almost exclusively during deodorization at high temperatures. A study by Matthäus et al. (2011) showed that the formation of 3-MCPDE is directly correlated to the temperature of deodorization. In this study, CPO containing 4.2% FFA and 4.75 ppm TC was subjected to deodorization temperature between 230 and 260 °C for 90 min. As anticipated, the highest levels of 3-MCPDE are found in oil samples that were deodorized at 260 °C. This finding is in good agreement with the low 3-MCPDE content in red palm olein. Typically, the production of the red palm olein undergoes mild refining with low deodorization temperature (<180 °C) and resulted in a very low amount of 3-MCPDE (Mayamol et al., 2007). Therefore, the deodorization temperature is recommended to be as low as possible to reduce the 3-MCPDE formation without compromising other quality parameters.

Chemical refining could also help to reduce 3-MCPDE formation. The review by Oey et al. (2019) showed that neutralization processes could result in 3-MCDPE reduction between 31 and 81% with different alkali solutions and concentrations. Similar to CPO washing, neutralization could reduce the TC content in the CPO, resulting in a low 3-MCPDE formation. Apart from lowering the 3-MCPDE precursor, the neutralization process reduces the acidity of the CPO before deodorization. During neutralization, most FFAs and excess acids (from the degumming process) are removed to produce neutral oil that is less susceptible to 3-MCPDE formation (Freudenstein et al., 2013). Furthermore, the neutralization process could enable mild deodorization and lead to refined oil production with low 3-MCPDE content. A recent study by Hori et al. (2021) showed that a combination of neutralization and steam sparging could reduce 80% of 3-MCPDE formation. Nevertheless, chemical refining resulted in high oil losses that could increase the production cost.

Another possible mitigation strategy is the adoption of frying applications in the refining process. A study showed that prolonged frying could reduce 3-MCPDE formation (Chew et al., 2021c). Therefore, prolonged heating during deodorization could help to decompose 3-MCPDE in the oil. However, it was worth noting that this method could result in the degradation of oil quality. Hence, an in-depth study of this method should be investigated. A prolonged deodorization time with a lower deodorization temperature could perhaps reduce 3-MCPDE without compromising the oil quality. Alternatively, the introduction of antioxidants during deodorization could minimize oil oxidation and slow down 3-MCPDE formation. A study showed that the addition of TBHQ (66 ppm) and alpha-tocopherol (172 ppm) into oil under heat (230 °C) for 30 min resulted in 44% and 22% of reduction of 3-MCPDE (Li et al., 2015). Typically, palm oil is rich in antioxidants, and most of these compounds are removed during refining (deodorization). As such, there is a possibility to explore the application of these abundantly available antioxidants in palm oil for 3-MCPDE mitigation. For instance, process modifications to retain these compounds in the oil or adding back the antioxidants into the oil could be a new strategy for 3-MCPDE mitigation.

It is well noted that not all mitigation strategies, especially the laboratory and pilot scales, could be successfully implemented commercially. Various aspects such as feasibility, cost, sustainability, scalability, and implementation risk should be considered for commercialization. The overall potential mitigation strategies for 3-MCPDE mitigation in the palm oil supply chain are summarized in Fig. 7. The mitigation should begin from upstream, in which the oil palm plantation estate could improve the quality of palm fruits in order to reduce the amount of solid impurities with chlorine sources. Subsequently, the mill could improve its processes to produce high-quality CPO with low DAG and TC contents. Finally, the reduction of 3-MCPDE formation can be made by improving the refining process conditions. With the integration of these mitigation strategies, the 3-MCPDE issue in the palm oil industry could be resolved. Additionally, high-quality oil products with exceptionally low 3-MCDPE (<0.35 ppm) could also be produced for speciality fat products such as infant formula.

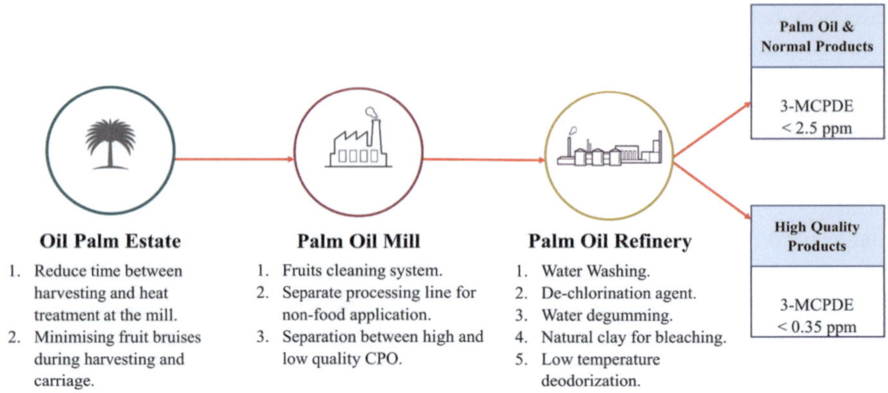

Fig. 7 Proposed mitigation strategies for 3-MCPDE throughout the palm oil supply chain

5 Conclusion

According to studies, DAG and chlorinated compounds in CPO promote the 3-MCPDE formation during the thermal treatment in the refining stage. Although the 3-MCPDE is formed during refining, chlorinated compounds are detected in palm fruits and CPO in the estate and oil mill. Therefore, an integrated mitigation strategy across palm oil processing could be a promising approach to address the 3-MCPDE issue. The mitigation strategy is proposed to start in estates and mills by addressing the precursors of 3-MCPDE. This approach is followed by the improvement of refinery process conditions focusing on minimizing 3-MCPDE formation. These findings lead to various possible mitigation strategies. By incorporating precursors reduction and process modification techniques, the overall content of 3-MCPDE in the palm oil could be reduced. This will help to produce food that complies with 3-MCPDE's dietary limit set by the authorities.

Acknowledgements The authors are grateful for the support of Sime Darby Plantation Research. Input from Nik Mohd Farid Mat Yasin, Norliza Saparin, and Syed Mohd Hadi Syed Hilmi are gratefully acknowledged.

References

Arisseto, A. P., Marcolino, P. F. C., Augusti, A. C., Scaranelo, G. R., Berbari, S. A. G., Miguel, A. M. R. O., Morgano, M. A., & Vicente, E. (2017). Contamination of fried foods by 3-monochloropropane-1, 2-diol fatty acid esters during frying. *Journal of the American Oil Chemists' Society, 94*(3), 449–455. https://doi.org/10.1007/s11746-017-2951-9

ASTM International. (2010). ASTM D4929-04: Standard test method for salt in crude oils (potentiometric method). *West Conshohocken (PA).*

Che Man, Y. B., Haryati, T., Ghazali, H. M., & Asbi, B. A. (1999). Composition and thermal profile of crude palm oil and its products. *Journal of the American Oil Chemists' Society, 76*(2), 237–242. https://doi.org/10.1007/s11746-999-0224-y

Chew, C. L., & Saparin, N. (2021). Principal formation and mitigation strategies for 3-mcpde in palm oil processing. *Journal of Oil Palm, Environment and Health (JOPEH), 12.* https://doi.org/10.5366/jope.2021.06

Chew, C. L., Ng, C. Y., Hong, W. O., Wu, T. Y., Lee, Y. Y., Low, L. E., Kong, P. S., & Chan, E. S. (2021a). Improving sustainability of palm oil production by increasing oil extraction rate: A review. *Food and Bioprocess Technology, 14*(4), 573–586. https://doi.org/10.1007/s11947-020-02555-1

Chew, C. L., Low, L. E., Chia, W. Y., Chew, K. W., Liew, Z. K., Chan, E. S., Chan, Y. J., Kong, P. S. & Show, P. L. (2021b). Prospects of palm fruit extraction technology: Palm oil recovery processes and quality enhancement. *Food Reviews International*, 1–28. https://doi.org/10.1080/87559129.2021.1890117

Chew, C. L, Ab Karim N. A., Quek, W. P., Wong, S. K., Lee, Y. Y., Chan, E. S. (2021c). Aerobic-liquor treatment improves the quality and deep-frying performance of refined palm oil. *Food Control, 108072*(126), 1–8. https://doi.org/10.1016/j.foodcont.2021.108072

Chew, C. L., Ab Karim, N. A., San Kong, P., Tang, S. Y., & Chan, E. S. (2021d). A sustainable in situ treatment method to improve the quality of crude palm oil by repurposing treated aerobic liquor. *Food and Bioprocess Technology, 14*(4), 679–691. https://doi.org/10.1007/s11947-021-02582-6

Chew, C. L., Kong, P. S., & Chan, E. S. (2021e). Aerobic liquor washing improves the quality of crude palm oil by reducing free fatty acids and chloride contents. *European Journal of Lipid Science and Technology, 2000347.* https://doi.org/10.1002/ejlt.202000347

Chew, C. L., S. Hilmi, S. M. H., Saparin, N., M. Hassan, N. S., M. Siran, Y., Asis, A. J., Chan, E. S., Tang, S. Y. (2018, November 21–23). Effect of steriliser condensate and empty fruit bunch's liquor restreaming on the physicochemical properties and stability of palm oil. *Monash Science Symposium 2018*, Monash University Malaysia.

Chew, C. L. (2021). Quality improvement of crude palm oil via in situ washing with treated aerobic liquor: Process development and product evaluation. Monash University. Thesis. https://doi.org/10.26180/14984700.v1

Chong, C. L. (2012). Measurement and maintenance of palm oil quality. In *Palm Oil* (pp. 431–470). Urbana: AOCS Press.

Craft, B. D., Nagy, K., Seefelder, W., Dubois, M., & Destaillats, F. (2012). Glycidyl esters in refined palm (Elaeis guineensis) oil and related fractions. Part II: Practical recommendations for effective mitigation. *Food Chemistry, 132*, 73–79. https://doi.org/10.1016/j.foodchem.2011.10.034

Crews, C., Brereton, P., & Davies, A. (2001). The effects of domestic cooking on the levels of 3-monochloropropanediol in foods. *Food Additives & Contaminants, 18*(4), 271–280. https://doi.org/10.1080/02652030120064

Destaillats, F., Craft, B. D., Sandoz, L., & Nagy, K. (2012). Formation mechanisms of monochloropropanediol (MCPD) fatty acid diesters in refined palm (Elaeis guineensis) oil and related fractions. *Food Additives & Contaminants: Part A, 29*(1), 29–37. https://doi.org/10.1080/19440049.2011.633493

Dingel, A., & Matissek, R. (2015). Esters of 3-monochloropropane-1, 2-diol and glycidol: No formation by deep frying during large-scale production of potato crisps. *European Food Research and Technology, 241*(5), 719–723. https://doi.org/10.1007/s00217-015-2491-1

Ebongue, G. N., Dhouib, R., Carriere, F., Zollo, P. H. A., & Arondel, V. (2006). Assaying lipase activity from oil palm fruit (Elaeis guineensis Jacq.) mesocarp. *Plant Physiology and Biochemistry, 44*(10), 611–617. https://doi.org/10.1016/j.plaphy.2006.09.006

Ermacora, A., & Hrncirik, K. (2014). Influence of oil composition on the formation of fatty acid esters of 2-chloropropane-1, 3-diol (2-MCPD) and 3-chloropropane-1, 2-diol (3-MCPD) under conditions simulating oil refining. *Food Chemistry, 161*, 383–389. https://doi.org/10.1016/j.foodchem.2014.03.130

European Food Safety Authority (EFSA). (2018). Update of the risk assessment on 3-monochloropropane diol and its fatty acid esters. *EFSA Journal, 16*(1), e5083. https://doi.org/10.2903/j.efsa.2018.5083. Accessed on 7 April 2021.

Freudenstein, A., Weking, J., & Matthäus, B. (2013). Influence of precursors on the formation of 3-MCPD and glycidyl esters in a model oil under simulated deodorization conditions. *European Journal of Lipid Science and Technology, 115*(3), 286–294. https://doi.org/10.1002/ejlt.201200226

Goh, K. M., Wong, Y. H., Abas, F., Lai, O. M., Mat Yusoff, M., Tan, T. B., Wang, Y., Nehdi, I. A., & Tan, C. P. (2020). Changes in 3-, 2-Monochloropropandiol and glycidyl esters during a conventional baking system with addition of antioxidants. *Foods, 9*(6), 739. https://doi.org/10.3390/foods9060739

Hori, K., Hashimoto, Y., Itani, A., Okada, T., & Tsumura, K. (2021). Effects of neutralization combined with steam distillation on the formation of monochloropropanediol esters and glycidyl esters in palm oil under laboratory-scale conditions. *LWT, 139*, 110783. https://doi.org/10.1016/j.lwt.2020.110783

Huang, Z., Xie, D., Cao, Z., Guo, Z., Chen, L., Jiang, L., Sui, X., & Wang, Z. (2021). The effects of chloride and the antioxidant capacity of fried foods on 3-chloro-1, 2-propanediol esters and glycidyl esters during long-term deep-frying. *LWT, 145*, 111511. https://doi.org/10.1016/j.lwt.2021.111511

JECFA. (2017). Evaluation of certain contaminants in food (Eighty-third report of the Joint FAO/WHO Expert Committee on Food Additives). WHO Technical Report Series, No.1002. Retrieved from https://apps.who.int/iris/handle/10665/254893. Accessed 25 December 2020.

Lakshmanan, S., & Yung, Y. L. (2021). Chloride reduction by water washing of crude palm oil to assist in 3-monochloropropane-1, 2 diol ester (3-MCPDE) mitigation. *Food Additives & Contaminants: Part A, 38*(3), 371–387. https://doi.org/10.1080/19440049.2020.1842516

Li, C., Jia, H., Shen, M., Wang, Y., Nie, S., Chen, Y., Zhou, Y., Wang, Y., & Xie, M. (2015). Antioxidants inhibit formation of 3-monochloropropane-1, 2-diol esters in model reactions. *Journal of Agricultural and Food Chemistry, 63*(44), 9850–9854. https://doi.org/10.1021/acs.jafc.5b03503

Lumanlan, J. C., Fernando, W. M., & Jayasena, V. (2019). Mechanisms of oil uptake during deep frying and applications of predrying and hydrocolloids in reducing fat content of chips. *International Journal of Food Science Technology, 55*(4), 1661–1670. https://doi.org/10.1111/ijfs.14435

Matthäus, B., Pudel, F., Fehling, P., Vosmann, K., & Freudenstein, A. (2011). Strategies for the reduction of 3-MCPD esters and related compounds in vegetable oils. *European Journal of Lipid Science and Technology, 113*(3), 380–386. https://doi.org/10.1002/ejlt.201000300

Matthäus, B., & Pudel, F. (2013). Mitigation of 3-MCPD and glycidyl esters within the production chain of vegetable oils especially palm oil. *Lipid Technology, 25*(7), 151–155. https://doi.org/10.1002/lite.201300288

Mayamol, P. N., Balachandran, C., Samuel, T., Sundaresan, A., & Arumughan, C. (2007). Process technology for the production of micronutrient rich red palm olein. *Journal of the American Oil Chemists' Society, 84*(6), 587–596. https://doi.org/10.1007/s11746-007-1078-9

MPOB, Malaysian Palm Oil Board Licensing and Enforcement Division. (2019). Enforcement of additional licencing conditions imposed on licensees of palm oil mill (mf), palm oil refinery (rf), palm oil products exporter (px) and palm oil products importer (pm) categories: food safety & good quality palm oil. Enforcement Circular (Licensing) MPOB, Pk (EL) MPOB 01/2019, 1–4.

Nagy, K., Sandoz, L., Craft, B. D., & Destaillats, F. (2011). Mass-defect filtering of isotope signatures to reveal the source of chlorinated palm oil contaminants. *Food Additives & Contaminants: Part A, 28*(11), 1492–1500. https://doi.org/10.1080/19440049.2011.618467

Oey, S. B., Van der Fels-Klerx, H. J., Fogliano, V., & van Leeuwen, S. P. (2019). Mitigation strategies for the reduction of 2-and 3-MCPD esters and glycidyl esters in the vegetable oil processing industry. *Comprehensive Reviews in Food Science and Food Safety, 18*(2), 349–361. https://doi.org/10.1111/1541-4337.12415

PORAM. PORAM Standard Specifications for Processed Palm Oil. (2000). http://poram.org.my/p/wp-content/uploads/2013/12/1.-PORAM-Standard-Specification.pdf. Accessed 29 November 2019.

Pudel, F., Benecke, P., Fehling, P., Freudenstein, A., Matthäus, B., & Schwaf, A. (2011). On the necessity of edible oil refining and possible sources of 3-MCPD and glycidyl esters. *European Journal of Lipid Science and Technology, 113*(3), 368–373. https://doi.org/10.1002/ejlt.201000460

Rahmat, N., Syed Mohd Hadi, S. H., Norliza, S., Syahril Anuar, M. R., Yosri, M. S., Mohammed Faisal, M. Y., Ahmadilfitri, M. N., & Ahmad Jaril, A. (2019). Production of high quality crude palm oil (CPO) and low 3-MCPD ester RBD palm oil. *Palm Oil Engineering Bulletin, 131*, 24–28.

Rahn, A. K. K., & Yaylayan, V. A. (2011). What do we know about the molecular mechanism of 3-MCPD ester formation. *European Journal of Lipid Science and Technology, 113*, 323–329. https://doi.org/10.1002/ejlt.201000310

Ramli, M. R., Siew, W. L., Ibrahim, N. A., Hussein, R., Kuntom, A., Abd. Razak, R. A., & Nesaretnam, K. (2011). Effects of degumming and bleaching on 3-MCPD esters formation during physical refining. *Journal of the American Oil Chemists' Society, 88*(11), 1839–1844. https://doi.org/10.1007/s11746-011-1858-0

Spaparin, N., Krishnan, A., Md Noor, A. (2018). Process for producing a refined vegetable oil. WO2018182396A1. Available at: https://patents.google.com/patent/WO2018182396A1/en. Accessed 5 February 2020.

Shimizu, M., Vosmann, K., & Matthäus, B. (2012). Generation of 3-monochloro-1, 2-propanediol and related materials from tri-, di-, and monoolein at deodorization temperature. *European Journal of Lipid Science and Technology, 114*(11), 1268–1273. https://doi.org/10.1002/ejlt.201200078

Šmidrkal, J., Tesařová, M., Hrádková, I., Berčíková, M., Adamčíková, A., & Filip, V. (2016). Mechanism of formation of 3-chloropropan-1,2-diol (3-MCPD) esters under conditions of the vegetable oil refining. *Food Chemistry, 211*, 124–129. https://doi.org/10.2903/j.foodchem.2016.05.039

Syed Hilmi, S. M. H., Othman, N. H., Saparin, N., Jahaya, S. S., Md Noor, A., & Asis, A. J. (2018). Process for producing a refined palm fruit oil having a reduced 3-mcpd content. WO2019027315. Available at: https://patentscope.wipo.int/search/en/detail.jsf?docId=WO2019027315. Accessed 5 February 2020.

Tagoe, S., Dickinson, M., & Apetorgbor, M. (2012). Factors influencing quality of palm oil produced at the cottage industry level in Ghana. *International Food Research Journal, 19*(1), 271–278.

Tiong, S. H., Saparin, N., The, H. F., Ng, T. L. M., Md Zain, M. Z. b., Neoh, B. K., Md Noor, A., Tan, C. P., Lai, O. M., & Appleton, D. R. (2018). Natural organochlorines as precursors of 3-monochloropropanediol esters in vegetable oils. *Journal of Agricultural and Food Chemistry, 66*, 999–1007. https://doi.org/10.1021/acs.jafc.7b04995

Touffet, M., Trystram, G., & Vitrac, O. (2020). Revisiting the mechanisms of oil uptake during deep-frying. *Food and Bioproducts Processing, 123*, 14–30. https://doi.org/10.1016/j.fbp.2020.06.007

United States Department of Agriculture (USDA). (2020). Production, supply and distribution. Available at: https://apps.fas.usda.gov/psdonline/circulars/oilseeds.pdf. Accessed 17 July 2021.

Union, E. (2020). COMMISSION REGULATION (EU) 2020/1322 of 23 September 2020 amending Regulation (EC) No 1881/2006 as regards maximum levels of 3-monochloropropanediol (3-MCPD), 3-MCPD fatty acid esters and glycidyl fatty acid esters in certain foods. *Official Journal of the European Union*. Retrieved from https://eur-lex.europa.eu/legal-content/EN/TXT/?uri=CELEX:32020R1322. Accessed 25 December 2021.

Walpole, S. C., Prieto-Merino, D., Edwards, P., et al. (2012). The weight of nations: An estimation of adult human biomass. *BMC Public Health, 12*(439), 1–6. https://doi.org/10.1186/1471-2458-12-439

Wong, Y. H., Muhamad, H., Abas, F., Lai, O. M., Nyam, K. L., & Tan, C. P. (2017). Effects of temperature and NaCl on the formation of 3-MCPD esters and glycidyl esters in refined, bleached and deodorized palm olein during deep-fat frying of potato chips. *Food Chemistry, 219*, 126–130. https://doi.org/10.1016/j.foodchem.2016.09.130

Wong, Y. H., Goh, K. M., Nyam, K. L., Nehdi, I. A., Sbihi, H. M., & Tan, C. P. (2019). Effects of natural and synthetic antioxidants on changes in 3-MCPD esters and glycidyl ester in palm olein during deep-fat frying. *Food Control, 96*, 488–493. https://doi.org/10.1016/j.foodcont.2018.10.006

Zhang, X., Gao, B., Qin, F., Shi, H., Jiang, Y., Xu, X., & Yu, L. (2013). Free radical mediated formation of 3-monochloropropanediol (3-MCPD) fatty acid diesters. *Journal of Agricultural and Food Chemistry, 61*(10), 2548–2555. https://doi.org/10.1021/jf501662y

Zhang, H., Jin, P., Zhang, M., Cheong, L. Z., Hu, P., Zhao, Y., Yu, L., Wang, Y., Jiang, Y., & Xu, X. (2016). Mitigation of 3-monochloro-1, 2-propanediol ester formation by radical scavengers. *Journal of Agricultural and Food Chemistry, 64*(29), 5887–5892. https://doi.org/10.1021/acs.jafc.6b02016

Zhou, H., Jin, Q., Wang, X., & Xu, X. (2014). Effects of temperature and water content on the formation of 3-chloropropane-1, 2-diol fatty acid esters in palm oil under conditions simulating deep fat frying. *European Food Research and Technology, 238*(3), 495–501. https://doi.org/10.1007/s00217-013-2126-3

Zulkurnain, M., Lai, O. M., Latip, R. A., Nehdi, I. A., Ling, T. C., & Tan, C. P. (2012). The effects of physical refining on the formation of 3-monochloropropane-1, 2-diol esters in relation to palm oil minor components. *Food Chemistry, 135*(2), 799–805. https://doi.org/10.1016/j.foodchem.2012.04.144

Zulkurnain, M., Lai, O. M., Tan, S. C., Abdul Latip, R., & Tan, C. P. (2013). Optimization of palm oil physical refining process for reduction of 3-monochloropropane-1,2-diol (3-MCPD) ester formation. *Journal of Agricultural and Food Chemistry, 61*(13), 3341–3349. https://doi.org/10.1021/jf4009185

Sustainable Practices of IOI Palm Oil and Palm Kernel Processing Complex in Sabah

Shyam Lakshmanan, Yen Li Yung, Boon San Chan, and Zhe Haw Chong

Abstract As part of the palm oil supply chain, IOI Edible Oils Sdn. Bhd. (IOIEO) plays a vital role in processing crude palm oil, and palm kernel. IOIEO strives to improve its operations and is committed to reducing its carbon footprint. Improvements are made by heat recovery, water reuse, implementing process changes and installing new equipment. The company has achieved 85% recovery of the treated effluent water while reducing its power consumption by 40%. In addition to becoming a greener process, additional benefits were observed in terms of improved product quality, and lower treatment chemical usage. The company decided to switch to LED lights and inverters were installed for large motors which were operating suboptimally. These two measures (LED lights and inverters) resulted in 2.5% reduction in energy consumption. Steam condensate and fugitive steam recovery resulted in reduction of 41% and 20% in the palm kernel dry fractionation plant and dry fractionation plant's steam consumption. Heat recovery in the refinery, resulted in energy savings of 30%. The refinery switched to burning liquified natural gas, which resulted in lower carbon footprint of 10%, when compared to its previous use of fuel oil for the same purpose.

Keywords Palm oil refinery · Palm dry fractionation · Palm kernel dry fractionation · Vent economizer · Heat recovery

Abbreviations

3-MCPDE	3-Mono-chloro-propane-diol ester
ASEAN	Association of South East Asian Nations
BPO	Bleached Palm Oil
BPKO	Bleached Palm Kernel Oil
BPKOL	Bleached Palm Kernel Olein

S. Lakshmanan · Y. L. Yung (✉) · B. S. Chan · Z. H. Chong
IOI Edible Oils Sdn Bhd, 12 KM, Sungai Mowtas, Jalan Jaya Chip, Off Jalan Batu Sapi, 90000 Sandakan, Sabah, Malaysia
e-mail: yung@ioigroup.com

D. C. Y. Foo et al. (eds.), *Sustainable Technologies for the Oil Palm Industry*,
https://doi.org/10.1007/978-981-19-4847-3_14

BPKST	Bleached Palm Kernel Stearin
CO_2	Carbon dioxide
CPO	Crude Palm Oil
CPKO	Crude Palm Kernel Oil
DF	Dry fractionation
EMC	Energy Management Committee
ERT	Emergency Response Teams
ESP	Electrostatic precipitator
ETP	Effluent Treatment Plant
EU	European Union
FFA	Free fatty acid
GC-MS	Gas Chromatograph–Mass Spectrometer
GE	Glycidyl Ester
HACCP	Hazard Analysis and Critical Control Points (Food Safety Management System)
IChemE	Institution of Chemical Engineers
ICP–OES	Inductively Coupled Plasma–Optical Emission Spectrometer
IE3	International Efficiency (Premium Efficiency)
IOIEO	IOI Edible Oils Sdn Bhd, Sandakan
ISCC	International Sustainability and Carbon Certification
ISO 9001	Quality Management System
ISO 14001	Environmental Management System
ISO 45001	Safety Management System
KCP	Kernel Crushing Plant
KeTSA	Ministry of Energy and Natural Resources
LED	Light Emitting Diode
LNG	Liquefied Natural Gas
MSPO	Malaysian Sustainable Palm Oil Certification
MPOB	Malaysian Palm Oil Board
PFAD	Palm Fatty Acid Distillate
PKDF	Palm Kernel Dry Fractionation
PKFAD	Palm Kernel Fatty Acid Distillate
RBDPO	Refined Bleached & Deodorized Palm Oil
RBDOL	Refined Bleached & Deodorized Palm Olein
RBDST	Refined Bleached & Deodorized Palm Stearin
RBDPKO	Refined Bleached & Deodorized Palm Kernel Oil
RBDPKOL	Refined Bleached & Deodorized Palm Kernel Olein
RBDPKST	Refined Bleached & Deodorized Palm Kernel Stearin
R&D	Research and Development
RO	Reverse Osmosis
RSPO	Roundtable on Sustainable Palm Oil Certification
UF	Ultra Filtration

1 Introduction

IOI Edible Oils Sdn Bhd (Sandakan Refinery) is one of the largest exporters of various palm oil products from Sabah. Greater than 99% of all the products produced on the site are destined for export. In order to accommodate these exports, the company has a tank farm with more than 60 tanks of varying sizes. This tank farm's pump house is connected to an export jetty which handles mainly IOI Edible Oils' (IOIEO) own products. In addition to these facilities, the complex houses three physical refineries, four dry fractionation plants, two palm kernel fractionation plants and three kernel crushing plants. The neighbouring IOI Bio-Energy (IOIBE) facility has biomass fired steam boilers and a renewable energy power plant. Both these boilers utilize biomass from oil palm as its fuel to supply steam and power to the IOIEO complex. The combined IOIEO and IOIBE complexes are built on 62 acres in Batu Sapi, Sandakan, Sabah (Fig. 1).

The plant commenced its operation on the site in year 1997, complete with its own power generating facility (diesel fired) and water treatment plants. This was necessary as there initially was no public supply of municipal water and electricity to the site.

As the complex invested in additional plants, more power generating equipment were installed to meet the increasing power consumption of the site. When public electricity supply was available in 2003, the company secured public power supply, and maintained the power generating equipment as standby generators.

Over the years, these powers generating equipment have been operated to help reduce the load consumed from the grid, as there was power shortage in the region from 2010 to 2017. The refineries require steam for their process, and the complex was

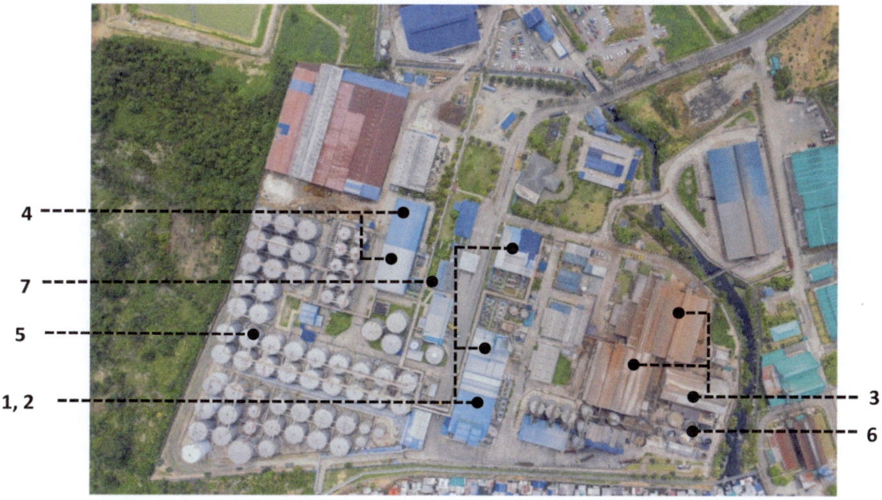

Fig. 1 Aerial view of IOIEO site in Sandakan, Sabah (Number description as shown in Table 1)

Table 1 Facilities on IOIEO site

No.	Plants	Quantity	Capacity
1	Palm oil refinery	3 plants	1,000,000 t/y
2	Dry fractionation	4 plants	1,000,000 t/y
3	Kernel crushing plant	3 plants	300,000 t/y
4	Palm kernel dry fractionation	2 plants	120,000 t/y
5	Tank farm/bulking installation	64 tanks	175,000 t
6	Effluent treatment	1 plant	365,000 t/y
7	Research and development	Facility	3-MCPDE & GE

originally equipped with fuel oil-fired boilers. Subsequently in 2004, the company decided to reduce its carbon footprint and worked on securing supply from a biomass fired boiler and retired these fuel–oil fired boilers. The company continued to grow over the years and is currently the largest electrical power consumer in the district. The company has a preference to employ locals and is currently operated entirely by Malaysians. In excess of 99% of its employees are locals from Sabah. As a food processing company, it is concerned about the quality of its products and it maintains high standards and has HACCP, ISCC, ISO 9001, 14001, 45001, MSPO, and RSPO among its many certifications.

Table 1 shows the production facilities on site.

The main products of this complex are refined palm oil and palm kernel oil products, which are mainly used directly in food applications. This complex processes both palm oils as well as palm kernel oils. The exports from this complex include crude and refined palm and palm kernel oils, as well as their fractions. Once supplied to our customers, the valuable ingredients in palm and palm kernel oil may be extracted for use. Palm and palm kernel oils are used in a wide variety of food, oleochemical and energy, biomass and others applications (Radzian et al., 2009), including cooking oils, margarine, non-dairy creamer, infant formula and pharmaceutical products such as tocotrienol and other supplements or vitamins and many food related applications as well as in anti-bacterial, anti-fungal and anti-viral applications. Under oleochemical category, there are surfactants, cosmetics, detergents, lubricants, skin lotions, soaps, toothpaste, polyurethane, polyols and other agrochemical products. Other areas of use include bio-diesel, bio-composites, fertilizer and for animal feed (Sue, 2017). Figure 2 shows a cut-out view of the palm fruit. While crude palm kernel oil (CPKO) is obtained from the kernel (Ibrahim et al., 2003), crude palm oil (CPO) is extracted from the mesocarp. It is at the palm oil mills that CPO is extracted from the mesocarp. It is also at the mills that palm kernels (PK) are segregated and pre-treated. This CPO and PK are supplied to IOIEO for further processing.

Fig. 2 Cut out of oil palm fruit showing mesocarp (flesh) and kernel

Figure 3 summarizes the processing of palm oil. The palm fruits which are harvested in the estates are quickly delivered to nearby palm oil mills for processing. At the mills, the bunches are graded and processed. The first step in the process is sterilization of the fruit bunches which also serves to deactivate enzymes. The mills employ various process steps to produce crude palm oil while also removing the palm kernels. Both CPO and PK are processed in palm oil mills, and then supplied to the refinery for further processing. Figure 3 shows the different handling of CPO and PK in the refining complex. While CPO is channelled directly to the refinery for processing, PK has to be sent to the kernel crushing plant (KCP) for oil extraction first. These processes are explained in greater detail in their respective sections.

IOIEO mainly supplies refined products to its customers who may use it directly for food or may further process it; e.g., oleochemicals plants extract various components of the oil. Edible oils are comprised mainly of triglycerides. These triglycerides are made up of a glycerol backbone that is connected to three fatty acid chains. CPO and CPKO have different fatty acid compositions as shown in Table 2.

The different fatty acid compositions of CPO and CPKO gives both these oils their distinct properties. CPKO is mainly comprised of shorter chain triglycerides (Goon et al., 2019) as compared to CPO. CPKO also has a higher level of saturated fatty acids, which then makes it a more stable oil as compared to CPO (Bahadi et al., 2019). The higher fraction of longer chain fatty acids in CPO also explains its higher melting point and deodorizing temperature required, when compared with CPKO. Therefore, there are two separate processing streams in IOIEO, for palm oil and palm kernel oil, as shown in Fig. 3.

The main processing sections of IOIEO complex are explained in the following sections.

(a) Physical Refinery

CPO from the mills is received after quality testing, and is subsequently supplied to the physical refinery to be processed. The physical refining process is shown in

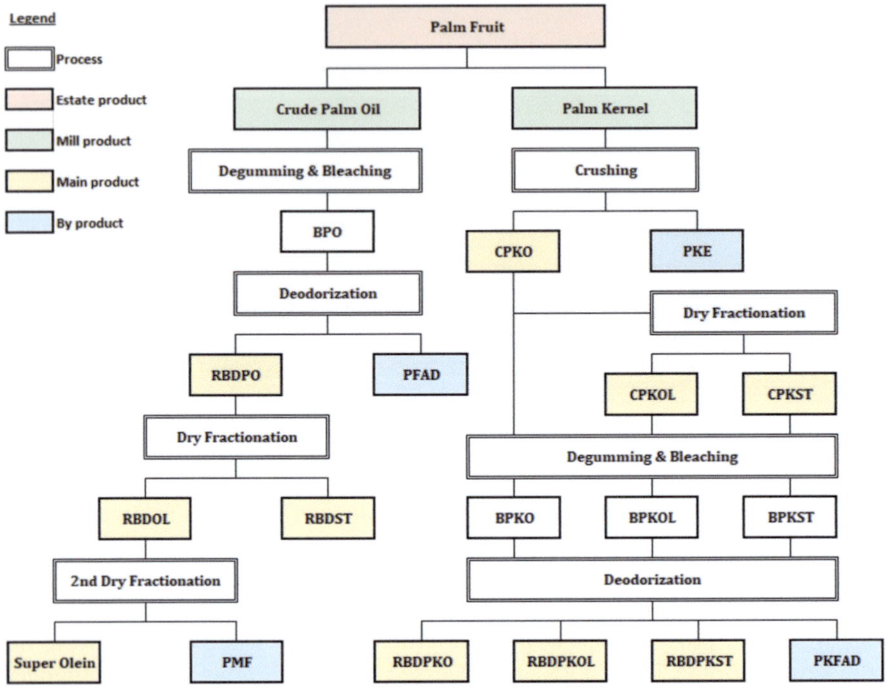

Fig. 3 Block diagram of palm refining and fractionation process and product

Table 2 Fatty acid composition of CPO and CPKO (Radzian et al., 2009)

Fatty acid composition (wt% as methyl esters)		CPO	CPKO
Caproic	C6	–	0.3
Caprylic	C8	–	4.4
Capric	C10	–	3.7
Lauric	C12	0.3	48.3
Myristic	C14	1.2	15.6
Palmitic	C16	42.5	7.8
Palmitoleic	C16:1	0.2	–
Stearic	C18	4.6	2
Oleic	C18:1	40.8	15.1
Linoleic	C18:2	10.6	2.7
Linolenic	C18:3	0.3	–
Arachidic	C20	0.3	0.2

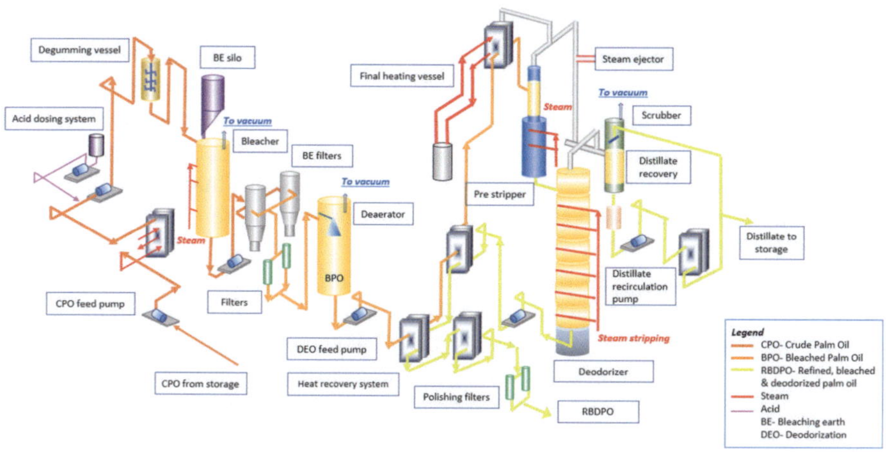

Fig. 4 Physical refinery degumming, bleaching and deodorizing processes

Fig. 4. The term physical refinery is used here to differentiate it from the chemical refining (Pan et al., 2020; Xie et al. 2019) process. Due to the low gums and impurities content in CPO, it is amenable to direct physical refining, without requiring the use of chemicals such as caustic soda. The physical refining process is more environment friendly as it avoids the use of chemicals, and does not generate an alkaline waste water stream that requires further chemical neutralization with sulphuric acid and treatment of the waste water produced from it. In the physical refinery, CPO is first filtered and then heated up to around 100 °C. The CPO is then conditioned with the addition of a small quantity of phosphoric or citric acid (See Fig. 4). The acid dosage would be adjusted to be proportionate to the gums contained in the incoming CPO. The acids serve the purpose of hydrolysing the gums (phospholipids) contained in the CPO. A short contact time along with vigorous mixing is carried out to ensure thorough mixing of the CPO with the acid to allow adequate reaction time for hydrolysis of the gums contained in the CPO.

After the CPO has been thoroughly mixed with the acid, this acid-CPO mixture is retained in an agitated vessel and allowed to remain in contact for 20 min. The acid hydrolysed CPO is next sent to the bleaching step. To aid in the removal of impurities contained in the CPO, an adsorbent (Hew et al., 2020) is added to the acid hydrolysed CPO. The process involved is adsorption and not absorption. Adsorption is mainly a surface phenomenon, whereby a particular substance may be attracted to and sticks to the surface of the adsorbent. This adsorbent is commonly called bleaching earth (BE) by the industry (Patterson, 1992). The main bleaching agents used are activated bentonites or attapulgites, (Brooks et al., n.d.; Cunha de Melo, et al., 2020; Łaska-Zieja et al., 2020; Matthaus, 2012) or a mixture of the two. In some cases, activated carbon is also added into this BE to enhance the bleaching performance. Bentonites and attapulgites are clay minerals which exhibit adsorption properties which are found to be beneficial for processing of edible oils. The BE is mixed with the acid hydrolysed CPO for around 40 min. This mixture is normally agitated using steam to improve the contact between the adsorbent and the impurities in CPO. The BE is next removed from CPO via a filtration process using vertical leaf filters (Fig. 4). During this filtration process, the impurities in the CPO are removed along with the hydrolysed gums by the adsorption action of the BE. This filtered CPO has a significantly lighter appearance and is less viscous than raw CPO.

The degumming acid and BE are used primarily for the removal of unwanted constituents found in CPO such as phosphatides, pigments, trace metal complexes, carbohydrate materials from fruit bunches, oxidation products or traces of soaps. Therefore, the degumming and bleaching stages are important to reduce the unwanted constituents to a minimal level to increase the efficiency of subsequent refining steps. However, some CPO sources contain higher levels of phosphatides or impurities which may require additional processing aids in order to obtain acceptable colour and stable refined edible oil that meet or exceed customers' expectations. In order not to compromise on quality, the refinery continuously monitors raw material and product quality and constantly pursues quality improvement initiatives.

At this stage, the filtered oil is called bleached palm oil (BPO). After filtration, the BPO goes through a drying step to reduce its moisture content. Moisture levels must be minimized as moisture will hydrolyse the triglycerides in the oil, resulting in greater formation of free fatty acid (FFA) and hence greater oil loss. After the drying step, BPO goes through a series of heating stages as shown in Fig. 4 as part of the heat recovery implemented in the process to reduce fuel consumption, before final heating to raise BPO to the deodorization temperature of 260 °C (Lakshmanan et al., 2020).

The heated BPO is then channelled to the pre-stripper. The high vacuum in the pre-stripper aids in the removal of volatile impurities and FFA. The stripped BPO exits from the bottom of the pre-stripper and enters the top of the deodorizer by gravity flow (Fig. 4). The deodorizer also operates under very high vacuum (Lakshmanan & Yung, 2021). Here, much of the remaining volatiles and other decomposition products (aldehydes, ketones, etc.) contained in the BPO are removed in the vapour stream leaving the deodorizer. Steam is supplied to each tray of the deodorizer. Steam bubbles are injected at the bottom of each tray and rapidly increase in volume due to the vacuum conditions in the deodorizer. This action by the steam gives intimate contact and mixing, while also increasing the contact surface area between steam and oil to aid in stripping of impurities contained in the BPO. The mixing action of the steam also raises volatile matter to the surface of the BPO in each tray, so they may be removed by the vacuum conditions in the deodorizer. The vapours and volatile matter leaving the deodorizer are channelled to a vapour scrubber that is irrigated with palm fatty acid distillate (PFAD), to help condense the volatile matter (Tan et al., 2020). The product leaving the vacuum scrubber is known as PFAD, although not the main product of the refinery, it is commonly used in oleochemical production, as renewable material in biofuels production and in animal feed. The deodorizer also serves another function in palm oil refining. It provides retention time at high temperature to break down many of the colour-causing compounds contained in the

BPO into smaller molecules which may then be distilled off, or removed by the vapour stream.

The refined oil exits from the bottom of the deodorizer. This product is termed as *refined, bleached, and deodorized palm oil* (RBDPO). The hot RBDPO leaving the deodorizer flows through a heat recovery system, where the outgoing RBDPO heats up the incoming BPO. This heat recovery greatly reduces the fuel consumption of the plant. As high temperatures rapidly deteriorate the quality of the RBDPO, it is rapidly cooled down to produce a bland, odourless and stable oil that is low in impurities content. As mentioned earlier, the pre-stripper and deodorizer operate under high vacuum conditions. This vacuum is generated using steam. The steam is generated from palm empty fruit bunch fibre and palm kernel shell (Lakshmanan et al., 2021) obtained from palm oil mills located within close proximity of the complex. This is another one of the company's many sustainability initiatives.

The complex houses three refineries from different technology providers with varying configurations. However, despite the differences, they serve the same function, which is to refine CPO. The smallest refinery in the complex is used mainly for refining palm kernel oils. The process steps involved in refining of palm kernel oils are similar to refining CPO, but with much lower dosages of degumming acid and BE being used, while deodorization is conducted at lower temperatures.

(b) Dry Fractionation

At room temperatures, RBDPO forms distinct solid and liquid fractions. This is due to the nature of the fatty acids in palm oil. Palm oil contains a large proportion of saturated fatty acids which solidify at room temperature. This solidification of the oil can be undesirable for certain applications. While some applications may require a liquid oil instead of a semi-solid one, others may be able to operate with a solid fat. In order to address this issue, RBDPO often undergoes an additional process, known as fractionation. The purpose of this process is mainly to separate the solid fat (at room temperature) from the liquid oils. While the liquid fraction is known as olein (RBDOL), the solid fraction is called stearin (RBDST). There are two main fractionation processes. The first process uses a solvent, while the second does not. The solvent free process is commonly known as the *dry fractionation* process. There is no waste generated by the dry fractionation process and no additives or supplementary post treatment of the end products is required. IOIEO employs the dry fractionation process as part of its sustainability initiatives (Fig. 5).

Fractionation is very much a batch operation, with crystallization followed by filtration. While cooling the oil following a desired cooling profile, crystals start to form and consolidate. At the end of the cooling cycle, this oil which is full of crystals is drained from the crystallizers. In the second stage of the fractionation process, the slurry from the crystallizer is pumped in batches to a filter-press for filtration (Fig. 5).

In the filter press, the slurry is pumped through purpose made filter cloth which retain the solid fats, while the liquid fraction flows through the filter cloth. The liquid fraction that flows through the filter cloth is called RBDOL. The solid fat that gets trapped on the filter cloth is next squeezed, to extract as much of the liquid fraction as possible. The RBDOL is then pumped to storage where it is now ready for export

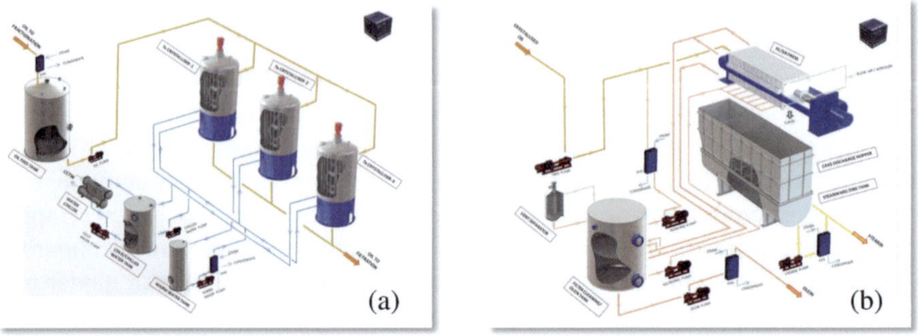

Fig. 5 Palm oil dry fractionation process **a** crystallization; **b** filtration (Ballestra, 2015)

and/or to be used directly as cooking oil. In the following step of the filter press, the solids are discharged into a heated trough, where this solid fraction called RBDST, is melted using hot water, and the melted stearin is pumped to storage tanks. The stearin fraction tends to solidify in storage, and it will need to be heated prior to shipment. In the palm oil dry fractionation process, the yield obtained is determined from the percentage of RBDOL obtained from a known RBDPO feed. This became the industrial norm as olein was sold at a premium to stearin. However, in recent years, there has been a change with stearin frequently becoming the more valuable of the two products. However, the practise of calculating the olein yield as the basis of the plant separation efficiency has been maintained. Plant operators adjust their operating temperatures and cycle times, depending on the prevailing market prices for olein relative to stearin.

(iii) Kernel Crushing Plant

The complex also receives palm kernel from the palm oil mills. These palm kernels are fed to specially designed presses. Here, the kernels are cracked and the oil is pressed out from the kernel. The oil content in palm kernel is typically around 50%. The KCP in IOIEO extracts the oil out from the kernel using mechanical presses. This is done through two-stages of screw pressing. The more environment friendly solvent-free screw presses utilized at IOIOE produces no waste at the KCP and thus contribute to a lower carbon footprint. The oil extracted through these presses is called CPKO. The remaining solids after pressing is called palm kernel expeller (PKE) and is currently exported to markets where it is found to be a superior quality animal feed. Besides processing normal PK, certified or segregated sustainable grade (SG) PK is processed at dedicated production lines to produce SG CPKO. This SG CPKO undergoes further processing downstream before being exported as SG grade. The segregation of the feed is so as to enable certified sustainable grades of oil, to be processed and stored and exported as sustainable products.

(iv) Palm Kernel Dry Fractionation

The CPKO that is extracted in the KCP is next sent to the palm kernel dry fractionation (PKDF) plant for further processing. The general fractionation method employed under dry fractionation is also used for PKDF, but the three-stage batch process employed here is carried out in a different plant that is dedicated only for CPKO. Temperatures and cycle times also do differ from that used for palm oil dry fractionation. The CPKO dry fractionation process as shown in Fig. 6 is also a batch process. The process is called dry fractionation as it does not employ the use of any solvents. The PKDF process is more environment friendly as it does not generate any effluent, does not use any solvent, and hence there are no releases to the environment. The three main steps in palm kernel fractionation are crystallization, followed by statolization (or static crystallization) and finally, filtration.

In terms of capacity, the PKDF plants are much smaller than the RBDPO dry fractionation plants, with the volume of CPKO produced being much lower than the volume of CPO. The process, operating parameters and yields of solid and liquid fractions differ from those of fractionation of RBDPO. As shown in Fig. 6, for CPKO dry fractionation, crystallization is followed by a step called statolization. Both of these processes are carried out under strict temperature control. Statolization is followed by a filtration operation to separate the oil into its solid and liquid fractions, namely crude palm kernel olein (CPKOL) and crude palm kernel stearin (CPKST). The semi-solid slurry is pumped into a filter-press where the fractions are separated at high pressures. For the PKDF plant, the solid fraction, or palm kernel stearin is the more valuable product. Hence for the PKDF plant, it is the stearin yield that is calculated. The palm kernel stearin fraction finds application in many areas including confectionery, making it the more valuable of the two products obtained from the plant. The CPKOL is supplied to the oleochemical industry which extracts basic oleochemicals (such as fatty acids, fatty alcohols, methyl esters and glycerine) or it may be further processed into oleochemical derivatives (such as fatty esters, fatty amines and personal care products), all of which takes it beyond the basic palm kernel oil.

2 Conservation of Energy

In compliance with the Energy Commission's (EC) regulations, the company set up an Energy Management Committee (EMC) which is comprised of the key plant executives. The committee meets once every two months to review the status of energy savings projects that have been undertaken, and to evaluate new energy savings projects. Although the EC's focus is mainly on electricity conservation, the company's EMC looked into potential areas for reducing the consumption of utilities such as electricity, fuel, steam and water usage. After its formation, the EMC has led the implementation of many energy saving projects, and monitors and reports the saving obtained after implementation of these projects.

(a) Electricity Conservation

For many of the larger motors, after studies being carried out to optimize their operation, inverters have been installed to control the power consumption. Before

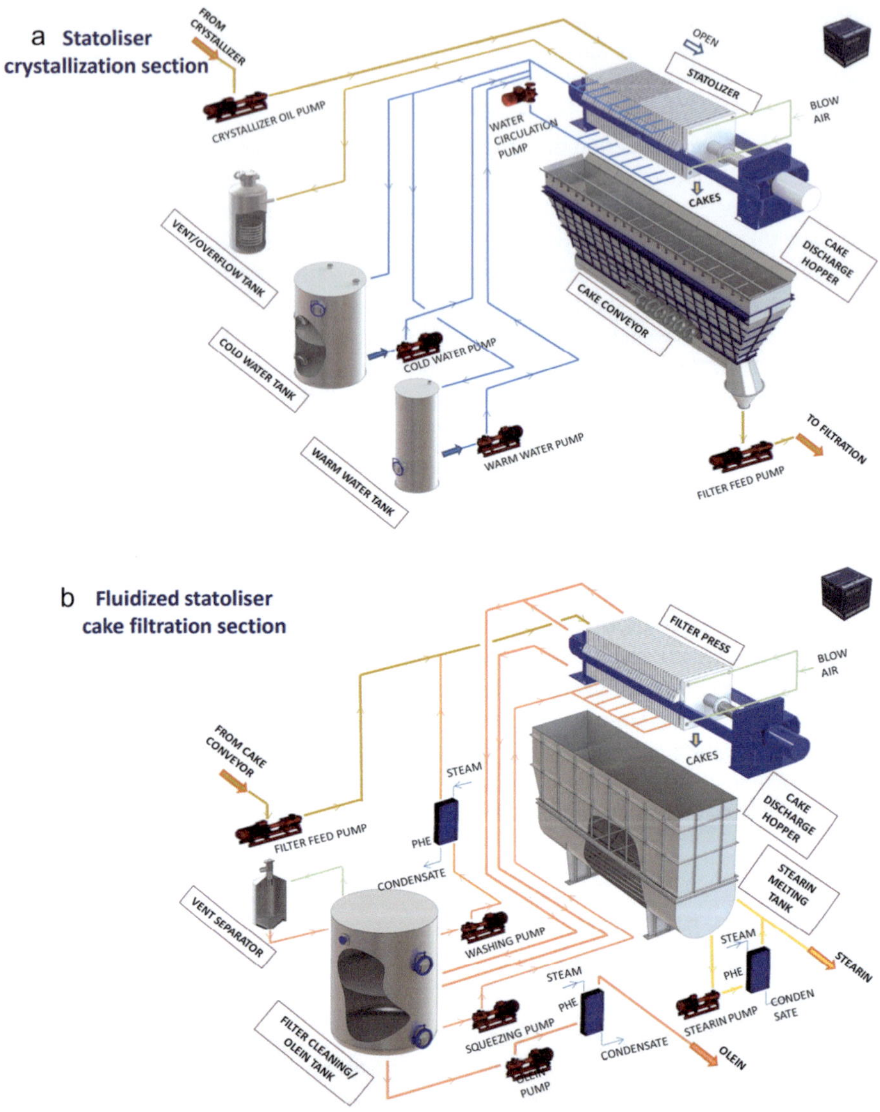

Fig. 6 Palm kernel oil dry fractionation **a** crystallization and statolization process; **b** filtration (Ballestra, 2015)

installation of these inverters, valves were used to throttle the fluid flow so as to supply the desired volume of flow. The placement of valves in the piping circuit, operated in a throttled position creates higher pressure drop for the piping system. This resistance to flow, causes more energy to be consumed by the motor, in order to make up for the pressure loss. The valves are now placed in the fully open position, after the inverters were installed. In the new operating method, the inverters vary the speed of the motor to deliver the desired liquid flowrate, hence resulting in higher efficiency in energy utilization. The EMC also encourages team members to evaluate whether their respective plants could benefit by utilizing high-efficiency pumps and motors as well as implementation of inverter drives to reduce electricity consumption. At each EMC meeting, energy saving analysis is presented for sharing, and to identify the plants with the greatest and lowest energy savings. The meeting also evaluates other areas or plants within the complex where these implemented energy savings projects may be duplicated because of evidenced significant saving.

Selected cooling tower fans were fitted with inverters. These inverters modulate the cooling tower fan speed in order to maintain the required water temperature. During low heat load periods, the cooling tower fan may be completely stopped, if the cooling tower water temperature is already very low. This control action also helps to reduce water evaporative losses at the cooling towers. The complex has also migrated to utilizing LED lights and spot lights as shown in Fig. 7 (Gayral, 2017) in replacement of conventional lights and spot lights to achieve energy saving.

Fig. 7 **a** Replacing light bulbs and spot lights with energy efficient LED lights, **b** Storage tank farm illuminated with LED lights

Fig. 8 Usage of inverter for cooling tower fans and pumps

Inverter systems regulate the speed of motor based on the required pump's capacity. Their usage in selected locations has improved energy efficiency and reduced electricity consumption. These efforts have shown positive results with savings being recorded in electricity consumption in the past few years as shown in Fig. 9. Implementing such measures has yielded a win–win situation; a win for the environment, and a win for the company through cost reductions achieved (Fig. 8).

The EMC has mandated that all new motors that are purchased for the complex must be either of Premium Efficiency (IE3) or higher efficiency rating. While the efficiency of an 80-kW motor rated as IE3 would be 95%, the IE2 rated motor would have a lower 94% efficiency, whereas an IE1 rated motor's efficiency would be even lower at 93%. The higher the IE rating of the motor, the power consumed will be lower for same amount of work. Although these higher efficiency motors are higher priced, the EMC supports their purchase as they are more environment friendly and in line with the sustainable practises embraced by the company.

Preventive and predictive maintenance are practised in the company. Tools such as ultrasonic detectors, thermal cameras, vibration probes and temperature probes are provided for the maintenance team to conduct periodic readings for operating equipment. The ultrasonic detectors can help identify leaking steam traps and areas with air leaks. These checks help to reduce wastage of these resources. Thermal cameras are used to detect hotspots on electrical and other equipment, to assist to

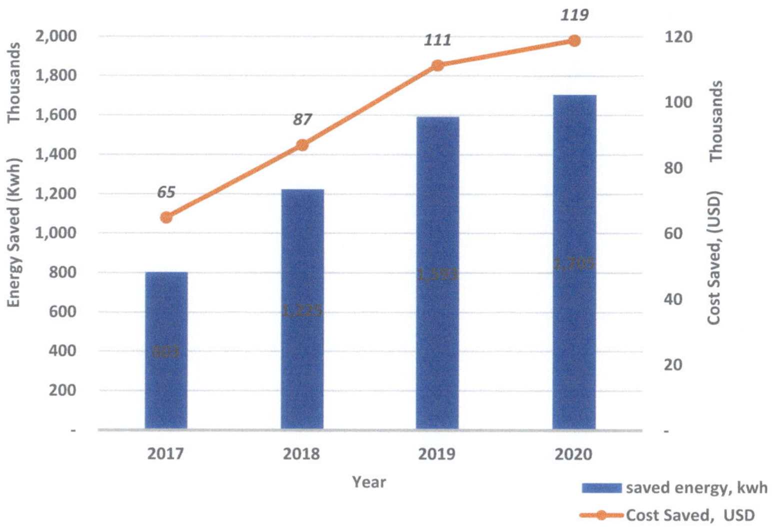

Fig. 9 Energy savings achieved in IOIEO between 2017 and 2020

predict failure, and to then take necessary action to prevent the failure. It is also used to identify areas where insulation may be insufficient, and to improve it to reduce energy losses. The vibration and temperature probes are used to check vibration and temperature of rotating equipment such as pumps, agitators and to determine if there is any deterioration in their operating conditions. The trend analysis conducted using these data, helps production personnel to determine when to schedule preventive maintenance work. When repeated deterioration is noticed for the same equipment, further analysis is conducted to study its installation and improve it.

For the air-compressors, internal leak test is conducted on a monthly basis to determine if there is deterioration in its efficiency, and preventive maintenance will be carried out when such leakage has been identified. These measures have helped to reduce electricity wastage of the air-compressors. The company's active pursuit of sustainable practises has been recognized as it was placed as the national champions for the Energy Award (see Awards & Recognition, Fig. 28a) organized by Ministry of Energy and Natural Resources (KeTSA) in 2018. The company was next selected to represent the country at the ASEAN level, and here it was placed as the 2nd runner up for energy management, in the large industry category (Fig. 28c).

The implementation of EC regulation of having a Registered Electrical Energy Manager and an EMC has helped the company focus efforts on reducing costs. There is a requirement to report the progress of projects to EC, and even reasons for any delays, because of which there is greater commitment to achieve reduction in utility consumptions.

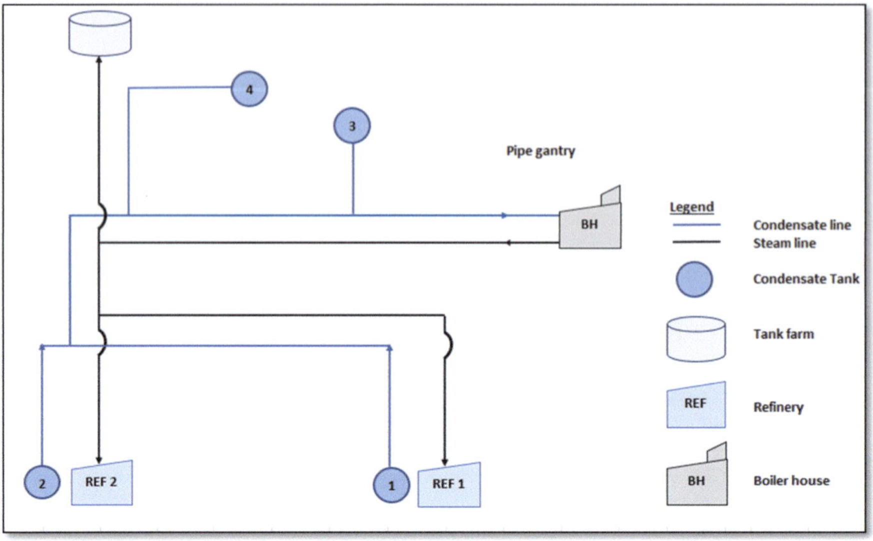

Fig. 10 Steam condensate recovery

(b) Heat Recovery and Utilisation—Vent Economizers

After the plant had been in operation for a few years, a steam condensate recovery system was implemented in the complex. The condensate recovery pipes travel past all the process plants and condensate is channelled to condensate tanks beside each plant, as shown in Fig. 10. Here, the hot water may be used for heating purposes, hence reducing the usage of steam. After the successful implementation and usage of these condensate streams, vapour could be observed exiting from these condensate tanks. The EMC recommended that a project be implemented to recover heat from these fugitive vapour emissions.

The evaluation carried out recommended the installation of a vent economizer to recover heat from the vapour stream. In the project that was implemented, heat was captured in a water stream, and the hot water was successfully utilized for melting stearin. Where previously steam was utilized, now the heating and melting of stearin was carried out using condensate and by condensing fugitive steam vapours. This vent economizer is basically a heat exchanger. It captures the heat that is escaping along with the vapour stream and condenses the vapours (to recover pure water). The vapours condense and some heat is absorbed. The vent economizer installation is shown in Fig. 11. This heat recovered here is used for melting stearin, which is the solid fraction of palm oil. The Heat Recovery Using Vent Economizer project won the Palm Oil Award of the Institution of Chemical Engineers (IChemE) annual awards in 2018 (see Awards & Recognition, Fig. 28d). More vent economizers have since been installed around the complex.

There was an additional benefit from implementing this project. The usage of hot water instead of steam for heating and melting the stearin cake has resulted

Fig. 11 View of the vent economizer

in improved product quality for our RBDST in terms of colour and stability. For this project, steam condensate is collected in individual condensate tanks located within each plant. When condensate is utilized, it was observed that water temperature is maintained at around 90 °C and it is recirculated for reuse to recover any remaining heat. However, when steam was used previously, the temperature was as high as 110 °C depending on the steam pressure. The higher temperature degrades product quality as a result of localized overheating of stearin. After implementing the vent economizer project, it resulted in savings in steam consumption for the dry fractionation plant (Table 3). The first vent economizer was installed in 2016, and a 13.7% reduction in steam consumption was observed. A second vent economizer was installed in 2019 which resulted in further reduction in steam consumption yielding cumulative reduction of steam consumption of 19.6%, when compared against the base year (see Table 3).

Upon observing the savings achieved in the dry fractionation plant, a similar project was implemented for the PKDF plant. Here the savings achieved were higher. Initially, low pressure steam consumption was around 110 kg/t. After implementing the project, steam consumption reduced by 41%, improving to 65 kg/t, as shown in Fig. 12. Efforts are ongoing to reduce this further by utilizing condensate streams that are available nearby.

Table 3 Reduction in steam consumed in dry fractionation plant

Year	2014	2015	2016	2017	2018	2019	2020
Steam used (kg/t)	25.5	24.9	22.0	20.5	21.1	20.5	17.8
Steam saving (%)	Base year	2.4	13.7	19.6	17.3	19.6	30.2

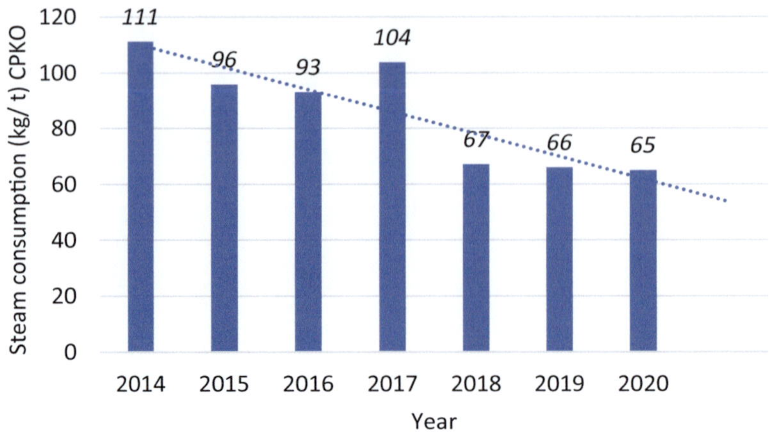

Fig. 12 Reduction in steam consumption in PKDF plant

(iii) Heat Recovery for Physical Refinery

The fuel consumption of Refinery 3 (the largest physical refinery in the complex) in 2014 was 5.73 L/t RBDPO (Table 4). This was much higher than the fuel consumption of the other physical refineries in the complex. After studies were carried out to better understand the process, the high fuel consumption was traced to the inefficient heat exchange in the falling film heat exchanger (Fig. 13a). Efforts were undertaken to clean the heat exchanger to enhance heat exchange. These efforts only yielded limited improvement in performance and that too only for short spans, before the heat exchange deteriorated. In 2016, the team decided to install two spiral heat exchangers to replace the falling film heat exchanger (Fig. 13b). This project was highly successful as it helped to reduce fuel consumption to 3.38 L/t RBDPO (see Table 4). In year 2019, a proposal was put forth to reintroduce the falling film heat exchanger (after conducting extensive chemical cleaning). It was reintroduced and placed in series with the spiral heat exchangers, as shown in Fig. 13c. This heat recovery project resulted in further fuel reduction to 2.89 L/t RBDPO (see Table 4). This refinery which previously had the worst fuel consumption record, now has the lowest fuel consumption when compared with the other refineries in the complex.

Table 4 Diesel consumption for Refinery high pressure boiler operation

Year	2014	2015	2016	2017	2018	2019	2020
Refinery 3 (Diesel + LNG) L/t or Sm^3/t RBDPO	5.73	5.17	3.93	3.38	3.65	3.65	2.89
Reduction (%)	Baseline	10	31	41	36	36	50
Configuration	Falling film (FF)		Change to Spiral Heat Exchanger (SHE)				SHE + FF

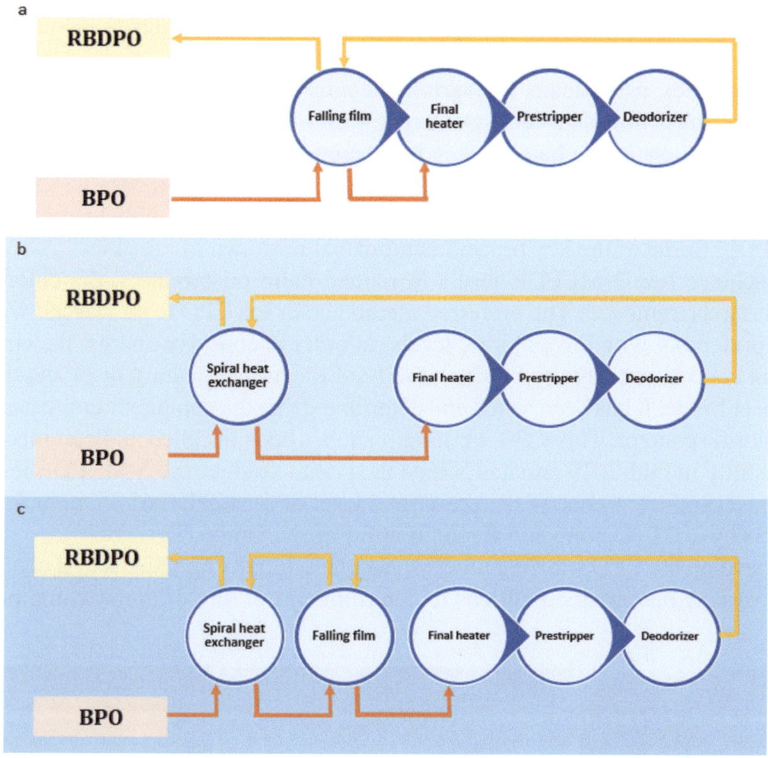

Fig. 13 Heat recovery system in physical refinery, **a** original design with falling film heat exchanger only; **b** after falling film heat exchanger is replaced by spiral heat exchanger; **c** combination of falling film and spiral heat exchangers in series

3 CPO Washing Process for Physical Refinery: 3-monochloropropanediol Ester (3-MCPDE) and Glycidyl Ester (GE) Mitigation Initiatives

The company installed a washing centrifuge along with associated equipment in its pursuit to reduce the formation of 3-monochloropropane-1,2-diol ester (3-MCPDE) in its refined products. The CPO washing process has helped the plant achieve its target to manufacture products with low levels of 3-MCPDE. The company has designed its own CPO washing process with the involvement of its dedicated team members. The CPO washing process has also resulted in lower BE consumption as well as lower steam and electricity consumption. The removal of impurities in the oil during the washing stage has resulted in improved quality of washed CPO being sent to the bleaching section. As some of these impurities have been removed during the washing stage, it reduces the BE dosages required. The ensuing higher

BPO quality also results in higher plant throughputs being achieved, which greatly contribute towards lower utility consumption.

The complex uses steam for various heating applications and the condensate is collected in condensate recovery tanks. From here the hot water is channelled for use where low-grade heating may be required. Some of this condensate is now channelled to be used for the in-house designed CPO washing process. CPO washing is implemented to remove chlorides to produce refined oils with lower levels of 3-MCPDE. Some of the key process equipment are shown in Fig. 14.

To achieve low 3-MCPDE levels in refined palm oil products, the chloride in CPO has to be removed. The preferred method is for the CPO to be washed with hot water, preferably with low chloride levels. In order to conserve energy, the complex uses hot condensate directly (without the need for further heating or processing) to remove chloride. It has been tested and determined that the complex's condensate has low chloride content. This CPO washing plant has been installed and commissioned successfully in mid-2019 and has helped the plant to reduce the 3-MCPDE levels in its refined products to below the prescribed level of 3-MCPDE of 2.5 ppm for food products by 2021 (Commision Regulation European Union (EU), 2018).

To develop the CPO washing process, the plant initially operated a trial centrifuge to test various operating conditions to determine the ability of the washing process

Fig. 14 CPO washing plant equipment, **a** Centrifuge; **b–d** Reaction vessels; **e** Settling tank

to remove chlorides. This centrifuge is the type now used in palm oil Mills for CPO washing. The company subsequently secured a grant from Malaysia Palm Oil Board (MPOB) for 3-MCPDE mitigation and invested in its own refinery scale centrifuge. The process which was developed in-house, has been highly successful in reducing 3-MCPDE by 74%, along with reduction in processing aids dosage and resulting in improved final product quality.

4 Research and Development Department

Realizing the importance of innovation, a Research and Development (R&D) department was set up in 2017. One of the first R&D efforts was the set-up of lab scale CPO degumming, bleaching and deodorization. From here experiments were conducted on the incoming CPO and incoming BE quality to guide plant operation. The experiments would be able to advice the plant on the degree of difficulty in refining the incoming CPO of various quality, and the required type and dosage of BE and phosphoric acid to be used. Due to the high plant capacities here and the rapid changeover of tanks from which CPO is consumed, this guide from the lab greatly assisted the refinery to achieve optimum dosage levels of both chemicals. The next instruments to be purchased were Elemental Analyzer and Inductively Coupled Plasma-Optical Emission Spectrometer (ICP-OES). The elemental analyser is used to determine chloride content and ICP-OES is used to analyse trace metals, in the oil and water samples. Chloride was identified as a precursor in 3-MCPDE formation (Shimizu et al., 2013; Tiong et al., 2018). This became a vital instrument in helping the R&D team to identify CPO sources with high chloride levels, and which may contribute to high 3-MCPDE in refined products. Next, a gas chromatography-mass spectrometer (GC–MS) was purchased. This equipment was procured to analyse the 3-MCPDE and GE levels formed in our refined products. Much research work was conducted with the production team to help design a suitable process to reduce the formation of 3-MCPDE and GE in refined palm oil and palm kernel oils.

R&D work was carried out to guide the plant on reduction of 3-MCPDE and GE. The plant has commenced GE reduction experiments, and these trials are showing promising results. Plant modification is being carried out to test these in actual plant operation. For palm kernel oil, after much R&D and process improvement, the plant is currently able to achieve less than 0.5 ppm 3-MCPDE and less than 0.5 ppm GE commercially. This was achieved after numerous experiments conducted at R&D and plant scale trials. Its commercial products are now below 0.5 ppm 3-MCPDE and GE which is well below the stipulated limit of 1.25 ppm for 3-MCPDE and 1 ppm for GE (Commision Regulation European Union (EU) 2018/290, 2018).

The elemental analyser and GC–MS instruments helped in developing the company's own process for low 3-MCPDE and GE products. The company now operates its very successful and internally developed CPO washing process. This process has the added advantage as it has helped reduce the refinery's operating cost by reducing the consumption of processing aids such as bleaching earth and acid, as

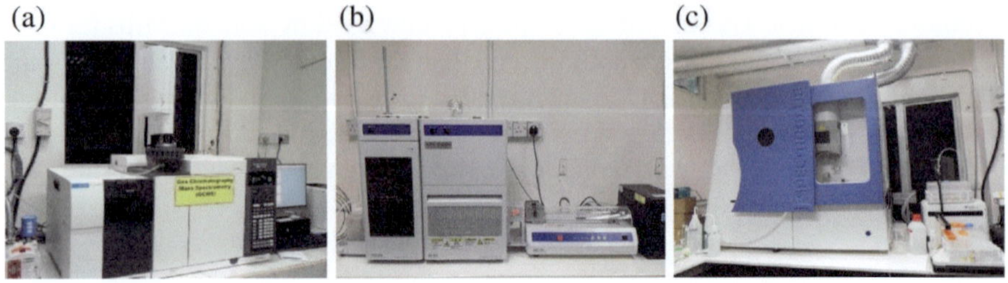

Fig. 15 **a** GC–MS; **b** Elemental analyzer; **c** ICP-OES

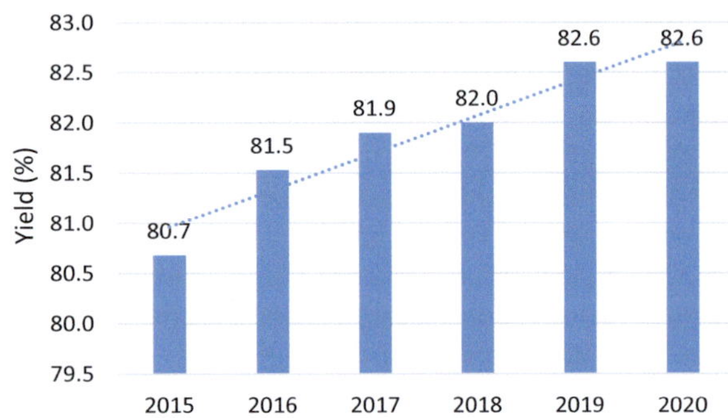

Fig. 16 Yield improvement in fractionation plant

well as reducing oil loss. Figure 15 shows some of these research and development instruments.

These equipment have been utilized in testing heavy metals in product, and the determination of 3-MCPDE and GE levels in processed oils. In addition to assisting the physical refinery, the R&D department has also helped the dry fractionation plant to achieve higher yield, rising from 78% initially, to 80%, and now at 82.6% (See Fig. 16). The next target set to be achieved is 83%.

5 Nitrogen Blowing—Quality Preservation

The company invested in a pressure swing adsorption (PSA) plant for on-site nitrogen generation. Nitrogen is bubbled into storage tanks to preserve the quality of its contents (Fig. 17). Nitrogen usage has been extended for use in blowing and pigging of production and shipments lines. Nitrogen is currently extensively used for blowing to prevent oxidation of the product. The upward migration of nitrogen bubbles displaces oxygen and enhances the preservation of oil quality. Blowing conducted using normal compressed air would oxidize palm oil products. The plant upgraded its infrastructure to ensure palm oil products quality is well preserved in storage, upon loading into ship storage tanks, and until it is delivered to destination port by

Fig. 17 **a** PSA nitrogen generation plant; **b**, **c** Nitrogen is pulse-bubbled into storage tanks to preserve the quality of its contents

using nitrogen blowing instead of compressed air (Chong, 2015). Nitrogen blowing is also used at the fractionation plant to further preserve the stearin quality when blowing to storage tank. Studies conducted by the research and development team has shown that the usage of nitrogen blowing to replace compressed air, has helped to decelerate the rise of peroxide value in stearin fractions, indicating oil oxidation has been retarded.

6 Resource Utilization and Zero Waste

The complex employs solvent free processes for fractionation as well as for its KCP, so these processes do not generate any waste. In the complex, only the physical refinery generates some effluent water and solid wastes. No other waste is generated from the other plants in the complex. The effluent water from the physical refinery comes from its vacuum generation system. Some fatty acid vapours do condense into this water. Due to the presence of fatty matter, this water has to be treated in the effluent treatment plant (ETP) to reduce its Biochemical oxygen demand (BOD) and Chemical oxygen demand (COD) levels before it can be released to the environment. In the IOIEO complex, after treating this water, around 85% of it is recirculated back to be used in the physical refinery's vacuum system and for cleaning purposes.

The ETP of the IOIEO complex uses aerobic digesters to treat the waste water. Studies were commissioned with the intention of boosting the ETP performance to handle the anticipated additional effluent load from the newly commissioned CPO washing process. A new effluent reactor was installed to cater for the additional waste water expected from CPO washing at refinery (Fig. 18a). This amounts to an additional 15% processing capacity. Mini pilot plant studies were conducted commencing on December 2019 to enhance aerobic activity of the effluent process. As a result of these plant improvement studies, some process improvements were implemented. Amongst some of these improvements included, proper aeration (as well as monitoring of dissolved oxygen levels) and addition of appropriate amounts of nutrients for the bacteria. These steps have helped to keep the bacterial colony healthy and at optimum performance levels for digesting organic matter in the waste water. Amongst other measures taken, the reactors were always kept operating to avoid starving the bacteria which would result in decay and loss in bacterial mass.

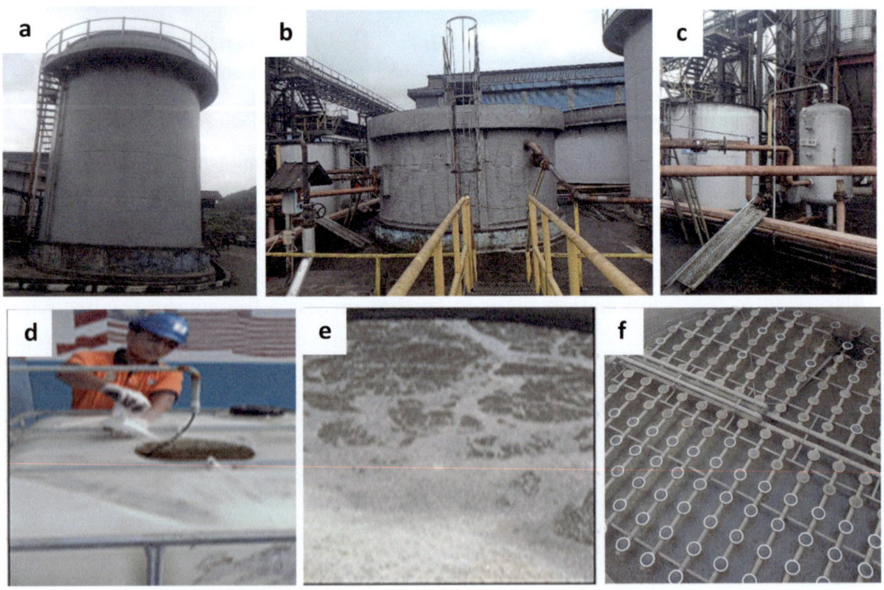

Fig. 18 Effluent treatment plant **a** new reactor; **b** DAF tank bypassed; **c** aeration vessel bypassed; **d** Urea preparation tank; **e** aeration in progress in reactor; **f** aeration nozzles in reactor

Along with operator retraining, these changes were implemented in the ETP, and it resulted in substantial energy savings. The treatment cycle time was successfully reduced from the original 17 h, to only 6 h and this has helped to reduce the electricity consumption by a significant 40%. In addition, the dissolved air flotation (DAF) section has been stopped (Fig. 18b). The monthly electricity consumption of ETP has been reduced by 10,000 kWh. The reduced operating time for the blowers has also helped to extend their life span along with that of the associated motors, and other moving parts. This would lead to reduced maintenance downtime and costs. Upon analysing and realizing that the refinery effluent water was mainly organic in nature, the ETP process was further modified from its original design. With the new changes made, the complete effluent streams are channelled to the digesters to be processed by the bacteria without any segregation or separation. As a result of this, polyaluminium chloride (PAC) usage for flocculation of solids has been discontinued. The successful reduction in chemicals usage and the shorter cycle times at the ETP have contributed to the reduction of greenhouse gas emissions from the complex. Any sludge generated is sent back to the reactors for digestion by the bacteria. Close to 85% of the treated water from the ETP is reused as industrial water in the complex. The quality of the treated water has improved in terms of its BOD and COD levels.

In parallel to the above study, a joint study was conducted with University of Nottingham (Malaysia Campus), to optimize the deodorization and fatty acid scrubber operation (Tan et al., 2020). This study was successfully conducted for Refinery 3, and was also applied for the other two refineries in the complex. The application of the findings from this study has resulted in higher FFA content in the PFAD, and lower oil loss into the vacuum condensing water stream, reducing the biological load to the effluent treatment plant.

When the plant was built in 1997, it had no municipal water supply, and hence required its own water source. The complex built and operates its own raw water treatment plant. In 2016, two additional water catchment ponds were constructed to increase the water catchment volume by 100,000 m^3. The treated water is supplied for the refinery complex as well as the biomass power plant complex. Rainwater harvesting projects were implemented at several large roof areas to increase water collection. The harvested water flows to the catchment ponds by gravity, requiring no additional energy input.

For the power plant, water of higher quality is required. Ultrafiltration (UF) and Reverse Osmosis (RO) plants first process the treated water before supplying the water to the demineralized water plant. Demineralized water is used as feed water by the power plant boiler. Both the UF and RO water treatment plants generate a waste water stream that is referred to as reject water. In the IOIBE complex, the reject water streams are collected and channelled to applications such as for fire-fighting, and general washing applications, which helps to minimise fresh water usage and wastage. During severe drought, the reject water has also been supplied as make up water to the cooling towers. The summary for water recycling and reuse is shown in Fig. 19. In 2019, as shown in Fig. 19, the total RO reject water that was generated was 75,000 m^3. The fraction of reject water recycled was 89%. This is almost double the reuse achieved in the previous year (46%).

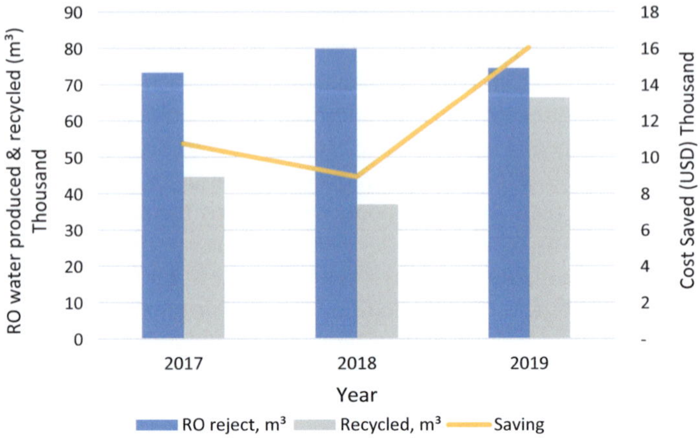

Fig. 19 Summary of reuse and recycle of reverse osmosis reject water in years 2017–2019[1]

7 Reduction in Atmospheric Emissions

The company was the first among the refineries in Sabah to approach and convince the Sabah State Energy Company (SEC) to supply liquified natural gas (LNG) to the refineries for their high-pressure boilers. After conducting several meetings, SEC was convinced of the viability of this project. SEC submitted proposals to its Board to support and supply sufficient volumes to the palm oil refineries, using a pricing mechanism that was at a discount to diesel supply prices. Considering the logistics challenge, most of the refineries in Sabah installed dual fired burners for this application. The use of dual fuel burners would allow the refineries to switch to diesel burning if there were interruptions to LNG supply. The IOIEO complex converted to LNG in July 2019, with the major Sabah refineries following suit. For the IOIEO complex, the switch from diesel burning to LNG has resulted in reduced CO_2 emission of 10% from the high-pressure boiler.

In efforts to further reduce fuel consumption, heat transfer surfaces of several critical heat exchangers for Refinery 2 were cleaned using specialised cleaning equipment (Fig. 20). After this cleaning exercise, the consumption of LNG was monitored. The average LNG consumption for the period from March to May 2020 was noticed to have reduced to 3.348 Sm^3/t RBDPO, which was equivalent to a 16.6% reduction. Given the annual throughput of 245,000 t (for Refinery 2), this results in a savings of 160,000 Sm^3 LNG, which amounts to a cost savings of USD 57,831, while the total CO_2 emission has been reduced by 460,000 kg/y.

[1] *USD/MYR exchange rate is 4.15 MYR, updated on 26 March 2021.*

Fig. 20 High pressure jet cleaning of spiral heat exchangers

8 Kernel Crushing Plant

The KCP has been actively improving its operation, with improved yields, higher throughput and lower maintenance costs. Much research and development work are conducted at the KCP to achieve this. The plant has been continuously improving its yield and throughput over the past few years (Fig. 21a), while also reducing its losses and power consumption (Fig. 21b), leading to lowering in greenhouse gas emissions. Due to its achievements the KCP was awarded MPOB's Best Kernel Crushing Plant in 2017/2018. (See Fig. 22).

The KCP invested in a dust mitigation system with the first unit installed in 2016 at KCP B to provide cleaner working environmental for staff. The second and improved dust plant was installed at KCP C in 2018, and an improved version was installed at KCP A in 2019. A further improvement noticed as a result of these efforts and improvements, was a steady decline in losses, as shown in Fig. 21c. The company's dust plant is featured on the cover of Asia Palm Oil magazine Vol. 8 No 1, Apr-Jun 2019 (see Awards & Recognition, Fig. 28h). IOIEO's KCP made it to the cover of the magazine in recognition of its achievements as the Best KCP in the country (Fig. 23).

9 Corporate Social Responsibility

The company's award-winning emergency response team has put out several fires in the adjacent hills and has also assisted the local fire and rescue department to put out fires in the nearby "Kampung Gas". Kampung Gas is a neighbouring residential community which has built houses on stilts in the nearby bay.

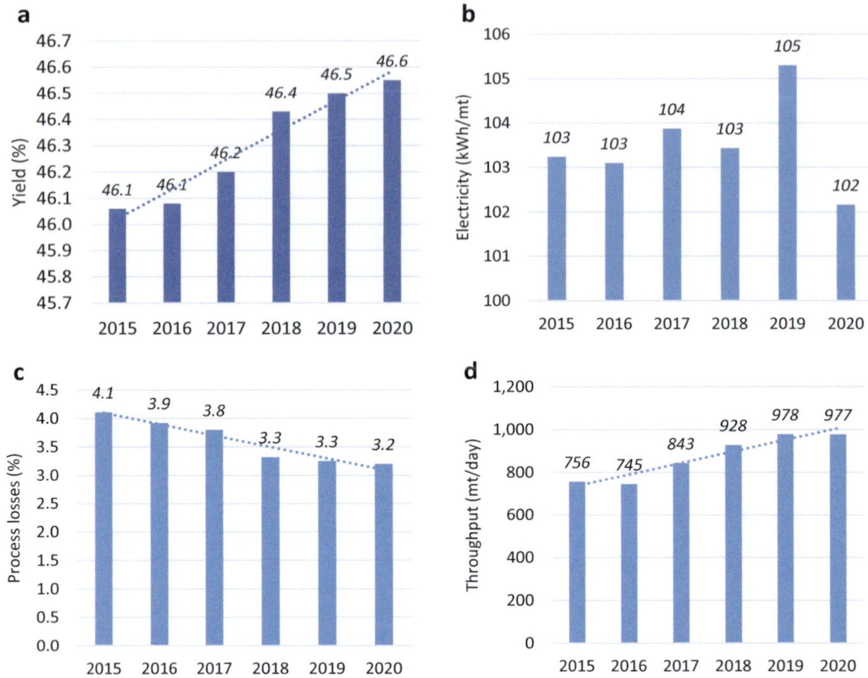

Fig. 21 Kernel crushing plant performance **a** CPKO Yield; **b** electricity consumption; **c** plant losses; **d** plant throughput

Fig. 22 MPOB Award for Best KCP 2017/2018

Fig. 23 **a–c** Dust plants in each of the three KCP in the complex

IOI is actively involved with its neighbouring communities. Local villages such as Kampong Gas and Kampong Bahagia are some of the beneficiaries. Land filling works, road repairs, activity sponsorship and donations for the village and schools are some of its contributions. During two fire incidents in December 2014 and June 2018, IOIEO's Emergency Response Team (ERT) provided on-site assistance to not only put out the fire, but also to help evacuate Kampung Gas villagers to safety and coordinate traffic to help firemen and other rescue teams (Malaysiaskini, 2018). The ERT also deployed IOI's fire hoses to the site to connect to the nearest water source to aid in firefighting (See Fig. 24). The speedy response and assistance provided by the ERT, helped to prevent more extensive fire damage, and was commended by the Head of Sandakan Fire and Rescue Department.

On several occasions in the past few years (with the latest being in March 2017 and April 2018), the company's employees have cleared landslides (see Fig. 25) on the nearby roads which would have isolated nearby villages. The company deployed its earth moving equipment and drivers to clear the blockages, and also sent its personnel to control and direct traffic during this time.

The company's ERT has been the champion for the state level ERT for several years (see Awards & Recognition, Fig. 29f). The company has trained several teams

Fig. 24 IOIEO's ERT assists the local Fire Brigade to put out fire on a floating village

Fig. 25 Landslide on nearby roads being cleared by IOIEO personnel and equipment

Fig. 26 Mock drills conducted during safety campaign (March 2019)

to be capable of attending and responding to emergencies. Many mock drills with varying scenarios are conducted annually to test the ability of the ERT (Fig. 26).

The emergency response centre is equipped with facilities such as oil boom, portable pump, breathing apparatus, confined space escape set, stretcher and fire proof jacket to attend to potential emergencies. Blood donation programs are conducted annually during IOI's Safety, Health and Environment Campaign Week in March every year. For the past few years, during the annual safety campaign week, the company has collected in excess of the targeted 101 pints of blood for Duchess of Kent Hospital's blood bank. This excludes the other instances during the year where the employees voluntarily donate blood.

The company annually trains internship students from Universities and Community College from various disciplines, at certificate level up to degree. Many of them gain employment at government or private sectors and some do join the IOI family in various departments as technicians, officers, chemists, engineers, etc. Many are grateful and commend the internship they experience at IOIEO. The company's

(a)

(b)

Fig. 27 **a** Visits by IOIEO's Sports and Recreation club to various homes around Sandakan; **b** Donation to charity events, school welfare, health department and fire rescue team

executives also do serve as industrial supervisor for undergraduate and postgraduate design projects.

The company has a sports and recreation club that contributes to charitable activities around Sandakan. The committee first visits the charity, to find out what essentials are required before proceeding to buy and supply these items. The photos below (Fig. 27) show several visits made to a nearby orphanage, old folks' home, home for the blind and spastic children's home.

10 Awards and Recognition

See Figs. 28 and 29.

Fig. 28 Awards and recognition for IOIEO **a** National champions award; **b** certificate—National Energy Awards (NEA) Category 1—energy efficiency (energy management in large buildings), organised by Ministry of Energy and Natural Resources (KeTSA); **c** 2nd runner-up award at ASEAN Level Energy Management in Buildings and Industries (Large Industry Category) 36th ASEAN Ministers on Energy meeting; **d, e** IChemE Palm Oil award "Vent Economizer", organised by Institution of Chemical Engineers; **f** Emergency Response Team Championship winner's award, State level; **g** Malaysia Palm Oil Board (MPOB) award for Best Kernel Crushing Plant (KCP) 2017/2018; **h** IOIEO's Best Kernel Crushing Plant makes the cover of Asia Palm Oil Magazine Vol. 8 No 1, Apr–Jun 2019

Fig. 29 Birds eye view of the plant showing greenery within and around it

The company maintains the greenery around its facility as can be seen in Fig. 30.

Fig. 30 Herbal garden and mini rest area next to the main office

The company's employees created a resting area adjacent to the main office. This area is surrounded by lush greenery with various medicinal herbs planted (see Fig. 30).

11 Conclusion

This chapter shows many areas where sustainable practises have been implemented in a palm refinery. The various energy saving practices implemented in the IOIEO complex over the past five years have resulted in power savings of 1,700,000 kWh, which led to savings amounting to USD 118, 807.[2]

The introduction of vent economizer system at fractionation plants has reduced steam consumption at DF plant by 20% and more significant reduction of 41% has been achieved in the PKDF plant. Together with steam saving implementation, both DF and PKDF plants had achieved high yields of 83% and 38.5%, respectively. The

[2] *USD/MYR exchange rate is 4.15 MYR, updated on 26 March 2021.*

conversion from diesel fuel to LNG has resulted in reduced CO_2 emission of 10% at refinery. The average LNG consumption has reduced by 17% giving a cost savings of USD 57,831 and CO_2 emission reduction of 460,000 kg/y. At KCP, the plant has been continuously improving its yield from 46.1 to 46.6% while also reducing its power consumption from average 104 to 102 kWh/t. Meanwhile, an air emission control system has been installed at each KCP to provide cleaner working environment for staff.

RO reject water utilization has increased from 37,000 to 66,000 m^3/y. The effluent plant process was successfully modified to remove chemical dosing as well as reduce air bubbling at ETP, resulting in electricity saving of 40%, while obtaining the same treated water quality of which 85% is recycled and consumed by the refinery. The R&D team has worked on 3-MCPDE and GE mitigation, and reported 74% reduction of 3-MCPDE in refined palm oil and 3-MCPDE and GE below 1 ppm for all refined palm kernel products. The complex has also conducted joint study with academia to optimize deodorization and fatty acid scrubber operation to obtain high RBDPO yield, at low cost, and with minimal waste or pollutions.

Sustainability is no longer just a buzzword but is a concept that is here to stay, and is gaining traction with more people opting for sustainably produced materials. IOIEO takes sustainability very seriously as evidenced by the various efforts taken in reducing fuel, electricity, gas, water and steam consumption.

Acknowledgements The authors acknowledge the strong encouragement and support given by the Group Managing Director of IOI, Dato' Yeow Chor Lee, in the company's pursuit of various sustainability initiatives.

References

Bahadi, M., Yusoff, M. F. M., Salimon, J., Jumaah, M. A., & Derawi, D. (2019). Physicochemical characteristics of Malaysian. *Malaysian Journal of Chemistry, 21*(2), 17–27.

Ballestra, D. (2015). *Specialty fats fractionation: New possibilities.* Statoliser Technologies. Accessed May 20, 2020. http://www.desmetballestra.com/oils-fats/fat-modification/dry-fractionation/palm-kernel-fractionation-statoliser

Brooks, D. D., Berbesi, R., & Hodgson, A. S. (2019). *Optimization of bleaching process.* Available online: https://lipidlibrary.aocs.org/edible-oil-processing/optimization-of-bleaching-process

Chong, C. L. (2015). Measurement and maintenance of palm oil quality. In Palm oil: Production, processing, characterization, and uses (p. 460). AOCS Press.

Commision Regulation European Union. (2018). *Commision Regulation European Union (EU) 2018/290.* http://data.europa.eu/eli/reg/2018/290/oj

Cunha de Melo, K., Silva de Oliveira, I., Helena de Oliveira Pires, L., Santos do Nascimento, L. A., Roberto Zamian, J., Narciso da Rocha Filho, G., Fonseca Passos, M., Santos Lopes, A., Converti, A., & de Costa, C. E. F. (2020). Study of the antioxidant power of the waste oil from palm oil bleaching clay. *Energies, 13*(4), 804. https://doi.org/10.3390/en13040804

Gayral, B. (2017). LEDs for lighting: Basic physics and prospects for energy savings. Demin Ienergie-Seminaire Daniel-Dautreppe, France, pp. 453-461. https://doi.org/10.1016/j.crhy.2017.09.001

Goon, D. E., Abdul Kadir, S. H. S., Ab Latip, N., Ab. Rahim, S., & Mazlan, M. (2019). Palm oil in lipid-based formulations and drug delivery system. *Biomolecules, 9*(64). https://doi.org/10.3390/biom9020064

Hew, K. S., Asis, A. J., Tan, T. B., Yusoff, M. M., Lai, O. M., Nehdi, I. A., & Tan, C. P. (2020). Revising degumming and bleaching processes of palm oil refining for the mitigation of 3-monochloropropane-1,2-diol esters (3-MCPDE) and glycidyl esters (GE) contents in refined palm oil. *Food Chemistry, 307*, 125545. ISSN: 0308-8146. https://doi.org/10.1016/j.foodchem.2019.125545

Ibrahim, N. A., Kuntom, A., Sue, T. T., & Lin, S. W. (2003). Current status of Malaysian crude palm kernel oil. *Oil Palm Bulletin, 47*, 15–27. Malaysia Palm Oil Board.

Lakshmanan, S., & Yung, Y. L. (2021).Chloride reduction by water washing of crude palm oil to assist in 3-monochloropropane-1, 2 diol ester (3-MCPDE) mitigation. *Journal of Food Additives & Contaminants: Part A*. https://doi.org/10.1080/19440049.2020.1842516.

Lakshmanan, S., Yung, Y. L., Chan, B. S., & Chong, T. H. (2020). Sustainable practice of and edible oils refining complex. *Journal of Oil Palm, Environment & Health, 11*, 42–56. https://doi.org/10.5366/jope.2020.05

Lakshmanan, S., Yung, Y. L., Palanisamy, K., & Ling, H. K. (2021). Lessons learnt from biomass-fueled power plant. *Journal of Oil Palm, Environment & Health, 11*, 6–10. https://doi.org/10.5366/jope.2020.02

Łaska-Zieja, B., Marcinkowski, D., Golimowski, W., Niedbała, G., & Wojciechowska, E. (2020). Low-cost investment with high quality performance. Bleaching earths for phosphorus reduction in the low-temperature bleaching process of Rapeseed oil. *Foods, 9*(5), 603. https://doi.org/10.3390/foods9050603

Malaysiaskini. (2018). *Kampung Gas fire razes 90 homes*. 27 Jun. https://www.malaysiakini.com/news/431586

Matthaus, B. (2012). Technological innovations in major world oil crops (S.K. Gupta (Ed.), Vol. 2). Springer New York. ISBN: 978-1-4614-0826-0.

Pan, F., Li, Y., Luo, X. et al. (2020). Effect of the chemical refining process on composition and oxidative stability of evening primrose oil. *Journal of Food Processing and Preservation, 44*, e14800. https://doi.org/10.1111/jfpp.14800

Patterson, H. B. W. (1992). *Bleaching and purifying fats and oils: Theory and practice*. American Oil Chemists' Society.

Radzian, R., Sue, T. T., Keat, O. C., & Khwan, W. S. (2009). Pocketbook of palm oil uses (6th ed.). Malaysia Palm Oil Board (MPOB), ISBN: 978-967-961-151-9.

Shimizu, M., Weikamp, P., Vosmann, K., & Matthaus, B. (2013). Influence of chloride and glycidyl-ester on the generation of 3-MCPD- and glycidyl-esters. *European Journal of Lipid Science and Technology, 115*(7), 735–739. https://doi.org/10.1002/ejlt.201200310

Sue, T. T. (2017). *Pocketbook of palm oil uses* (7th ed.). Malaysia Palm Oil Board (MPOB), ISBN: 978-967-961-228-8.

Tan, M. C., Foo, D. C. Y., & Lakshmanan, S. (2020). An integrated simulation-optimisation approach for free fatty acid (FFA) removal in palm oil deodorization process. *Asia-Pacific Journal of Chemical Engineering*. https://doi.org/10.1002/apj.2602

Tiong, S. H., Saparin, N., Teh, H. F., Ng, T. L. M., Zain, M. Z. B. M., Neoh, B. K., Noor, A. M., Tan, C. P., Lai, O. M., & Appleton, D. R. (2018). Natural Organochlorines as precursors of 3-Monochloropropanediol Esters in Vegetable Oils. *Journal of Agriculture and Food Chemistry, 66*(4), 999–1007. https://doi.org/10.1021/acs.jafc.7b04995

Xie, D., Zhou, H., & Jiang, X. (2019). Effect of chemical refining on the levels of bioactive components and hazardous substances in soybean oil. *Food Measure, 13*, 1423–1430. https://doi.org/10.1007/s11694-019-00058-y

Correction to: 3-MCPDE in Palm Oil Processing: Formation Factors, Transference to Food and Mitigation Approaches

Chien Lye Chew, Amirul Al Hafiz Abdul Hamid, Hemavathi Silvamany, and Soon Huat Tiong

Correction to:
Chapter "3-MCPDE in Palm Oil Processing: Formation Factors, Transference to Food and Mitigation Approaches" in: D. C. Y. Foo et al. (eds.), *Sustainable Technologies for the Oil Palm Industry*, https://doi.org/10.1007/978-981-19-4847-3_13

In the original version of the chapter, the text on page 336 where the text in the bracket has been missed, this has now been updated from "and 3-MCPDE content by 50% (). The studies…" to "and 3-MCPDE content by 50% (Chew et al., 2021). The studies…". The correction chapter and the book has been updated with the change.

The updated original version of this chapter can be found at https://doi.org/10.1007/978-981-19-4847-3_13

D. C. Y. Foo et al. (eds.), *Sustainable Technologies for the Oil Palm Industry*,
https://doi.org/10.1007/978-981-19-4847-3_15

Index

D. C. Y. Foo et al. (eds.), *Sustainable Technologies for the Oil Palm Industry*,
https://doi.org/10.1007/978-981-19-4847-3

MIX
Papier aus verantwortungsvollen Quellen
Paper from responsible sources
FSC® C105338

If you have any concerns about our products,
you can contact us on
ProductSafety@springernature.com

In case Publisher is established outside the EU,
the EU authorized representative is:
Springer Nature Customer Service Center GmbH
Europaplatz 3, 69115 Heidelberg, Germany

Printed by Libri Plureos GmbH
in Hamburg, Germany